The Sustainability of Rice Farming

The Sustainability of Rice Farming

D.J. Greenland
Visiting Professor
Department of Soil Science
University of Reading, UK

Former Director of Scientific Services
CAB INTERNATIONAL

Former Deputy Director General (Research)
International Rice Research Institute
Philippines

CAB INTERNATIONAL
in association with the
International Rice Research Institute

CAB INTERNATIONAL
Wallingford
Oxon OX10 8DE
UK

Tel: +44 (0)1491 832111
Fax: +44 (0)1491 833508
E-mail: cabi@cabi.org

CAB INTERNATIONAL
198 Madison Avenue
New York, NY 10016-4341
USA

Tel: +1 212 726 6490
Fax: +1 212 686 7993
E-mail: cabi-nao@cabi.org

Published in association with:
International Rice Research Institute
PO Box 933
1099 Manila
The Philippines

A catalogue record for this book is available from the British Library, London, UK

ISBN 0 85199 163 7

Library of Congress Cataloging-in-Publication Data
Greenland, D.J.
The sustainability of rice farming/D.J. Greenland.
p. cm.
Includes bibliographical references and index.
ISBN 0-85199-163-7 (alk. paper)
1. Rice. I. Title.
SB191.RSG724 1997
338.1′7318—dc21

96-46785
CIP

Typeset in Photina by AMA Graphics Ltd
Printed and bound in the UK by Biddles Ltd, Guildford and King's Lynn

Contents

Preface

Rice is the world's most important crop. It has supported more people for more years than any other cereal. The great civilizations of Asia emerged in the broad river deltas of China, Southeast Asia and the Indian subcontinent because high yields of rice sufficient to support more than the food demands of those who produced it could be sustained. The importance of rice in Asia is such that it has become deeply entwined with the cultures of the region. The terraced systems by which water is channelled to the small fields in which rice is grown have characterized the Asian landscapes for many years.

The reasons why rice has been able to support so many for so long are due to the physical environment in which rice is grown. The high rainfall of the monsoon lands, and the fact that nutrients and fertile sediments are carried with the floodwaters that seasonally flow into these areas, provided the essential requirements of the crop from the time that it was first cultivated several thousand years ago until recently. But now the burgeoning population of Asia has outstripped the natural capacity of the rice areas to produce the flow of nutrients and water that are the essential requirements of the crop. Nutrients now have to be supplied using heavy dressings of inorganic fertilizers, and flood waters stored behind huge dams for later release to the rice fields. Rice varieties able to produce greater yields than any grown before have been bred. These changes have averted the famines which afflicted India, China and other densely populated parts of Asia in the past and which were predicted to do so on an even wider scale in the 20th century. While the earlier methods of rice production proved sustainable for millennia the sustainability of the new methods of production, giving much greater yields, has still to be established.

The International Rice Research Institute (IRRI) in the Philippines has been at the forefront of the changes which have prevented famine. While IRRI is most famous for having bred the high-yielding, semi-dwarf rice varieties which were the basis of the green revolution, the Institute has always worked to increase the knowledge base on which the sustainable production of any crop depends. Not only has the rice plant been studied, but also the environments in which rice is produced, the farming systems in which rice is grown, and the social and economic factors which determine how and where it can be produced profitably.

Over the past thirty years, a series of symposia have been held at IRRI, in which the knowledge of rice scientists from Asia and elsewhere, has been assembled and published. Much of the literature concerning rice published in Chinese and Japanese books and journals, not readily accessible to workers outside of those countries, has been brought together in those publications. Thus the IRRI symposia have helped to make the knowledge of the Asian scientists available to a much wider audience.

From 1978 to 1987 I had the privilege of directing the research activities at IRRI. The knowledge base expanded tremendously during that period, but the publications produced each had a specific focus, and did not place the sustainability of rice production systems in either a historical or environmental context, although much of the work of IRRI and the national rice research programmes with which it was associated were directed to that end. This book is an attempt to analyse the biophysical and socioeconomic factors which determine the productivity and sustainability of rice farming, and place them in a historical context. It is hoped that this will help to ensure that in future sufficient rice will be produced to meet the further growth of the population, and that it will not be produced at the expense of the natural resources on which future generations depend.

The book is based on the knowledge of many rice scientists, not only those with whom I worked at IRRI from 1978 to 1987, but also those who were on the staff before and since. A great debt of gratitude is owed to them. In particular I would like to acknowledge the help I received from Drs Bhuiyan, Doberman, Cassman, Denning, Fischer, Huggins, Jackson, Khush, Kirk, Olk and Pingali, all of whom read parts of the manuscript and provided much helpful advice. The staff of the library at IRRI, and Lina Vergara and her successor as librarian, Dr Wallace, were industrious in finding and obtaining references for me, as was the Head of Library Services at CAB INTERNATIONAL, Chris Hamilton. Dr E.T. Craswell and the staff of the library at the Australian Centre for International Agricultural Research in Canberra were also most helpful. My inadequate knowledge of the archaeology of Asia, and the history of rice farming, was greatly improved by Professor Charles Higham of the University of Otago, New Zealand, who introduced me not only to his own work in Southeast Asia but guided my faltering footsteps to much other work on the history of rice farming. Hugh Brammer in England helped me with information about Bangladesh, and Professor Kazutake Kyuma with information about early rice farming in Japan and China. Dr Lloyd Evans, of the CSIRO

Division of Plant Industry, Canberra, Australia, read all of the draft manuscript and provided much helpful and constructive criticism. They should in no way be held responsible for any inadequacies in the text, which are my responsibility. The views expressed are also my own, and not attributable to IRRI, or to CAB INTERNATIONAL.

The writing of the book was made possible by the award of an Emeritus Fellowship by the Leverhulme Foundation. This enabled me to return to IRRI, where with the permission and encouragement of the Director General, Dr George Rothschild, I was able to consult with scientists and use the invaluable collection of rice literature in the library. The generous support provided by the Leverhulme Foundation is gratefully acknowledged. The Overseas Development Administration, UK, provided a grant to assist in the production of illustrations for the book, and its assistance is gratefully acknowledged.

Permission from the following to reproduce figures and tables from their publications is also acknowledged: Annual Reviews Inc., for Fig. 4.10; Butterworth-Heinemann Ltd, for Tables 6.9 and 7.6; Chinese Academy of Agricultural Science, West Suburb, Beijing, China, for Appendix 1; Ecological Reviews, Tohoku University, Sendai, Japan, for Fig. 2.2; The Food and Agriculture Organization of the United Nations, for Figs 1.1 and 8.1; Institut Mondial du Phosphate, Casablanca, Morocco, for Fig. 5.6; Journal of Irrigation Engineering and Rural Planning, Tokyo, Japan, for Fig. 6.1; Kluwer Academic Publishers for Tables 4.3, 7.8 and 8.1; New Phytologist, for Fig. 6.6; Resources for the Future, for Fig. 7.2; The Royal Society, London, for Fig. 6.7; Springer-Verlag, Heidelberg, Germany and Science Press, Beijing, for Fig. 4.2; Society of Chemical Industry, for Table 4.2; University of Hawaii Press, Hawaii, USA, for Fig. 7.3; John Wiley and Sons, Chichester, UK, for Figs 8.2 and 8.4. Many others have been taken from publications of IRRI and CAB INTERNATIONAL, and permission to include them is also gratefully acknowledged.

Dedication

To Mary, without whose patient support this book and much other work would never have been possible.

1 The Importance of the Sustainability of Rice Farming

1.1 The Historical Importance of Rice

Rice has supported a greater number of people for a longer period of time than any other crop. Somewhere around 8000–10,000 years ago it was first domesticated and cultivated. The ability to produce a surplus beyond the immediate needs of the producers, which followed perhaps 1000 years later, made possible the initial development of the communities from which the great population centres of Asia arose. Kingdoms and empires based on rice subsequently flourished in China, Indo-China and India. Rice cultivation spread east into Japan, southwards through the Southeast Asian archipelago and westwards to Madagascar and the Middle East.

As methods to control the supply of water to the rice paddies were developed, so that production was not dependent on the vagaries of weather and the floods from the rivers, and as the farmers selected the better plants for the seed they would plant, yields increased and crop failures became less frequent. The importance of rice was increasingly recognized in religious and social ceremonies.

Elsewhere in the world the relatively short lives of many kingdoms and empires can be traced to their dependence on agricultural systems which proved not to be sustainable. In the Middle East salinization of soils was most often the cause of their demise (Artzy and Hillel, 1988; Tainter, 1988). Around the Mediterranean and in Latin America, and probably in Africa, soil erosion and desertification have been the cause (Hyams, 1976; Dregne, 1983; Hillel, 1991). In Asia pestilence and wars and the rise and fall of rivers and the sea have destroyed societies, but soil degradation has seldom been the cause. The stability of soil productivity under wetland rice farming has made rice production from the wetlands the world's most sustainable and productive farming system. It has been sustained on the same land for millennia. Today rice farming still feeds more people than any other crop, and the price of rice still controls the fate of governments.

1.2 The Dimensions of Sustainability

Sustainability may be defined in several ways. For a subsistence farmer it means survival. If a farming system cannot be sustained the people supported by it will either discover a new and more sustainable system or perish. This is an irrevocable truth. If any species, human or otherwise, loses access to the food required to sustain itself it cannot survive. At low population densities concerns about sustainability will not extend beyond the concerns of the clan or group, whose interests will lie almost entirely in the preservation of the biological and physical attributes of the land on which they live. As population increases, the problems that arise when the interests of one group impinge on those of another soon start to become important. This is particularly true of territorial interests. Thus in addition to the time dimension of sustainability – how long can the soil continue to produce a crop at a given place? – the spatial dimension becomes important – is there enough land to provide food for all? As society becomes more sophisticated so the social and economic dimensions of sustainability arise – are the efforts made in crop production impinging on other interests of the group, such as the amenity values of forests or the other needs for water?

An economist may define sustainability as living on interest and not capital, but this fails to emphasize adequately the importance of preservation of natural resources and the environmental aspects of sustainability. A more comprehensive definition is needed, such as that given by Dumanski (1993). He refers specifically to land management:

> Sustainable land management combines technologies, policies and activities aimed at integrating socioeconomic principles with environmental concerns so as to . . .
> (1) maintain or enhance production/services
> (2) reduce the level of production risk
> (3) protect the potential of natural resources and prevent degradation of soil and water quality
> (4) be economically viable
> (5) be socially acceptable.

This is the concept of sustainability that will be used here.

Taking crops from the soil removes nutrients and causes other changes. Most farming systems have evolved to maintain the ability of a soil to provide nutrients, to hold water and to avoid soil degradation. There are historically only three sustainable farming systems: (i) animal-based mixed farming mostly found in the semi-arid to sub-humid grasslands of the savannas and prairies. In this system animals graze the grasslands and concentrate nutrients in manure which can be used to maintain the fertility of cropped land; (ii) tree-based systems, mostly found in the humid forest regions. In this system trees gather nutrients from subsoils and concentrate them at the soil surface, so that an alternation of forest and cropping maintains soil nutrient levels as long as the forest has sufficient time to complete its task of nutrient replenishment; and (iii) water-based wetland rice production, in which the sediments deposited by floodwaters and the nutrients directly absorbed from the water maintain the productivity of land on which rice is grown.

In the course of the past 150 years a fourth system has been added to the animal-, tree- and water-based systems. This is the inorganic fertilizer-based system. The fertilizer system was developed for those areas in Europe where soil degradation other than that due to nutrient removal was seldom a problem. There are of course other factors besides nutrient depletion which affect sustainability. For instance, in areas where serious erosion occurs methods for its control have to be developed, and in systems where non-food crops are produced the appropriate marketing facilities must be economically viable. Nevertheless it is no exaggeration to say that the future of mankind depends on the sustainability of fertilizer-based farming systems. Without the use of fertilizers the present population of the world could not be fed, let alone the additional two billion who will be added in the next 25 years. There is now insufficient land for non-fertilizer-based systems to support the existing world population, and certainly not the additional billions who will be added in the coming century. Although some may question this fact and urge a return to 'organic' farming systems, such systems can do no more than recycle the nutrient elements that exist in the soil. They may maintain present productivity but cannot raise productivity to the extent demanded unless more land is used. In poor areas such systems will do no more than recycle poverty. For the better areas, they will force the exploitation of marginal lands and increase the pace of land degradation.

The fertilizer system has been remarkably successful, particularly in the countries of the temperate zone where it was developed. Nevertheless there have been many concerns expressed about its sustainability in those regions. Some of the concerns have been based on the ecological effects of the intensive farming techniques which the use of fertilizers and pesticides encourages (Carson, 1966) and some have been more directly concerned with associated soil degradation (MAFF, 1970). While it has been found that some of the fears were groundless, they have led to increasing awareness of possibly damaging effects, and of the need to give due attention to the sustainability of the system. Concern about the environmental effects of the fertilizer-based system has also led to renewed interest in alternative, less intensive agricultural systems (NRC, 1989).

As the number of people living in tropical regions has increased and the demand for greater food production has intensified there have been many attempts to boost yields by transfer of the technology of the fertilizer-based system to the tropics. The first important step was to breed fertilizer responsive crop varieties. This was done for rice and wheat at the International Agricultural Research Centers in the Philippines and Mexico, initially funded by the Ford and Rockefeller Foundations. The success of the technology in the irrigated areas with good soils led to the so-called 'green revolution'. Elsewhere when fertilizers have been used to complement indigenous tree-based farming systems modest gains have been achieved (Greenland, 1976). When attempts have been made to replace rather than complement the indigenous systems of the tropics by intensive fertilizer-based systems the results have often been dismal failure. The earliest of these was the 'Groundnuts Scheme' in Tanzania initiated immediately after World War II, with the laudable intention of developing lightly used land in Tanzania for intensive production of groundnuts by mechanized methods used in Europe. Without an understanding of the soils and climate and the biological stresses to which the crop would be subject the result was dismal failure (Wood, 1950). Many similar attempts to transfer the intensive, mechanized production systems of the temperate zone to the tropics have been made subsequently, with equally unsuccessful results.

It is not only attempts to transfer technology directly from temperate to tropical countries that have failed. Population pressure is also causing the inherent stability of animal- and tree-based cultivation systems to break down, as the extent of grazing land becomes insufficient to feed the number of animals necessary to support the population of the savanna regions, and as the forest disappears or is no longer allowed sufficient time to regenerate and restore the fertility of the soil in the lands of shifting cultivation (Nye and Greenland, 1960; Sanchez and Benites, 1987). The productivity of rice in the wetlands and the sustainability of that productivity has been much greater than that of the animal- and tree-based systems of the forest and savanna areas. In the subsequent chapters of this book the historical evidence for the sustainability of rice farming is examined, and the problems of maintaining the system under the additional stresses arising as rice soils are required to support more and more people analysed.

1.3 Rice and Population Momentum

At the present time there are about 2700 million people in the world whose major staple is rice, almost half of the total population of the globe. In the course of the next two decades the population of the world will increase by about 2000 million (Fig. 1.1) and over half of these will live in Asia and be dependent on rice. By 2015 sufficient rice will need to be produced to provide the major staple for 4000 million people. That requires an increase in rice production of 50%, to maintain present nutritional standards. As there are many people in the major rice consuming countries living at sub-optimal

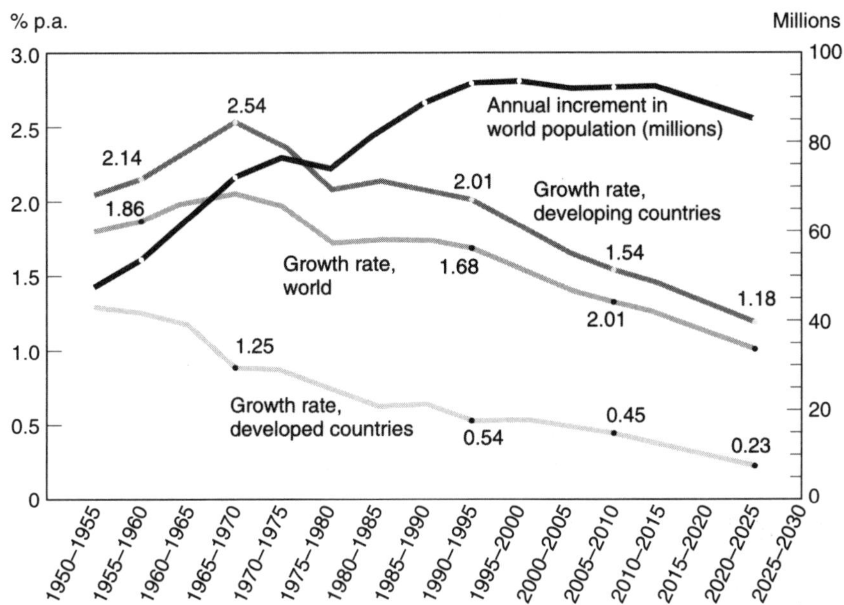

Fig. 1.1. Population grown rates and annual increments, 1930 to 1990 and projections to 2030 (Alexandratos, 1995).

nutritional levels the aim must be to increase rice production by up to 70% to raise nutritional levels to satisfactory standards.

Demand for rice from less advantaged areas in Asia, and from Africa where the coarse grains, sorghum and millet, and the starchy root crops such as taro and cassava, are currently the major sources of dietary calories, is certain to increase. Demand for rice has in fact been increasing rapidly in much of Africa and parts of Latin America (Table 1.1). Although demand is falling in some of the more affluent areas of Asia, as improvements in living standards enable diets to become more varied, IRRI (1993a) projects that the net effect will be an increase in average per caput consumption of milled rice from 56.4 kg per caput in 1990 to 58.6 kg per caput by 2000. The demand for rice in Asia in 1980 was 368 million tons (Mt) and was predicted to reach at least 494 Mt by the year 2000 (Barker *et al.*, 1985). Actual consumption in 1990 was already 405 Mt, and predicted demand in 2000 has been raised to 511 Mt for Asia and a global consumption of 560 Mt (Fig. 1.2). Crosson and Anderson (1992) in a careful and critical study of cereal requirements to 2030 also found that the global need for rice will continue to increase at over 2% per year until about 2005, but would then decrease gradually to a rate of 1.5% by about 2030. IRRI's projection of global demand in 2025 is 758 Mt.

The figures used above for population change over the next 20 years are the medium variant projection of the UN (UN, 1991). A little less than half of

Table 1.1. Changes in consumption of milled rice (kg) per person per year.

	1988	2000	(%)
Developing countries	70.4	70.8	+1
Africa	13.9	16.2	+16
Latin America	26.1	29.1	+11
Middle East	20.4	20.1	–1
Oceania	4.5	6.5	+45
Far East	92.9	95.0	+2
Bangladesh	133.9	130.3	–3
China	92.8	97.8	+5
India	72.8	74.9	+3
Indonesia	143.6	145.0	+1
Korea (South)	93.7	84.8	–10
Myanmar	188.8	192.2	+2
Pakistan	17.0	14.0	–8
Thailand	137.2	135.5	–2
Vietnam	155.2	165.1	+6
Developed countries	12.3	12.9	+4
North America	7.7	12.4	+61
Europe	4.0	4.9	+22
European Union	4.3	5.6	+30
Japan	70.6	59.8	–19
Ex-USSR	7.5	7.9	+8
World	56.4	58.6	+4

Source: IRRI (1994).

the global population is currently dependent on rice as their major staple food. An increase in the proportion dependent on rice is expected to arise not only because of the shifts in dietary preferences, but also because over half of the growth in the global population will be in Asia. The proportion of the additional population of two billion who will be dependent on rice is likely to be as high as 60%. These estimates are necessarily rather crude as it is difficult to draw a line between a person or family who are rice dependent and a family whose diet is becoming more varied but which still includes rice as the main cereal. However crude the figures, what is certain is that there will be a substantial increase in the demand for rice which will continue for at least the next 30 years.

The world population did not reach one billion until about 1825. In the past 10 years (1985–1995) it has added only a little less than one billion. Birth rates are declining, but a population momentum has been created due to the high number of women of child-bearing age living at present, and the decreasing death rate. Thus in spite of the reduction in the birth rate, the annual additions to the world population will only be a little below 100 million for the next 20 years (Fig. 1.1). In the coming decade the addition may well exceed one

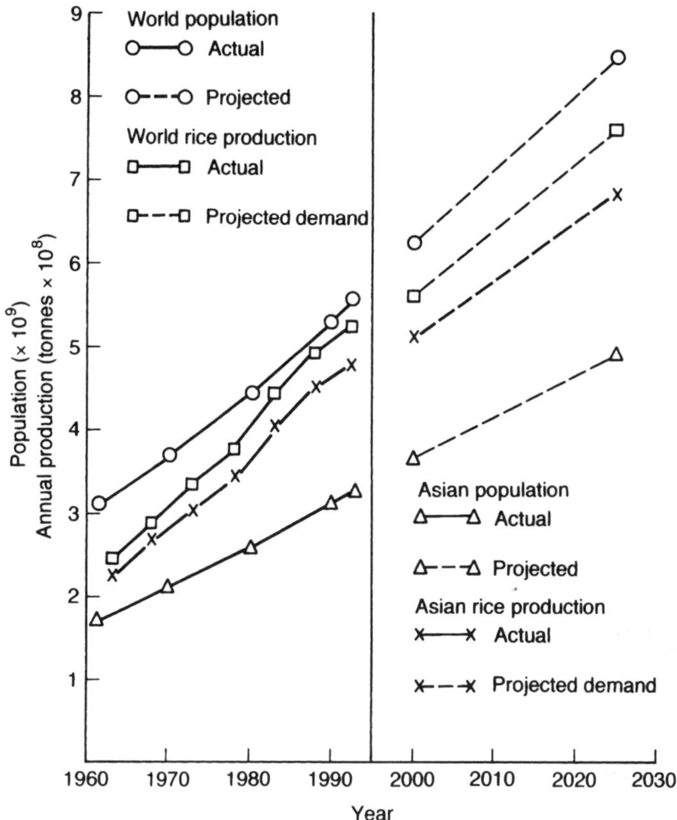

Fig. 1.2. World and Asian population and rice production, actual and projected; the projections for rice are of demand, not production. The production figures are annual averages, for 5-year periods, 1961–1990, and the annual average, 1991/93. (Data from IRRI, 1993a, 1995c; FAO)

billion (Graham-Smith, 1994). It has often been argued that given the continuing demand on resources it is more desirable to restrict population growth than to struggle to feed and meet the other requirements of a growing population. There should be little argument about that. The problem lies in the population momentum which has already been created. The rate of increase is not predicted to fall until 2015 to 2025; the increase in that decade will still be of the order of 800 million. A numerically stable population is predicted to be reached about the middle of the coming century, when the total population of the world will exceed 8 billion, and could well be as much as 11 billion, or double its present level.

Failure to meet the food demands of this growing population will result in misery for hundreds of millions, and almost certainly provoke continuing conflicts, such as those occurring now in parts of Africa where the ability to create either income or food for a growing population has already been failing. There needs to be a momentum in the production of rice and other foodstuffs to match the population momentum. Is there a realistic possibility of doing so, as happened in the past, or are we at last on the brink of the Malthusian precipice?

The question was raised at the time of the Bengali famines in the 1940s and 1950s, and much earlier concerned many others besides the Reverend Thomas Malthus. The Jesuit priests who informed Malthus about the population of China recognized not only the effects of 'pestilence and famine' on the size of the population but also the ingenuity which had led to the development of the irrigation systems responsible then for nurturing the rice crop to support the already considerable population. At the time of the Bengali famines half a century ago it was widely feared that they were the forerunners of many others. Fortunately, continuing scientific and engineering ingenuity have so far averted further major disasters in Asia.

Malthus (1798) expressed his concern about the relationship between food and population shortly before the global population had reached one billion. The population of China alone passed the one billion mark in 1980. Malthus gave an estimate of the population of China around 1790 of 333 million. This agrees remarkably well with the figures for the population of China between 1780 and 1790 given by Ho Ping-ti (1959). The data compiled from various official records and carefully reviewed by Ho Ping-ti reveal some catastrophic changes in the population in the 18th century of the order of 30–40 million. The disasters which caused the population to fall may well have encouraged Hung Liang-chi, described by Ho Ping-ti as 'the Chinese Malthus', to write his two essays, 'Reign of Peace' and 'Livelihood', setting forth a similar gloomy prognostication regarding the constraints to growth of the population of China. According to Ho Ping-ti, Hung's essays were published in 1793. The reference he gives is 'Chuan-shih-ko wen-chi (SPTK ed.), Series A, pp. 8a–10b'. The publication date is five years earlier than the appearance of the first version of Malthus' 'Essay on the Principles of Population'. The first version of Malthus' essay was published anonymously; only in 1803 when the second version appeared was Malthus' name attached to it. The sequence of disasters in China may have coloured not only the views of the Jesuit priests from whose reports Malthus derived his figures, but also Malthus' general views regarding the struggle between population growth and the causes of population decline.

As well as praising the 'persevering industry of the Chinese, in manuring, cultivating, and watering their lands', and noting that the land is well endowed with excellent soils and has 'a great number of lakes, rivers, brooks and canals wherewith the country is watered', Malthus quotes letters between Jesuit priests working in China in the 17th century. These letters show clearly that rice production in some parts of China was then insufficient to maintain the existing population. The letters describe graphically efforts to keep the numbers

Table 1.2. Calamities in Hubei province, China, 1650–1900.

Years	Famine	Drought	Floods, typhoons, etc.	Epidemic	Locusts, pests, etc.	Total counties affected in period	Average number counties affected per year
1650–1700	54	48	174	6	8	290	5.8
1700–1750	13	49	223	16	5	306	6.1
1750–1800	26	112	159	1	2	300	6.0
1800–1850	45	67	212	19	14	357	7.1
1850–1900	23	48	206	11	8	296	5.9

Source: Ho Ping-ti (1959).

of people in check, as well as the effects of wars, hurricanes, famines due to floods and droughts, and periodic locust and pest epidemics.

Ho Ping-ti (1959) lists the number of counties in Hubei Province affected by various calamities for the years from 1644 to 1911. Somewhere in the Province there was a calamity in at least one county in 237 years of the 267 between 1644 and 1911. On average between 1650 and 1900 there were six to seven counties affected by one sort of calamity or another every year (Table 1.2). The number and frequency of the calamities in Hubei in the central Yangzi Valley reveal how precarious life in China must have been throughout that period. Many of the problems continue, and help to explain the determination of China to proceed with the giant 'Three Gorges Dam Project' to control the waters of the Yangzi.

In addition to the natural disasters there were many conflicts and local wars causing death and destruction, notably in the second quarter of the 17th century and between 1840 and 1950. Ho Ping-ti (1959) describes many of these and records the large number of deaths that occurred. The population of Hubei is given as 34 million in 1850, and 28 million in 1953. The pressure of population in the Yangzi Valley had a major influence on the rebellion which grew to become the Taiping War, often considered the worst civil war that has ever occurred. Between 1850 and 1867 more than 30 million died. Both sides indulged in an extermination policy and destroyed much of the infrastructure created to support rice production (Ho Ping-ti, 1959). Further death and destruction occurred as the Chinese Republic was created, followed by the war with Japan and the establishment of Communist rule.

Ho Ping-ti quotes a diary written in 1855 which graphically depicts the relation between overpopulation and land pressure which helped to provoke the Taiping rebellion.

> The harm of overpopulation is that people are forced to plant cereals on mountain tops, and to reclaim sandbanks and islets. All the ancient forest of Szechwan has been cut down and the virgin timberland of the aboriginal regions turned into farmland. Yet there is still not enough for everybody. This proves that the resources of Heaven and Earth are exhausted (Ho Ping-ti, 1959, p.274.)

The remedies proposed by the writer of the diary were infanticide, at least of female children, compulsory sterilization, and heavy taxation of families having more than one child.

The slow increase in the Chinese population from 300 million to 550 million during the period of intense and bloody civil wars between 1800 and 1950 partly fulfilled the more gloomy prognostications of Malthus and Hung. The extraordinarily rapid expansion from 550 million to 1200 million between 1950 and 1995 contradicts their hypothesis. Although available data for population and rice production in China between 1900 and 1965 are rather unreliable, those available indicate that the rate of growth in rice production in China was below the rate of growth of the population until 1965. The growth rates were then closely matched until 1982, when the rice production rate passed the population rate. The improvement has depended not only on a degree of political stability and strong government, but also on increases in the food supply sufficient to match the growth of the population (Fig. 1.3). Without continuing increases in agricultural production the population increase certainly could not have been sustained. Since 1975 the government has curtailed the rate of growth of population by implementation of strict controls on family numbers, so that it is now below 1% per year.

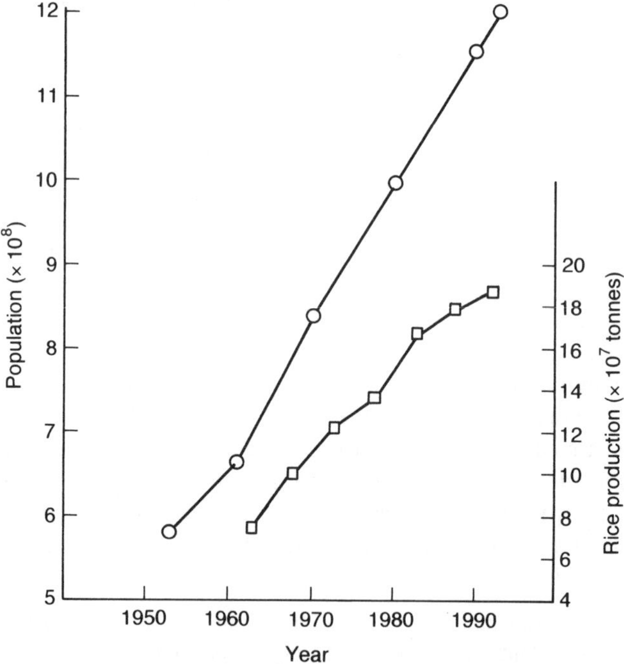

Fig. 1.3. China: population (○) and rice production (□), 1950–1993. Production figures are averages for 5-year periods, except 1991/93, which is an average of 3 years. (Data from IRRI, 1995c and Yu and Buckwell, 1991).

The increases in rice production which were needed to match population growth arose partly from the expansion of the rice area, partly from the continuing efforts to manage floodwaters and develop controlled irrigation systems, and from the improvements and discoveries in rice production methods. As rice production has reached levels where the majority of the population have sufficient to meet immediate needs more emphasis has been placed on the economics of rice production, and an increasing area has been transferred from production of the very high yielding hybrid rice varieties to lower yielding japonica varieties, which command a stronger market price.

The historical records of population and rice production before 1850 are more complete for China than for any other Asian country. Nevertheless it is well known that population and rice production have also grown rapidly in most other countries in tropical and subtropical Asia. Until about 1950 population numbers and rice production increased roughly in parallel with each other. Between 1950 and 1985 the rate of increase in production exceeded that of population growth (Fig. 1.2) but since 1985 there has been a fall in the rate at which production has been growing (Table 1.3).

Present trends in population growth rates indicate that outside of China the rate of decline will be much slower. The high rates of increase in production achieved between 1970 and 1990 need to be restored if increasing malnutrition and undernutrition are to be avoided, and the price of rice kept to a level that the urban poor can afford. The question that this book addresses is whether the present high levels of rice yields and production can be sustained and further increased, so that sufficient is produced to meet the demands of an increasing population. The projections in Fig. 1.2 are of increases in demand, not production. As production has generally kept pace with demand in the past, many economists have assumed that demand can be used as a proxy for production. It is in fact an open question as to whether the increases in production can continue to match demand, even in the short term.

1.4 Increasing Rice Production

The increases in rice production in the past three decades have been almost entirely due to increased yields; land area used to produce rice has increased only very slightly (Fig. 1.4). Much has been made of the importance of the contribution of the scientific community to the so-called 'green revolution' in rice which led to the higher yields and the development and popularization of the high-yielding, semi-dwarf tropical rice varieties. The contribution of the new varieties to the ability of Asia to feed itself during the past 30 years of rapid population growth has certainly been important, but it has been as part of a much wider effort involving many others, scientists and non-scientists (IRRI, 1985a).

Table 1.3. Rice: annual growth rates of area, production and yield, as per cent of the mean for the periods between the triennia indicated, for Asia, China and India, and other South Asian countries (Bangladesh, Nepal, Pakistan and Sri Lanka) and Southeast Asian countries (Cambodia, Indonesia, Laos, Malaysia, Myanmar, Philippines, Thailand and Vietnam).

Regions/ countries	1957/59–1965/67	1965/67–1973/75	1973/75–1981/83	1981/83–1985/87	1985/87–1991/93
Asia					
Area	0.85	1.09	0.24	0.19	0.38
Production	2.60	3.37	3.09	2.25	1.70
Yield	1.74	2.27	2.86	2.06	1.32
China					
Area	–0.58	2.25	–1.07	–0.90	–0.15
Production	2.62	3.92	2.98	2.00	0.90
Yield	3.21	1.68	4.06	3.10	1.05
India					
Area	1.21	0.74	0.46	0.25	0.40
Production	1.95	2.90	2.02	3.24	2.62
Yield	0.74	2.15	1.57	3.00	2.22
Other South Asia					
Area	1.26	0.61	0.88	0.34	0.00
Production	3.13	1.63	2.57	1.48	2.70
Yield	1.89	1.02	1.71	1.15	2.70
Southeast Asia					
Area	1.73	0.35	1.51	0.73	1.12
Production	3.17	3.29	4.29	2.80	2.12
Yield	1.46	2.94	2.83	2.06	1.00

Source: IRRI (1995c).

The story of the development of the high-yielding tropical rice varieties at the International Rice Research Institute (IRRI) has been well told by its first director, Robert Chandler (1985). The breeding of the first of the tropical high-yielding, semi-dwarf rice varieties, IR8, (Fig. 1.5) was due to the combined efforts of an international team of scientists recruited by Chandler. Most directly involved were a Chinese geneticist (T.T. Chang), a Japanese physiologist (Akiro Tanaka) and from the United States of America a plant breeder (Hank Beachell) and plant pathologist (Peter Jennings).

IRRI was established in the Philippines in 1960 as part of the response of the world scientific community to the threat of famine in Asia. Its fame was established first by the interest in and release of IR8 by the Philippine Government and then by the courageous decision taken in India to allow the entry of large quantities of seed of the high-yielding semi-dwarfs for immediate distribution to farmers, on the basis of the results obtained in multilocation field trials conducted by the Indian Council of Agricultural Research. The varieties were

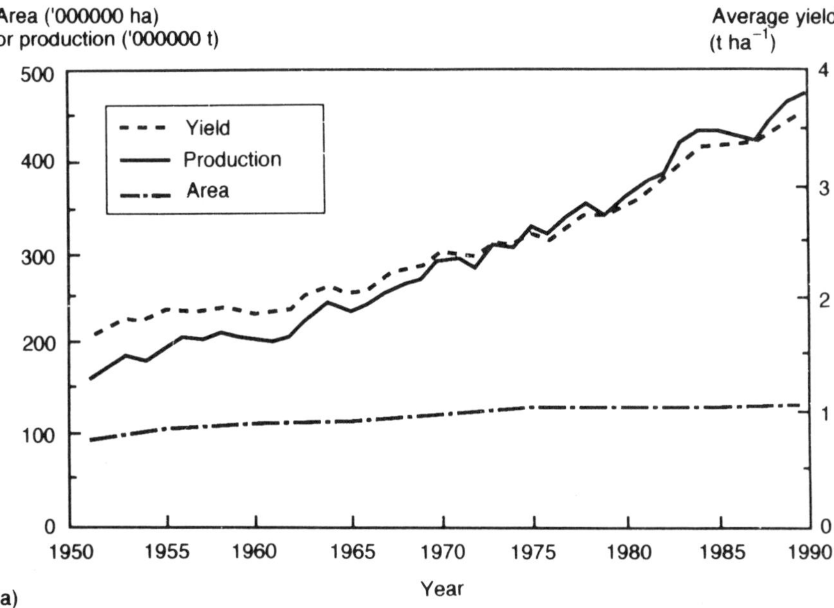

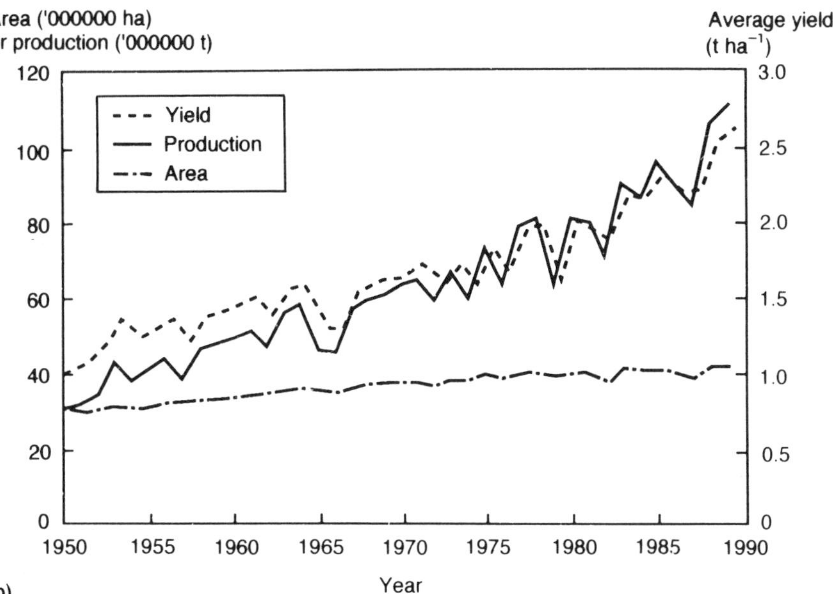

Fig. 1.4. Historical rice production statistics showing the area sown, average yields and annual production. (a) Asia, (b) India. Continued on pp.14-15

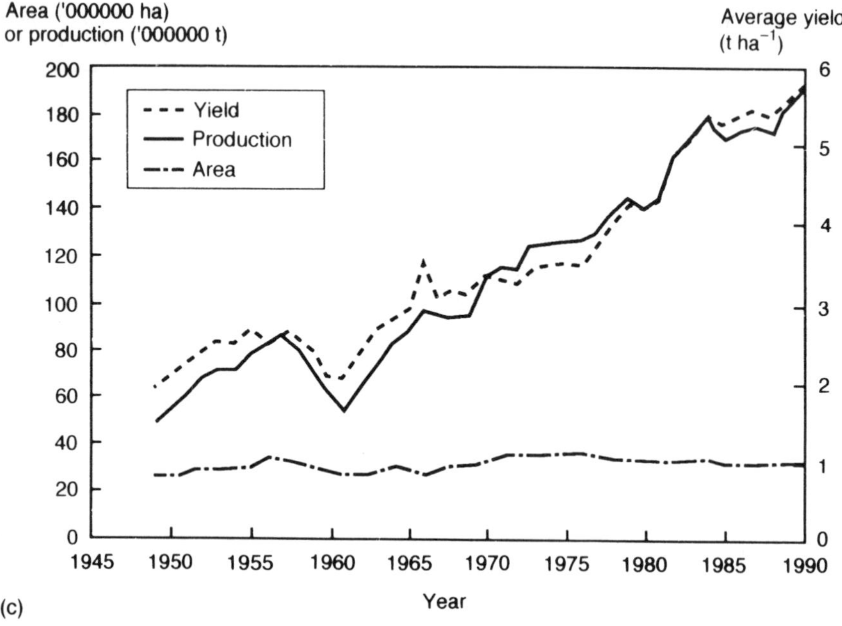

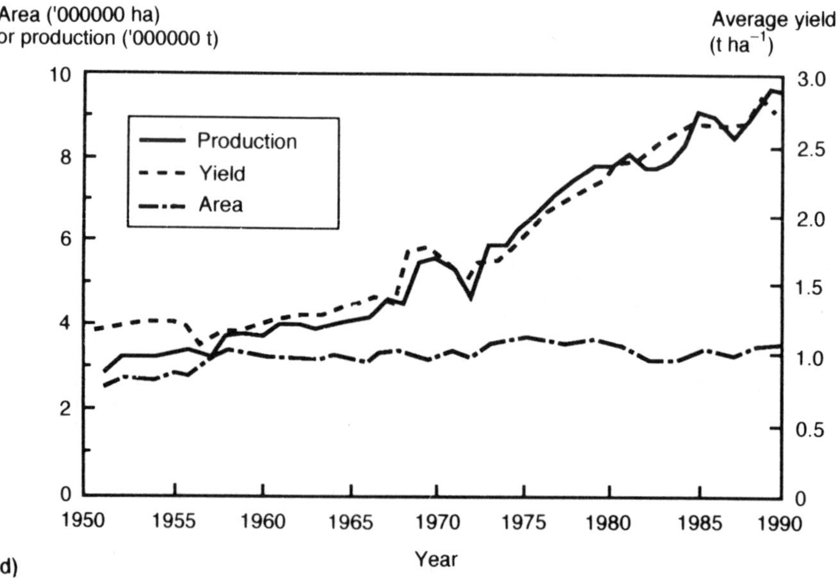

Figure 1.4. *continued* (c) China, (d) the Philippines.

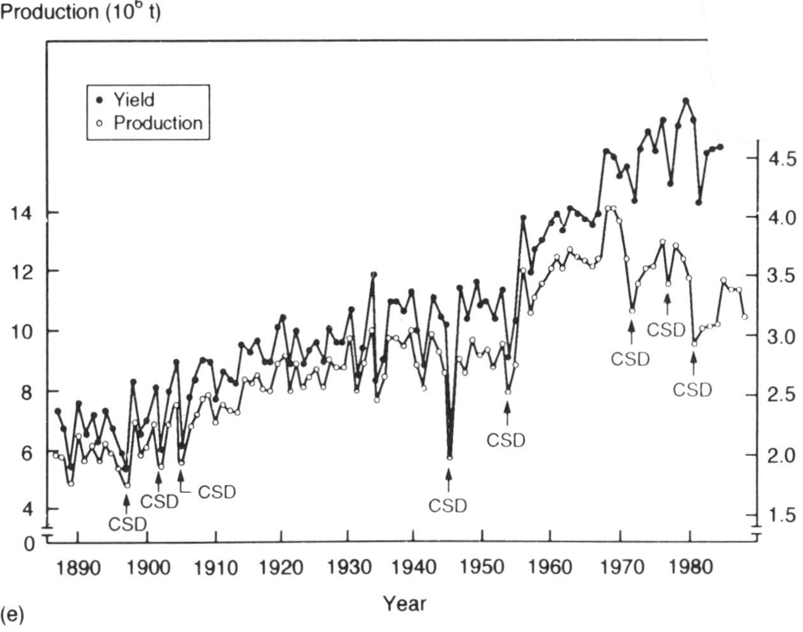

Figure 1.4. *continued* (e) Japan. CSD, cool summer damage.

Fig. 1.5. Peta, a traditional tall variety, and IR8, the first of the widely-used high-yielding, semi-dwarf tropical rice varieties.

an immediate success and undoubtedly had an effect in increasing food supplies in India at a critical time. The Director of the Indian Agricultural Research Institute who organized the rapid testing of the high-yielding varieties of rice and wheat in farmers' fields was M.S. Swaminathan, who later became Director General of IRRI.

The National Demonstration Trials organized in the fields of small farmers were the source of the great popularity which the high-yielding varieties gained in a very short space of time. Their release undoubtedly did much to stimulate rice research and extension, not least by increasing the confidence of Asian governments in the value of research. The new varieties enabled the advantages of improved irrigation availability and greater fertilizer use to be realized, and made possible the rapid increases in rice production which occurred between 1960 and 1985 (Fig. 1.2). Increasing production led to the declining trend in the world price of rice (Fig. 1.6).

The origin of the semi-dwarf varieties appears to have been a natural mutant discovered in south-central China about 1900. Seeds of the mutant were taken to Taiwan and used in breeding programmes from which the variety Dee-geo-woo-gen was produced, and then the high-yielding semi-dwarf Taichung Native 1 (Barker *et al.*, 1985). High-yielding japonica varieties were produced in Japan also early in this century, probably from the same mutant. The less leafy semi-dwarf plant type with large panicles produced grain rather than excess leaf and straw when grown under favourable conditions. As the price of fertilizers relative to milled rice fell from over three to less

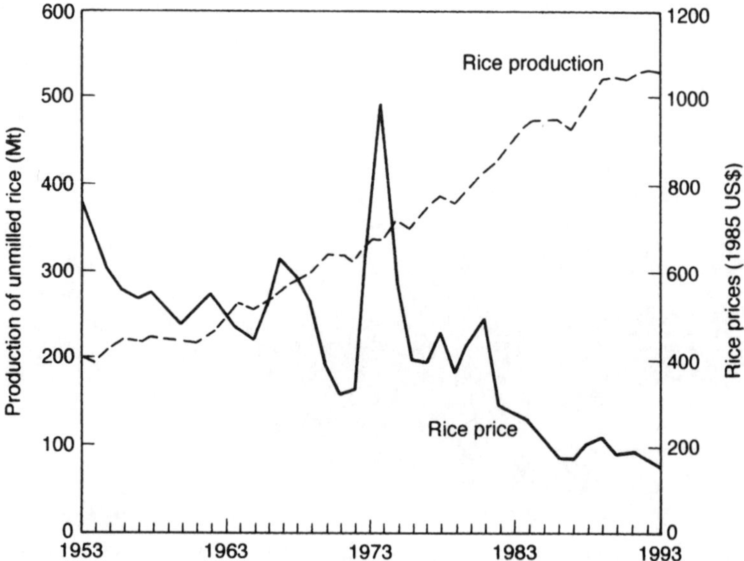

Fig. 1.6. Trends in world rice production and price, 1953–1993 (from IRRI World Rice Statistics Database).

than one during the 1930s it became advantageous to exploit the potential of these new varieties (Hayami and Otsuka, 1994). Similar conditions prevailed elsewhere in Asia during the 1960s and 70s, enabling the green revolution to occur.

Herdt and Capule (1983) analysed the relative contributions of plant characteristics ('modern varieties'), irrigation, and 'fertilizers' to rice production in several Asian countries in the 1960s and 70s (Fig. 1.7). They used national data for rice yields and production, area sown to modern varieties and changes in irrigated area and fertilizer use as factors in a modified production function. They found a similar contribution to be made by each factor, with a considerable residual which they interpreted as the interaction between the factors.

Green revolution technology is generally considered to include chemical control of pests as well as use of fertilizers and modern varieties. Chemical insecticides were necessary to protect some of the first high-yielding semi-dwarf varieties against insect pests, but the recent varieties, which have been much more widely grown than IR8 and other early semi-dwarf varieties, have multigenic resistance to most of the major pests of rice. This stable resistance has enabled varieties such as IR36 and more recently IR64 and IR72 to be extremely widely grown, and in the case of IR36 already for 20 years, without serious widespread pest outbreaks having occurred. The scientist primarily responsible for the breeding of the high-yielding and pest-resistant varieties was Dr Ghurdev Khush of IRRI.

Where irrigation water is available, and its delivery to and removal from the rice paddy can be controlled, where high-yielding varieties are grown,

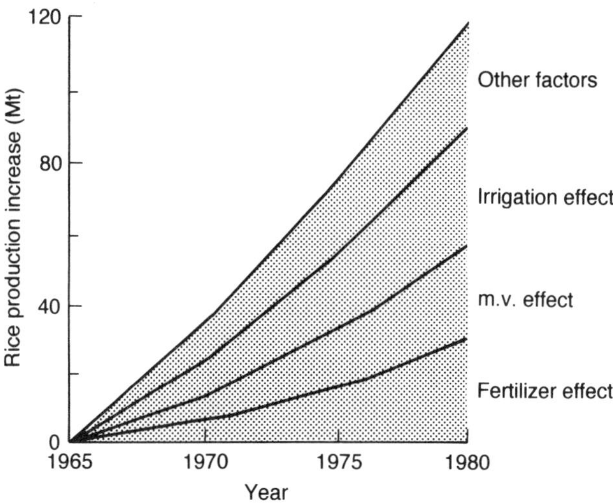

Fig. 1.7. Estimated contribution of fertilizers, modern varieties (m.v.), irrigation and other factors (including interactions and additional land factors) to rice production increases in eight Asian countries from 1965 to 1980 (Herdt and Capule, 1983).

and adequately fertilized, and where pests – insects, diseases and weeds – are controlled, much higher yields of rice can be obtained than in the past. Yields of 10–15 t ha^{-1} for a single crop are possible and sometimes achieved in dry season production in several tropical and subtropical countries. National average yields in China have recently approached 6 t ha^{-1}, as they have been for the irrigated crop in Indonesia. In Australia and California average yields have been over 8 t, in Egypt they reached 7.7 t in 1993, and the Province of Ninxi in China achieved an average yield of 9.18 t ha^{-1} in 1991.

Yoshida (1983) quotes maximum yields for a single crop of unhusked paddy rice of 17.8 t ha^{-1} recorded in India in 1974, and for Japan of 13.2 t ha^{-1}. In southern China and other areas where double and triple cropping of rice is practised the annual yield from two or three crops has often exceeded 20 t. De Datta (1970) reported a yield from Mindanao in the Philippines from three crops grown in one year of 23.6 t ha^{-1}, and Yoshida *et al.* (1972) of 25.7 t ha^{-1} from four crops grown in 335 days. At IRRI the continuous rice garden production system yielded over 25 t ha^{-1} $year^{-1}$ for 3 years (IRRI, 1979a, 1981a) but the system was abandoned when it was found difficult to sustain this yield level.

1.5 Potential Productivity of Rice and the Yield Gaps

Theoretical yield limits determined by radiation intensity, temperature and carbon dioxide content of the atmosphere may be calculated assuming no limitations to yield due to soil factors or pests. The present levels of efficiency attained by plants in converting radiation into the carbon compounds which constitute plant material could be improved but as yet progress in this area is slow (Coombs, 1984; Evans, 1990; Horton, 1994). At the present time photosynthetic efficiency rates of 2.5–3% have to be accepted as realistic. On this basis and using current atmospheric carbon dioxide concentrations, mean daily temperatures, and assuming no other external limitations, potential yields for rice at IRRI in the Philippines were calculated in 1982 to be a little more than 8 t ha^{-1} in the wet season and 12.8 t ha^{-1} in the dry season (Fig. 1.8).

Evans (1993) notes that estimators of potential yields have to sail between the Scylla of being conservative in their assumptions so as not to raise false hopes among policy makers and global planners, and the Charybdis of being perceptive enough not to be trapped by received wisdom and unreal limitations. IRRI in its planning for its third decade (1985–1995) preferred to sail closer to the Scylla of not wanting to raise false hopes, than the Charybdis of not being trapped by current perceptions. However Yoshida and Oka (1982) and Yoshida (1983) in discussing rice yield potential noted that if the plant type could be altered to raise the efficiency with which incoming radiation is used to 3.5%, then the yield potentials became 9.5 t ha^{-1} in the wet season and

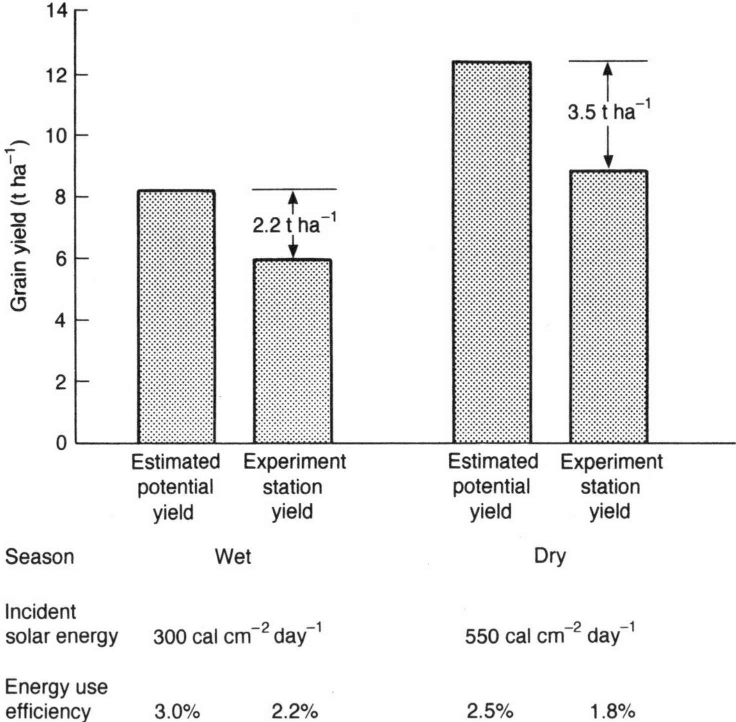

Fig. 1.8. The yield gap between potential and actual (1981) experiment station rice yields (IRRI, 1982c).

15.9 t ha^{-1} in the dry season. Shen Lian (1990) used a photosynthetic efficiency rate of 3.3% to arrive at a potential maximum yield of 14.4 t ha^{-1} for the first season rice crop in Taiwan, and Kropff *et al.* (1994) have produced higher estimates of potential yield at IRRI.

The maximum yield recorded at IRRI is 11.0 t ha^{-1} (Yoshida, 1983) in the dry season, and on several occasions close to 8 t ha^{-1} in the wet season. The highest yields attained in recent years at IRRI have been 10.7 t ha^{-1} for a hybrid rice variety, and 9.6 t ha^{-1} for IR72 (Cassman, 1994; Virmani, 1994; Khush, 1995). Thus the potential yields used by IRRI in planning for its third decade appear to have been realistic. IRRI as it enters its fourth decade has made a more rigorous analysis of the problem of adapting the plant type so that the potential of 15 t ha^{-1} under Philippine conditions may become a reality (IRRI, 1993b; Peng *et al.*, 1994; Setter *et al.* 1994).

There are two other yield gaps which are important as well as the gap between ultimate potential yield and that currently attained on the research station: (i) that between present farm yield levels and those attainable on favoured sites using established best practice; and (ii) that between the

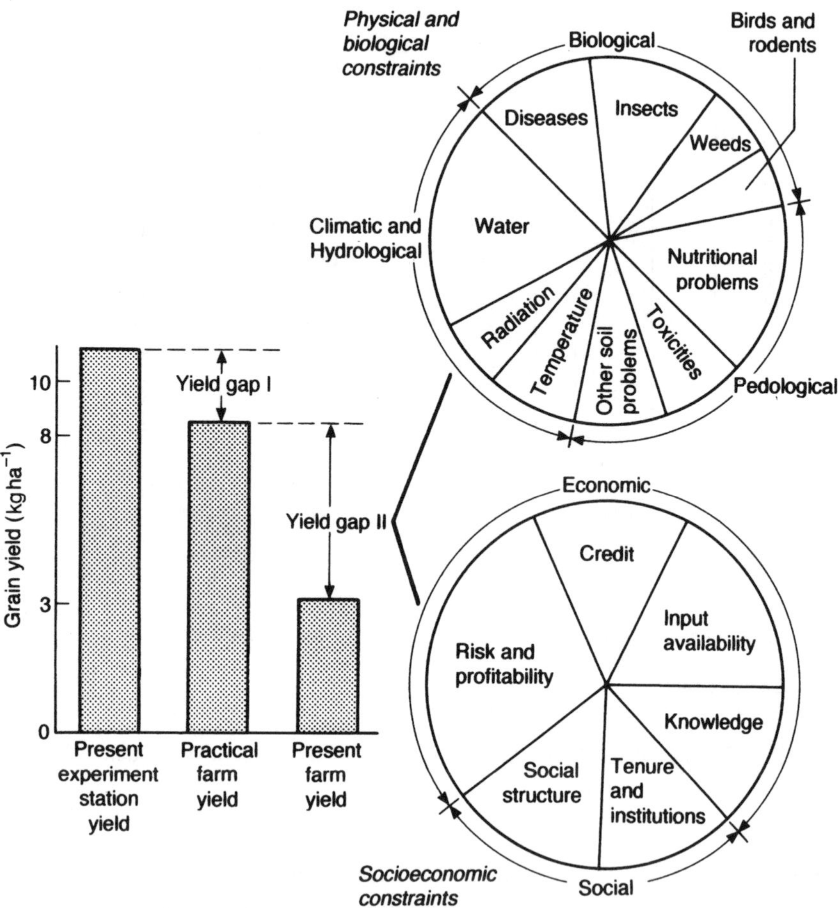

Fig. 1.9. Factors contributing to the gap between actual (1981) and practical farm yields (from IRRI, 1982c).

established best practice yields and those attained at experiment stations. The various constraints which give rise to the gaps are both biophysical and socio-economic. The pie-charts in Fig. 1.9 indicate the relative importance of the various constraints in the Philippines around 1980.

Studies of yields attained by rice farmers in Laguna Province of the Philippines, and in Nueva Ecija where the Philippine Rice Research Institute (Philrice) has been established, have been continued for many years (Pingali, 1994). They show that the yields attained by the top one-third of farmers now exceed those of the 'Highest Yielding Entries' (HYE) in the trials in both provinces (Fig. 1.10). They further show a continuing decline in the yields of the HYE. Thus not only has the yield gap between best farmers and best

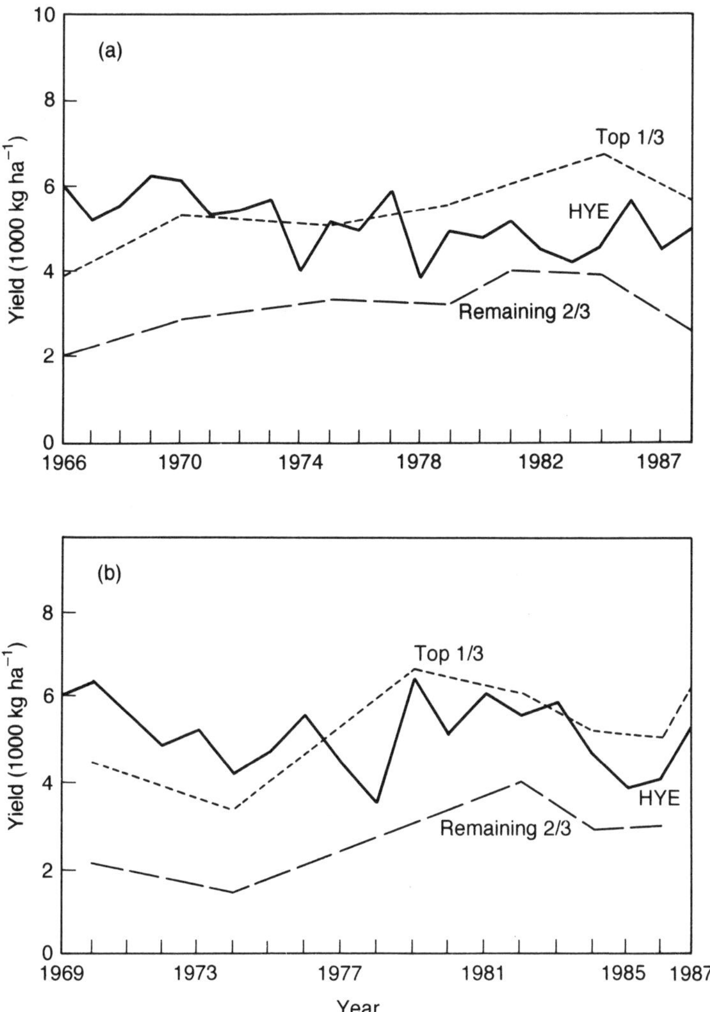

Fig. 1.10. Trends in yields obtained by top 1/3 of rice farmers and other farmers, compared with highest yielding entry (HYE) of rice varieties grown at experiment stations (a) in Laguna province (b) in Nueva Ecija province, the Philippines (Pingali, 1994).

practice been closed, but there must also be concern about the failure to sustain yields of the highest yielding entries.

Hybrid rice which has been important to the large yield increases in China may offer one avenue by which the yield plateau can be raised (IRRI, 1989; Virmani, 1994). Redesign of the rice plant to produce a plant with fewer tillers but more grains per panicle (Khush, 1990, 1995) is another. A further improved plant type should help to bring research station yields closer to climatically determined potential yields, and so reopen a gap between the yields of better farmers and those that researchers can achieve. Until there is a new yield plateau established it is difficult to see how the demand for a 50–70% increase in global rice production by 2015 will be satisfied (Pingali and Rosegrant, 1993; Cassman, 1994; IRRI, 1995a).

While there is promise for the future there is no doubt about the current slowing down of the momentum in production, not only of rice but of other food crops. Several recent publications have drawn attention to this problem (Crosson and Anderson, 1992; FAO, 1993; Pinstrup-Andersen, 1993; Plucknett, 1993; Tribe, 1994). There has been concern not only about yield levels and production potential, but also about environmental problems and the sustainability of the methods used to attain current yields. Environmental problems include pollution of water supplies by chemicals used to boost agricultural production, degradation of water supply systems due to soil erosion which causes siltation of reservoirs and canal systems, and the loss of biodiversity associated with the widening spread of intensive farming methods.

There must also be concern about the economic aspects of sustainability, particularly declining factor productivity. As yields increase the response obtained per unit of input tends to decrease (Cassman and Pingali, 1995a). This can be seen in declining responses obtained to successively higher rates of addition of nitrogen fertilizers used at any one site, and in the increasing costs of water as retention and distribution systems have to be constructed and developed in more difficult areas.

Declining factor productivity means that the costs of rice production increase. Rice production can then only be sustained if consumers have alternative sources of income and can afford to pay the higher prices. For the past three decades green revolution technology has led to lower costs of rice production as yields of rice have increased (Barker *et al.*, 1985). The price of rice to the consumer and on the world market has fallen steadily, apart from the 'blip' caused by the oil crisis of 1978 (Fig. 1.6). However since 1990 this trend has slowed, reflecting declining factor productivity. Without changes in technology which will enable rice to be produced more cheaply, or reductions in demand, it is likely that the trend will be reversed, and real prices will increase.

In the following chapters an attempt is made to place the sustainability of rice farming in historical perspective, and to examine more fully the biophysical basis for the sustainability of rice farming in Asia over the past several thousand years. Social and economic factors related to the past sustainability of rice farming are also considered, before conclusions are drawn about the future. The great Asian civilizations of the past were supported by rice farming systems which yielded 1 or 2 t $ha^{-1}$of rice. Can the present rapid development of Asian societies be sustained by a system in which rice yields will need to approach and often exceed 10 t ha^{-1}?

The Origins and History of Rice Farming

2

2.1 The Wild Species of Rice

Many millions of years ago the genus from which rice developed must have evolved on the great land mass of Gondwanaland (Chang, 1976a, b; 1985). When it divided to form the continents that became Asia, Australia, Africa, America and Antarctica most of the progenitors were concentrated in the area of the southeastern Himalayas, with scattered relatives in West Africa, northern Australia and New Guinea and Central and South America. Between 10 thousand and 20 thousand years ago, and perhaps earlier, humans started to gather the seed of the wild species as food. This was followed by domestication and planting, initially in wetlands, and later by clearing and cultivating fields where rainfall was retained by bunding, and later still flooded by diverting water to them from streams and rivers.

Wild rices continue to be harvested and used as food in several parts of the world (Higham, 1996). In Asia the practice is found for instance in the Jeypore tract of Orissa State in India and in Batticoloa District of Sri Lanka (Chang, 1976b). There are 19th century reports from West Africa of the collection of wild rice, probably *Oryza barthii*, for use as as a foodstuff (Harlan, 1989) and in some localities the indigenous *O. glaberrima* continues to be harvested as a wild species, in parallel with the use of domesticated varieties. Bancroft (1884) first recorded that the aboriginal people of Australia included wild rice as part of their food supply. More recently Higham (1996) has quoted other references to the use of the Australian wild rice *O. meridionalis*, as food by the Gidjingarli and Anbarra people of Arnhem Land in northern Australia. In North America *Zizania palustris*, a distant relative of *Oryza*, is now cultivated and sold as 'wild rice', but has been used as a foodstuff by the Ojibway and Menomini people of Wisconsin for many years (Higham, 1996). Grain from the wild floating rice varieties of *Oryza* spp. are harvested in the same way as in Cambodia and other parts of Asia, by using boats to visit the wild stands, and shaking the seed into the boat (Delvert, 1961).

The importance of the wild rices in relation to the sustainability of rice farming is considerable. They are an essential source of genes for diversifying the characteristics of *O. sativa* to sustain its productivity under new physical and biological stresses (Vaughan and Stich, 1991). The outstanding example is the use of *O. nivara* to confer resistance to the grassy stunt virus, but wild species can be valuable in conferring resistance not only to diseases and insect pests but also to biophysical stresses (Khush, 1977; Chang *et al.*, 1982; Shao *et al.*, 1986; Jackson, 1995).

2.2 The Domestication of Rice

Before cultivation of rice, when the land is worked to prepare a seedbed and the seed is sown, wild rice was domesticated by selection and planting of the wild species in favoured places from which other plants had been removed. It is still not clear exactly when and where this process first occurred and *O. sativa* first evolved. Carbonized rice grains carbon-dated to 4000–8500 years before the present time (BP) have been recovered from various archaeological sites in China, India and Southeast Asia (Fig. 2.1). These dates establish the time when species of rice were starting to be used as food, the initial step in the domestication of the rice plant. *O. sativa* with its many varieties evolved in the course of domestication. Some of the early specimens of rice belong to this species, but many are wild species, or at least closer to the wild species than to any known variety of *O. sativa*. To establish the date of the transition from the collection of seed of wild rices by people who were hunters and gatherers to the cultivation of *O. sativa* it is necessary to be able to distinguish the remains of the wild rices from those of *O. sativa*. This is not simple. It has mostly been done by comparison of the striations and reticulations on the surfaces of lemmas and paleas (Buth and Saraswath, 1972; Chang, 1976a) but Thompson (1992, 1995, 1996) has shown that glume surface patterning can be unreliable. She prefers to use abscission scars, supported by comparisons of the size and morphology of rice remains.

The location of early sites with rice remains, and the analysis of the physical evidence regarding the morphology of the remains, suggest that in Asia there was a transition from gathering of wild rice species in wetlands adjacent to rivers and marshland, to selection of seed and deliberate planting in favourable sites that were unlikely to be flooded too deeply, but which would remain water saturated (Higham, 1989). An intermediate stage may also have occurred, in which the wild rices were encouraged by clearing of other vegetation from where the wild rice was growing, as Evans (1993) has suggested for other domesticated species.

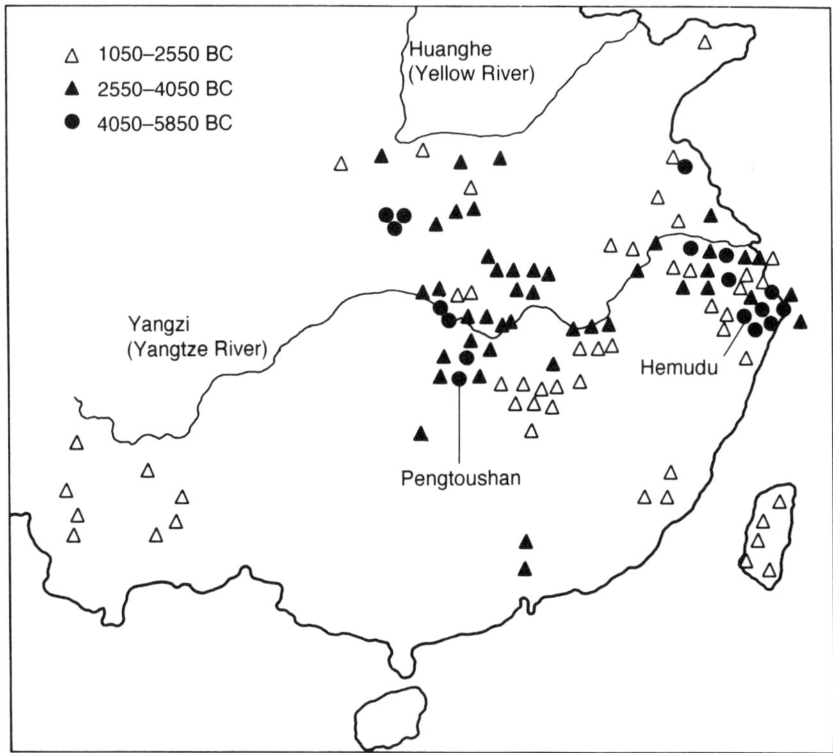

Fig. 2.1. Sites of early rice datings in China (Tang Shen-xiang, 1994). Note: some carbon datings of rice from Pengtoushan and Hemudu give dates earlier than 5850 BC.

2.3 From Domestication to Cultivation

To establish that the rice was deliberately planted rather than obtained from wild rice growing in the neighbourhood of settlements, other evidence has to be sought – from archaeological studies, by palynological analyses, and from references to rice cultivation and ceremonies which extol the importance of rice mentioned in early texts such as the Vedan (Solheim, 1972; Chang, 1976b; Higham, 1984, 1989; Maloney *et al.*, 1989; Maloney, 1991; Evans, 1993). Most evidence has come from China, mainland Southeast Asia and South Asia.

2.3.1 China

The oldest remains of what is believed to be cultivated rice have been found at Pengtoushan, Hunan Province in the central Yangzi valley (Yan, 1991; Tang Shen-xiang, 1994). The carbonized rice grains recovered have been carbon-dated to 7500–8500 years BP. It had previously been thought that the earliest 'true domestication' of rice occurred at Hemudu in the lower Yangzi region of southeast China, on the Ningshao Plain in Zhejiang Province (Chang, 1983).

Material from the Hemudu site certainly provides conclusive evidence of cultivation of *O. sativa*. The distribution and concentration of the relics within the soil and the characteristics of the soil profile and the soil properties indicate that the rice was grown in marshy places chosen so that they would not have been subjected to other than irregular and shallow flooding (Gong Zi-tong and Liu Liang-wu, 1994 and personal communication). Importantly, at Hemudu, rice husks, stalks and leaves were recovered in the same soil layers as the carbonized grains, together with 76 si-spades formed from animal bones with wooden handles attached. The spades were well polished indicating that they had been used over a long period. The carbon dates obtained from the rice husks are given as 6895 ± 180 BP and 6740 ± 130 BP (Liu Liang-wu, 1990, quoted by Gong Zitong and Liu Liang-wu, 1994) and the date of the grain as 7003 ± 117 BP (Chang, 1983). The site thus provides ample confirmation not only of domestication but also of the cultivation of rice.

The inhabitants of Hemudu cultivated their rice crop in ways which did not differ greatly from that in which much rainfed rice is grown today. In the floodplains of many major river systems, the marshy meadows (backswamps) behind the levees along the river banks become water saturated some weeks after the start of the rainy season. As the level of water in the river rises, the soil becomes soft and wet and easy to cultivate. Implements such as the si-spade would be quite adequate for working the wet soil. Rice seeds could be hand-sown into the cultivated soil. The farmers of Hemudu would have selected sites as far as possible free from flooding so that the young rice plants did not drown, but where the water saturation was sufficient to ensure little competition with the majority of weeds which prefer unsaturated soils.

The first rice cultivators must have come from the people who inhabited the Yangzi valley some 7000–9000 years ago. They were almost certainly responsible for the spread of cultivation to other parts of Asia, as population growth encouraged the expansion of human settlements, and as changing sea level and river meanderings, and oppression by Chinese warlords from the north forced further movement.

2.3.2 Southeast Asia

Higham (1989, p.86) has suggested that rice domestication may have occurred independently in mainland Southeast Asia through collection of wild rice, followed by encouragement of the favoured wild strains by removal of competing vegetation in the swampy areas where the wild species were found, and then the selection of seed and deliberate seeding of the wild species. Rice cultivation in naturally flooded areas without water control was certainly being practised in mainland Southeast Asia from at least 4000 years ago. At Khok Phanom Di (Higham, 1984; Maloney *et al.*, 1989; Thompson, 1995), about 70 km southeast of Bangkok, rice became part of the diet from about 4000 BP. It is not clear whether the rice was cultivated in the sense that the land was cleared and worked to prepare a seedbed and remove weeds, or whether the land was chosen as suitable and plants other than wild rice removed to allow the wild rice to develop. The perennial wild species, *O. rufipogon* (Chang, 1976a), or the species complex, termed by Oka and Morishima (1971) *O. perennis*, but which may have been *O. rufipogon*, were undoubtedly important progenitors of 'domesticated rice' (Morishima, 1986). Simple removal of competitors would have been adequate to obtain much more seed than collecting from plants subject to more severe competition and which had to regrow from seed each year.

The investigations at Khok Phanom Di (Maloney *et al.*, 1989; Thompson, 1992, 1995) show clearly in the record left by pollen deposits that the site was initially at the boundary between fresh and sea water, with mangrove vegetation, but that the vegetation later changed during the period of human occupation to freshwater vegetation which has persisted, perhaps for 4000 years until the present day. Maloney *et al.* (1989) found evidence that at Khok Phanom Di 7000 years ago burning of the vegetation was occurring. This may have been due to natural causes, to the work of hunter-gatherers, or to deliberate attempts to encourage graminaceous species, including wild rice. By 4000 to 3500 years ago the people of Kok Phanom Di were probably cultivating rice, as in addition to the pollen sequence there are abundant rice remains, together with hoes, reaping knives and ceramics.

The soil surface at Khok Phanom Di is now about 3 m above mean sea level and 12 m above the level of the rice fields in the area. The carbonized rice remains and artefacts used for dating all confirm occupation and the existence of a rice–fish culture from about 4000 BP, with the sea retreating from the site about 3600 BP. Careful analysis of the information available (Thompson, 1992, 1995) has failed to establish whether the retreat of the sea was associated with eustatic changes in sea level, a major flood, or general channel instability. Other evidence from East Asia indicates that mean sea level rose almost 4 m to about the present level between 4500 and 3500 BP, and then fell by 2.5 m between 3500 and 3000 BP (Yang Huai-jen and Xie Zhiren, 1984).

Although there is no record of any implements corresponding to the si-spades having been found in Southeast Asia, putative slate harvest knives have been found in the Thai–Myanmar border region and dated to 9000 BP. These may well have been used to collect panicles from wild species, but there is no evidence to show whether or not these included relatives of rice. The earliest convincing evidence for rice cultivation by sedentary village communities comes with the Phung Nguyen culture of the Red River Valley, dating from the fifth millennium BP, a date differing little from that of the start of cultivation at Khok Phanom Di (Higham, 1989; Maloney, 1991). Glover and Higham (1996) conclude that on present evidence rice cultivation in Southeast Asia was initiated some 4000 years later than in the middle Yangzi region, and 2000 years later than in the lower Yangzi region.

There is very little evidence to show when rice cultivation was initiated in insular Southeast Asia. Bellwood (1992, p.105) cites evidence that northern Luzon in the Philippines was settled by people from Taiwan about 4800 BP, and that these people brought with them their rice culture. Further movement to Java may have occurred about 4000 BP.

2.3.3 India

There were centres in South as well as Southeast Asia where rice became a major item in the diet around five millennia ago. Again, it has still to be finally resolved whether domestication of rice spread to these areas from southern China and Southeast Asia or arose independently. Much evidence has been presented which may indicate an early origin of rice farming in these areas (Vishnu-Mittre, 1974; Sharma *et al.*, 1980). Archaeological evidence from three sites in Uttar Pradesh near the Ganges and Belan Rivers is believed to indicate that rice farming there may have started as long ago as at Hemudu. Sharma *et al.* state (p.25) that the evidence reveals 'selective gathering' of wild rice from as long ago as 10,000 and perhaps even 20,000 years BP, with evidence of domestication in the 'earliest levels of the Neolithic'. They report carbon dates for rice husks as well as grain which indicate that cultivation of rice may have succeeded gathering of wild rice between 8500 and 6500 BP, but their evidence has not been widely accepted. Kumar (1988) also believes rice to have been cultivated in India since the Neolithic, but cautions about the difficulty of distinguishing wild rice and domesticated rice. Chang (1976b) first considered that the earliest domestication of rice occurred in India, but revised his views in the light of later archaeological findings from the lower Yangzi region (Chang, 1983).

There is much archaeological evidence for rice cultivation in the upper and middle Ganges, Rajasthan and Gujarat from 4500 BP onwards, and the Vedan and other literary sources are quoted which 'suggest a deep rooted antiquity of rice in India'. Some authorities have questioned the early dates for rice cultivation in India, but Hutchinson (1976) in reviewing the earlier archaeological evidence regarding rice farming in the upper Ganges floodplain concluded that the practices were already so sophisticated by 4500 BP that domestication of rice must have already been practised in the region for some time. Glover and Higham (1996), from a critical review of the available data, have concluded that towards the end of the third millennium BC (4000 to 4300 BP) rice was domesticated among small scale neolithic farming communities in the central and eastern parts of the Ganges Valley, and then spread to the socially and agriculturally more advanced communities in the Indus Valley, where it became cultivated as a summer crop. On present evidence it appears to be most likely that rice cultivation was initiated in India sometime between 4000 and 5000 BP.

2.3.4 Africa

A system of rice cultivation without bunds and transplanting is used in the upper reaches of the Niger in West Africa, where it has certainly been practised for many centuries and probably for much longer (Portères, 1950, 1956, 1976; Carpenter, 1978). The indigenous *O. glaberrima* is grown, a species which shows considerable diversity. Two major ecotypes have been distinguished (Jackson, 1995) with tolerance to flooding and to drought conditions. Yields as high as 3 t ha^{-1} have been reported from northern Nigeria (Carpenter, 1978). The tolerance of glaberrimas to poor conditions makes them useful in modern rice breeding programmes for crossing with *O. sativa* but their use as a crop is dying out. Where they persist they are often mixed with varieties of *O. sativa* so that interspecific crosses are now quite commonly found.

2.4 The Development of Water Management

From the above account it is clear that between 10,000 and 4000 years ago, in one or several sites between the Ganges Valley in India and the Yangzi in China, gathering of rice as food was replaced by sowing and cultivation. The methods of cultivation spread into other parts of China, to insular Southeast Asia, to Sri Lanka and to India, although the question of whether rice domestication and cultivation also developed independently in India and elsewhere remains open. The success of the methods of rice production fostered the emergence of well organized village communities. The development of more highly organized societies with large towns and sophisticated social systems had to await better controlled water distribution and management systems (Oka, 1988; Higham, 1989).

The earlier naturally flooded systems were undoubtedly affected by changing river courses and eustatic changes in sea level, as well as by sedimentation and other factors affecting the relative levels of land and sea. The evidence from Hemudu (Gong Zi-tong and Liu Liang-wu, 1994), for which the altitude is given as 3–4 m, is that rice cultivation may have persisted there for two or three millennia, but that there was then a major change in the river sedimentation pattern. Hui Lin-li (1983) states on the basis of archaeological evidence that the Hemudu site was occupied until 3710 BP. This agrees with the evidence in the soil profile which shows four superimposed sedimentary layers, the rice and other relics occurring in the lowest layer, which is 40–50 cm thick. The three higher layers correspond to later depositional patterns laid down by the river. The river changes must have been accompanied by major flooding, destroying homes and causing those who survived to find new sites for their habitation.

Before the development of methods for water control the production of rice was at the mercy of the river floods when the impact of the monsoon was felt, and of droughts if the arrival of the monsoon was staggered or delayed. There is no evidence at present that any attempts were made to control the spread and depth of floodwaters before about 2700 BP. The lack of water control and large storage or diversion systems would have made life precarious in the monsoonal areas close to the great rivers, and exerted a strong control on the growth of the population and the development of societal organization.

The unpredictability of the depth to which the river would rise, and the sensitivity of young rice seedlings to submergence, would at some stage have shown the early rice farmers the advantage to be obtained from sowing the seeds in a seed bed not subject to the vagaries of the river floods, and of transplanting the seedlings when they were sufficiently tall and growing quickly so that they would not drown. Chang (1976b) has suggested that transplanting began during the Eastern Han dynasty between AD 23 and 220 when other advanced agricultural methods were introduced to rice farming. However as transplanting was far more widely practised for rice than for the other major crops grown in northern China at that time it is possible that it evolved independently in southern China and probably at an earlier date, although the first literature references to transplanting appeared during the Han Dynasty.

Bray (1984) believes that farmers living during the Zhou Dynasty (3000–2500 BP) were already bunding and irrigating their fields. Advantages to be obtained by controlling water depth by use of bunds around the fields would have been easily appreciated by the early farmers who were growing rice in the backswamps of the Yangzi and other river systems. It is likely that they learnt to manage the rice crop using shallow bunds to retain the floodwater and rains falling on the fields. Such a system is still in use throughout South and Southeast Asia (IRRI, 1979b; Garrity *et al.*, 1986). Lambert (1985) for instance has described the farming system of the Pahang Malay in detail, and Higham (1984) discusses its close relation to the swamp farming of rice which would have been practised in the swampy areas of the middle and lower Yangzi, perhaps as long as 7000 years ago. In the delta areas, where water control was more difficult, selection of rice varieties able to withstand submergence and elongate rapidly as floodwater rose was taking place. This enabled communities in the flood-prone areas to develop as well as those in the more favoured lands above the delta areas.

Thus by 2500 BP systems of water control had been developed in which water from a stream or river, or held in a pond dug for the purpose, could be fed into the rice fields. Chaudhuri (1991) mentions that systems of water management may have been introduced to northern India from the Middle East at about that time, but such systems were almost certainly developed independently in China and possibly in Southeast Asia. Very simple diversion systems may well have been used as soon as wooden or bone tools were available to dig ditches. Opening and closing the bunds around the fields enabled the depth of water in the rice field to be controlled so that the young rice seedlings were not submerged, nor allowed to suffer from drought (Higham, 1989). Siting the rice fields in relation to the point of diversion from the river or reservoir allowed the water to flow from paddy to paddy under gravity, and by building terraces between successive paddy fields at shallow intervals the rate of flow could be controlled.

Between about 2500 and 2000 BP a major development took place in northern China. Iron implements were produced which made earth movement much easier. The more advanced methods of rice cultivation probably date from the time that the Han Chinese brought these iron implements, including the animal-drawn plough, from northern China to the lower Yangzi. With the iron plough and sophisticated understanding of canal building and water control methods rapid extension of rice cultivation was possible. Sufficient rice could be produced to enable large city communities to be supported. The plough with beam, iron ploughshare, and handle is mentioned in China under the ruler Wudi (140–187 BC) and was certainly used much earlier for the cultivation of millet. Bray (1984) gives the date of the invention in China of the plough and seeder with iron used in their construction as 2600 BP.

The construction of canals provided water supplies to the growing settlements, a means of communication and transport, and the opportunity to raise crops under irrigation. Xing Jia-ming (1988) describes the construction of canals in the period of the warring states (475–221 BC). These diverted water from the Juma He and enabled some 70,000 ha to be irrigated. Others built under the Eastern Han Dynasty (AD 23–220) were extensive enough for other areas of similar size to be irrigated. Construction of major canal systems on the lower Huanghe (Yellow River) was also well developed by 2000 BP (Greer, 1979).

Aerial photography has revealed that a canal system existed in the Mekong Delta dated by ground reconnaissance to about 1600 years ago (Higham, 1984). Recent work in which aerial photography has been linked with studies of satellite imagery (Parry, 1992) has shown how extensive these systems of reservoirs and canals were in Southeast Asia.

The time at which rice cultivation with controlled water management was initiated in India is not well established. Hutchinson (1976) has suggested that the alternation of a rice and a wheat crop in northern India commenced as long ago as 4000 BP. The sophisticated rice–wheat system requires some degree of water control and suitable implements for land preparation. Kumar (1988) and Randhawa (1980) both indicate that rice was being grown rather extensively, and by implication with some degree of water control, between 4500 and 3500 BP. It is quite possible that the water management systems brought to the Indus region from Persia and the Middle East were used for irrigating both the winter wheat crop and the summer rice crop, and that this management system was developed rather earlier than Chaudhuri (1991) suggests.

Storage systems with reservoirs known as tanks in India and Sri Lanka, barays in Cambodia and Vietnam, and ponds in Japan, were being widely constructed from 2000 BP and probably much earlier. The barays at Angkor (Yasodharapura) in Cambodia, which are able to hold several tens of millions of cubic metres of water, are perhaps the best known examples. They were constructed between AD 800 and 1200 (Chandler, 1983). Construction of tanks in southern India and Sri Lanka may well have preceded their use in Cambodia (Brohier, 1975). Randhawa (1980) states that tanks existed in India in 3000 BP, and Brohier (1975) suggests that the construction of tanks in Sri Lanka was introduced from southern India about 2400 BP. By the end of the 12th century King Parakrama the Great, ruler of Sri Lanka, could pronounce 'In my kingdom are many paddy fields cultivated by means of rainwater, some cultivated by means of perennial streams, and some by means of tanks both small and big' (Moormann and van Breemen, 1978). Many of these tanks are still in use in the dry zone of Sri Lanka, and well managed rainfed systems operate in the wet zone, with many small valleys carefully terraced and the flow of water controlled as it moves from higher to lower positions.

The *japonica* race of rice was introduced to Japan from China, and might be more accurately referred to as *sinica* (Chang, 1976b). The introduction occurred more than two millennia ago. Thus there is a long record of rice growing in Japan and many careful studies of its origin and distribution have been made by Japanese scientists. Among these is a study made at Tsukumino Moor, Aomori Prefecture, in northeast Japan (Yamanaka, 1979). The palynological record (Fig. 2.2) shows that large trees were growing at the site in the period 3000–2000 BP. The site may then have been used for shifting cultivation as there is also evidence of occupancy. A small amount of relic *Oryza* material is present before 1800 BP, suggesting that some form of upland rice may have been included among the crops grown. After 1800 BP there is much evidence of regular rice cultivation, and from 1500 BP it is clear that methods were introduced to control the depth of the river flood. From that time onwards rice was regularly cultivated at the site, although a brief gap occurs around 900 BP. The site provides firm evidence for rice production having been sustained under controlled water management at the same location for at least 1500 years.

The advent of canals to enable water to be diverted from streams and rivers, and of pond and tank storage systems made another innovation in rice farming possible, the development of double rice cropping in which a second rice crop is taken after the crop grown during the monsoon rains. Chang (1976b, 1987) suggests that double cropping became common in the lower Yangzi region about 1000 years ago, following the introduction of the short duration Champa (indica) varieties from Fujien Province (Evans, 1978),

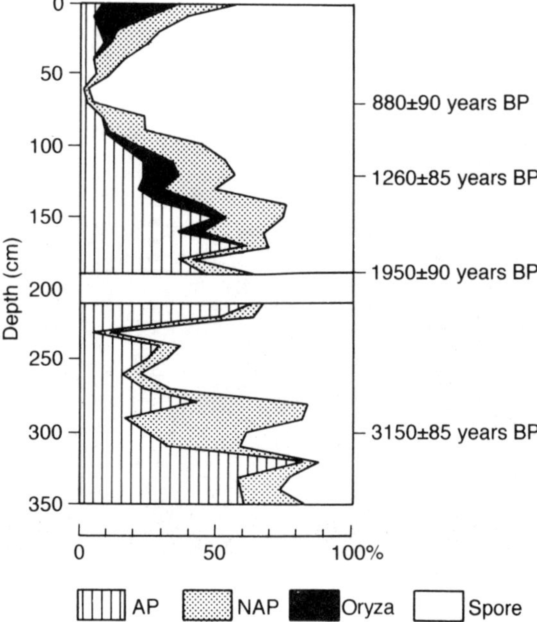

Fig. 2.2. Summarized pollen diagram of the deposits from Tsukimino Moor, Japan. Percentage based on total arboreal pollen (AP) (excluding *Pinus* and *Alnus*), non-arboreal pollen (NAP) and spores (Yamanaka, 1979).

Depth (cm)	Age (year BP)
70	880 ± 90
120	1260 ± 85
190	1950 ± 90
300	3150 ± 85

Corresponding mean sedimentation rates, mm year^{-1}.

880 BP to present 0.80	1260 BP to 880 BP 1.31
1260 BP to present 0.95	1950 BP to 1260 BP 1.01
1950 BP to present 0.97	3150 BP to 1950 BP 0.92
3150 BP to present 0.95	

whence they had been brought several centuries earlier from Vietnam. Multiple cropping of rice may have developed independently in India, using irrigation methods brought from the Middle East. Rice was noted to be planted three times each year by Ibn Battuta, writing in the 14th century (Gibb, 1971).

When the methods of constructing bunds to control water movement had been learnt, and the importance of levelling fields to ensure even distribution of water and prevent rill formation in the fields recognized, the opportunity to grow flooded rice at higher elevations arose. Wherever there were streams or rivers from which the water could be diverted, a terrace and a paddy field could

Fig. 2.3. Montane rice terraces, Ifugao, the Philippines.

be constructed. This would have led initially to development of the higher terraces of the major rivers and their tributaries, but spread rapidly into more hilly and mountainous regions such as the foothills of the Himalayas. As rice varieties were developed which were tolerant of lower temperatures and able to mature more quickly so the cultivation moved into the higher mountainous areas (Chang, 1987) until rice can now be found growing at elevations above 2500 m (Uhlig, 1978).

The date when terraces were first constructed in China is again hard to determine but there is no evidence that they were being built before 2220 BP, and were not common in China before the 9th century AD (Bray, 1984). Most of the higher montane terraces, which the Chinese refer to as 'ladder fields', are probably less than 1000 years old. The mountain terrace systems where rice is grown are often seen as the image of long-continued rice production in Asia. Relatively few if any of such systems are in fact as old as 1000 years. For example, the Ifugao terraces in the northern Philippines (Fig. 2.3) are often described as 'age old' and said to have existed for thousands of years. In fact most of these terraces were built within the past 500 years, as the Ifugao moved to the hills to escape the Spanish invaders, and the earliest were almost certainly constructed less than 2000 years ago, when the Ifugao were driven from the more fertile and easily cultivated valley lands as a result of population pressure and conflict with neighbouring tribes (Conklin, 1980).

An attempt to illustrate how the different innovations in rice production and water management systems have influenced rice yields is presented in Fig. 2.4. Bray (1986) has suggested that the changes in rice farming which

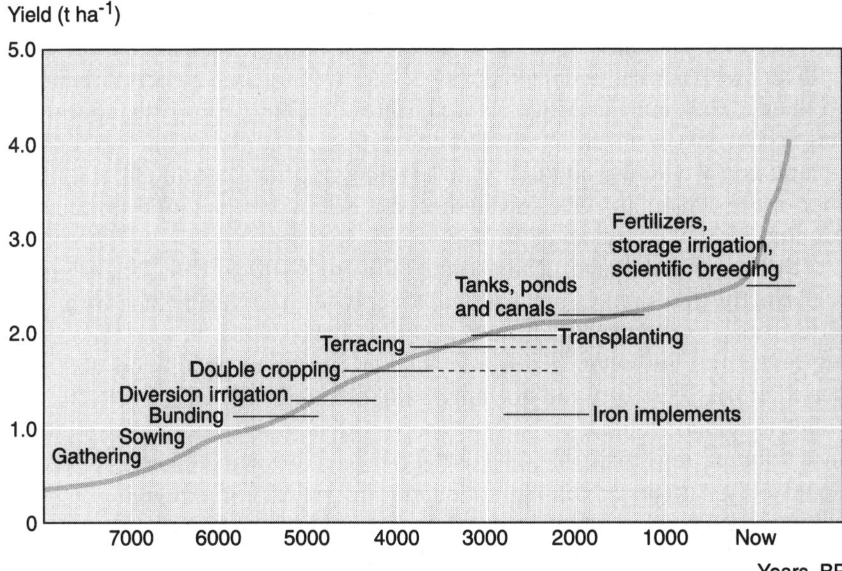

Fig. 2.4. Technical innovation and rice yields, 10,000 BP to the present.

occurred around 1000 years ago might be termed the first green revolution. In fact the changes during the Han dynasty laid the foundations for the increase in rice yields from about 1 t ha^{-1} to 2 t ha^{-1}, and the first green revolution, or perhaps evolution, should date from 2200 years BP.

2.5 The Origins of Upland Rice

There was at one time considerable speculation that rice originated as an upland rather than a wetland crop. Chang (1976a, b) discusses these speculations as far as Southeast Asia is concerned. He notes that the more advanced and less diverse plant and grain features of the upland varieties indicate a rather later evolution. The weight of recent evidence also supports the conclusion that upland rice developed as a domesticated species by adaptation from wetland precursors. Chang (1987) suggests that it followed a similar evolutionary path in both China and Southeast Asia. He states that 'Land use is from swidden agriculture (shifting cultivation) to unbunded permanent fields to bunded fields – first rainfed, later irrigated'. It is unlikely that this is in fact a general chronological sequence. More probably upland rice was introduced into swidden agriculture as a special crop rather than as a major staple, and that it evolved independently of the wetland rice sequence. Seeds of *O. sativa* have been found in early upland sites in Southeast Asia remote from water (Zeven, 1973; Grigg, 1974). It is most likely that some wetland species were found which were somewhat tolerant to upland conditions and a process of selection and adaptation over many years led to the development of the upland cultivars with thicker and stronger roots than wetland rice. Selection from the wetland glaberrimas in Africa, as well as from the sativas in Asia, appears to have occurred (Portères, 1956). The cultivation of the glaberrimas as both an upland and wetland crop in West Africa may not have continued as long as that of the sativas in Asia, as the diversity is less than is found in *O. sativa* (Chang, 1976b; Katayama, 1987).

During the Makalian Phase from 7500 to 4500 BP the climate in the region of the upper reaches of the Niger River where the glaberrimas are grown today was far wetter than it is now. It gradually became drier over the following millennia, but nevertheless a far higher density of population was supported in the years between one to two millennia ago (Davidson, 1972). The Sahelian region would then have provided a greater array of suitable wetland sites than are now available. The use of rice as a prestige crop in many traditional West African ceremonial functions may reflect a much greater availability of the crop in earlier times when it would have enjoyed a position of importance corresponding to that which it continues to enjoy in Asia.

Although rice is widely grown by cultivators practising shifting cultivation in many parts of the world, it is often sown as a special crop with taro and vegetables in wet or otherwise more favoured spots, and not rotated with the main subsistence crop.

2.6 The Further Spread of Cultivated Rice

Following the initiation of rice cultivation in China between 9000 and 7000 years BP, and its subsequent development in Southeast and South Asia, it spread gradually around the world, southwards to insular Asia about 4000 years ago, westwards from India to the Middle East about 3000 years ago, eastwards from China to Korea and Japan about 2500 years ago. Rice moved from the Middle East to Egypt and the Mediterranean region 1000–2000 years after it had first been grown in 'the fertile crescent' of the Middle East (Fig. 2.5).

In Africa *O. sativa* spread to Madagascar from insular Southeast Asia at least 1000 years ago. From Madagascar it spread slowly to East and Central Africa. Only in Madagascar did it become a major staple. It seems probable that the spread of rice westwards across the Indian Ocean took place in the course of the great expansion of the Moslem Empire around 1000 AD, and that a knowledge of water management for the development of inland rivers accompanied the rice plants (Rabesa Zafera Antoine, 1985). Many national Madagascan customs and rice cultivation practices reflect those now used in insular Southeast Asia. The languages of the Malagasy people are Austronesian, and indicate settlement from insular Southeast Asia. Linguistic and archaeological evidence support the view that settlement occurred about 2000 years ago.

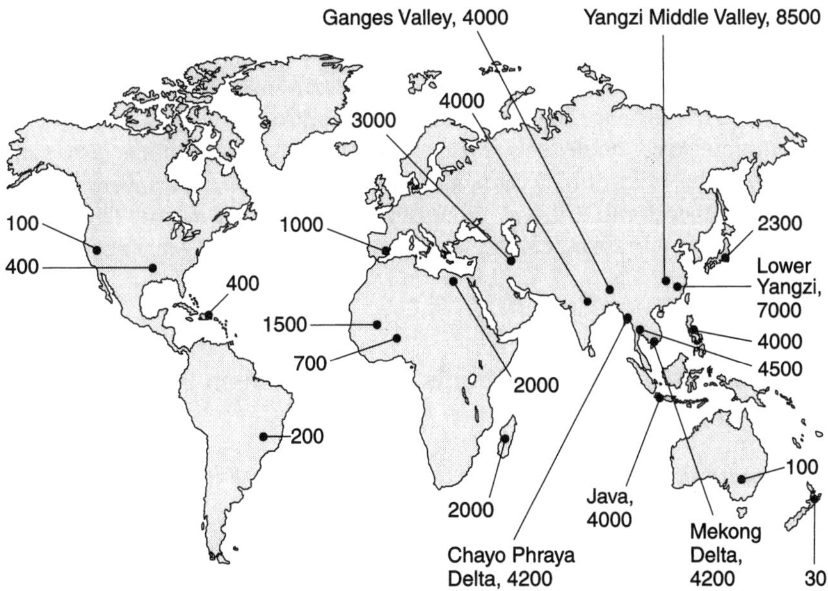

Fig. 2.5. The origin and spread of rice cultivation. Numbers indicate years BP when rice was first known to be cultivated.

Year BP	Location
8500	Yangzi Valley
5000	S.E. Asia, mainland
5000	Central India
4000	Island S.E. Asia
3800	Indus Valley
3000	Iran
2000	Madagascar and Egypt
1500	West Africa – glaberrima rice
1000	Southern Europe
700	West Africa – sativa rice
400	N. America
200	West Indies
100	Australia
30	New Zealand

Rice also reached the east coast of Africa from India, perhaps via Oman, and probably at about the same time as it was taken to Madagascar, but was not accompanied by migration of rice cultivators and so has not been as widely developed as in Madagascar. Although domestication of the glaberrima rices occurred independently of developments in Asia, the sativas also reached mainland Africa from Egypt and the Middle East and were cultivated perhaps 1000 years ago.

The arrival of rice in the Americas followed some time after Columbus, reaching the Mississippi delta region perhaps 300 years ago, and later spreading to California and South America, although it was probably also introduced to Brazil by the Portuguese rather earlier.

Attempts were made to grow rice in northern Australia almost 200 years ago, but it was not until 70 years ago when irrigated plantings were made in southern Australia that it was grown to any significant extent. The first attempts to grow it in New Zealand were only made about 30 years ago.

2.7 The History of Rice Yields – From a Green Evolution to a Green Revolution

Early information about rice yields is very limited and difficult to interpret. Evans (1993) has assembled such evidence as exists on the productivity of wild species. Based on that evidence an informed guess might be that early varieties of rice would have produced the equivalent of around 0.5 t ha^{-1} of grain, had the rice plants been growing in reasonably close proximity. By 2000 BP it is probable that yields were close to this level.

Yu Ye-fei (1980) and CAAS (1986) give 0.54 dan mu^{-1} or 402 kg ha^{-1} for the yield obtained during the Han dynasty of 206 BC to AD 222. Wu Hui (1985) and Bray (1984, p.128) give much higher figures of 2.64 dan mu^{-1} and 2 piculs mu^{-1} (1980 and 2550 kg ha^{-1} respectively). The difficulties of interpreting the older Chinese data are many. Not only is it seldom clear to what area the data refer, but the units were by no means fixed. Thus at one time a mu was 'one hundred steps' but it later became 240 steps – and a step was at one time 6 chi but later became 5 chi. A dan ranged from 0.2 to 1.035 times its presently accepted weight of 50 kg. Some uncertainty also attaches to the fact that yields were sometimes quoted as unmilled grain, and sometimes as brown rice after removal of the husk. More importantly, as Bray (1984) points out, yields were commonly quoted on an annual basis so that from the time of the introduction of the champa rices around AD 1000 and probably earlier some of the yields quoted referred to two or three crops of rice. These factors certainly help to account for the wide range of values given for yields in China during the Song, Ching and Ming dynasties by Min Zong-dian (1984) reproduced in Appendix 1. The range is from 34 to 1375 jin mu^{-1}, or probably 255 to 10,300 kg ha^{-1} if the Chinese units are converted to their probable modern equivalents. The very high yields are from Huzhou in the Lake Taihu region in the early Qing dynasty, perhaps about AD 1700. The reference given is 'Supplement to Agricultural Book' and it is presumably referring to yields from three crops which were commonly being grown in this region at that time. Bray (1984) reports several yields (probably annual, and including some for a single crop and some for double or triple crops) from the Yangzi region and southern

China obtained between AD 1000 and 1800. These range from 1000 to 7000 kg ha^{-1}.

To achieve the rapid turn around necessary to grow several crops in one year, harvesting, ploughing and harrowing have to be completed in as little time as possible. The Chinese devised many simple implements for harvesting and used water buffaloes to expedite the land cultivation. 'Three colours in a day' – the gold of the ripened rice, brown of the soil, and green of the transplanted rice – was an honourable objective then widely attempted. Interplanting a second rice crop as the first approached harvest was also practised (Bray, 1984).

Some of the references quoted by Min Zong-dian (1984) are reports from provincial governors to the emperor, and some exaggeration might be expected. The references which accompany the data show the wide range of sources used to gather these records. Average yields change little from the Song to the Qing dynasty.

You (1995) has reviewed much of the early data. He believes that the figures given by Yu Ye-fei (1980) are most likely to be correct. They also correspond closely with the data given by CAAS (1986). Data from Japan for the period from the 9th century AD to the end of the 19th century (Table 2.1a) are similar to those given for China in the corresponding periods by Yu Ye-fei. There is considerably less uncertainty about the early Japanese data than the Chinese, and the agreement between them lends further credence to Yu Ye-fei's data.

While there was a wide range in the annual yields obtained in different regions of China, and some uncertainty remains regarding actual yields, it appears likely that national average annual yields increased slowly to reach about 1 t ha^{-1} by AD 1000, and 2 t ha^{-1} by about 1850. Thereafter they increased more rapidly in Japan than in China, particularly after the introduction of scientific plant breeding methods in Japan in 1910, and following the improvement in methods for the manufacture of nitrogen fertilizers which allowed the price of nitrogen fertilizer to fall compared with the price of rice (Hayami and Otsuka, 1994).

The general turbulence in China which prevailed from the time of the start of the opium wars in 1840 and continued until about 1950 caused great disruption, not least to farming. Yields of rice and other crops declined. Ho Ping-ti (1959) notes that the population of the four Yangzi Provinces declined by 30 million between 1850 and 1953. He discounts the possibility that the apparent decline of the population could be due to exaggeration of the numbers pre-1850 and attributes the decline to the Taiping and other wars, destruction of farmland, famines, epidemics and the evils of opium – the problems that Malthus foresaw in 1798.

The rice yields in the Yangzi Provinces were the highest in China, but nevertheless there was a drastic fall in the population. The people at that time were encouraged to grow 'the new 30 day and 40 day miracle rices' (Ho

Table 2.1. Rice yields, (a) China and Japan, pre-1912 (b) China, Japan, India, the Philippines and Thailand, 1913–1991.

(a) Pre-1912.

1. China										
Dynasty		Han	Wei	Western Jin	Eastern Jin	Sui-Tang	Sung	Yuan	Ming	Qing
Years	Before 206BC	206BC–AD202	AD220–265	265–317	317–420	581–906	960–1279	1260–1368	1368–1644	1644–1911
Yield, t ha^{-1}	0.34	0.40	0.59	0.74	0.83	0.85	1.04	1.45	1.95	(1.61)
2. Japan										
Years, AD				800–900	1550	1720	1840	1878–1887	1893–1897	1903–1907
Yield, t ha^{-1}				1.01	1.65	1.92	1.92	1.85	2.60	3.10

(b) 1912–1991.

1. China									
Years	1915–1921	1931–1940	1946–1955	1956–1965	1966–1975	1976–1980	1980–1985	1985–1987	1991–1993
Yield, t ha^{-1}	1.27	1.77	2.50	2.94	3.63	4.01	5.00	5.30	5.80
2. Japan									
Years	1913–1917	1923–1927	1933–1937	1953–1957	1963–1967	1973–1977		1983–1989	1990–1993
Yield, t ha^{-1}	3.5	3.6	3.8	4.3	5.1	5.8		6.2	5.8
3. India									
Years	1913–1917	1923–1927	1931–1940	1955–1957	1963–1965	1975–1977		1987–1989	1990–1993
Yield, t ha^{-1}	1.01	0.92	0.96	1.3	1.5	1.8		2.4	2.6
4. Philippines									
Years	1913–1917	1923–1927	1931–1940	1955–1957	1963–1965	1975–1977		1987–1989	1990–1993
Yield, t ha^{-1}	0.78	1.06	1.11	1.1	1.3	1.9		2.7	2.8
5. Thailand									
Years	1913–1917	1923–1927	1931–1940	1955–1957	1963–1965	1975–1977		1987–1989	1990–1993
Yield, t ha^{-1}	1.6	1.65	1.56	1.4	1.8	1.8		2.1	2.1

Data for China from Yu Ye-fei, 1980; CAAS, 1986; IRRI, 1995c.
Data for Japan from Hayami and Otsuka, 1994; Ishizuka, 1969, quoted by Plucknett, 1993.
Other data from Barker *et al.*, 1985; IRRI, 1993a, 1995c.

Ping-ti, 1959, p. 285). There were probably no varieties with these characteristics available, and none are known now, although Bray (1984, pp. 493–494) also quotes in considerable detail data referring to 50-day rices. Annual as well as individual crop yields had probably reached the limit of the technology then available, so that the existing population of the Yangzi delta region could no longer be supported.

While the national average yield in Japan reached 3 t ha^{-1} by 1903, that of China did not do so until 1966 (Table 2.1b) after communist rule had re-established conditions in which agriculture could flourish. Elsewhere in Asia yields of rice at the beginning of the 20th century lagged well behind those of Japan. The green revolution taking place since 1960 has seen the global average rice yield increase from 2 t ha^{-1} to 3.6 t ha^{-1} in 1993, and it may not yet be over. Rose (1985) has compiled data for yields obtained by many Asian countries from 1913 to 1980, and IRRI (1995c) data for the years to 1993. The data for India, the Philippines and Thailand (Table 2.1b) are typical of many countries in South and Southeast Asia; yields were below 1.8 t ha^{-1} until after 1960, when the green revolution began to make its impact.

Rice Farming Today

3

There are at the time of writing probably more than 200 million rice farmers in the world. Most of them live in Asia and very few farm more than 2 ha of rice. In China alone there are probably 75 million rice farmers – a figure which may be compared with the figure of 11 million families in China around 1730 (Duhalde, 1738, quoted by Malthus, 1798) of whom rather more than half might have been supported on rice farms. The Chinese rice farmers of that time would have been practising farming methods similar to those that had then been in use for more than a thousand years. They would have been only too aware of the extent to which their livelihood, and indeed their lives, depended on the monsoon rains and the behaviour of the great rivers, although the knowledge that their farming system had been sustained over many generations may have given them some sense of security. As the Jesuit records show, an awareness already existed that the system was not so sustainable that it could support great increases in population numbers, and considerable movements of population had already been enforced in many places as population outstripped the ability of the land to provide sufficient food (Ho Ping-ti, 1959; Bray, 1984).

The range of conditions in which rice farming is practised in China and elsewhere has continued to grow, and there have been many advances in rice farming practices which have greatly increased the productive potential of rice, and the availability and sustainability of water supplies. In this chapter the environments in which rice is grown are described. These are used to help define the present range of rice farming systems and their relative extent and productivity, as an essential background to understanding the sustainability of present day rice farming.

3.1 The Environments in Which Rice is Grown

Rice evolved as a wetland species, and although varieties suited to upland conditions have developed, it remains more sensitive to water deficiency than most other crops. Critical factors in rice productivity are the supply of water to the soil, from rain, river, reservoir or groundwater, and the ability of the soil to retain water. Landscape position largely determines the way in which water can be supplied to the soil, and soil texture and drainage characteristics the extent to which it is retained.

Most rice is produced, and the highest yields obtained, on the alluvial deposits formed where major rivers emerge from hilly and mountainous areas. Eroded material from the upland soils is washed into streams and rivers. When the rate of flow decreases as the river emerges from the hills this material can no longer be carried along in the river water. The coarse sandy materials will first be deposited, and finer silty and clayey material carried further until the rate of flow becomes very slow, or the river or stream overflows its banks and the fine particles settle in the backswamp areas (Fig. 3.1). The quality of the deposited material in terms of future crop production will depend on the soils and rocks from which it is derived. The coarse sandy material is seldom suitable for rice production because the light textured soils formed will not retain water

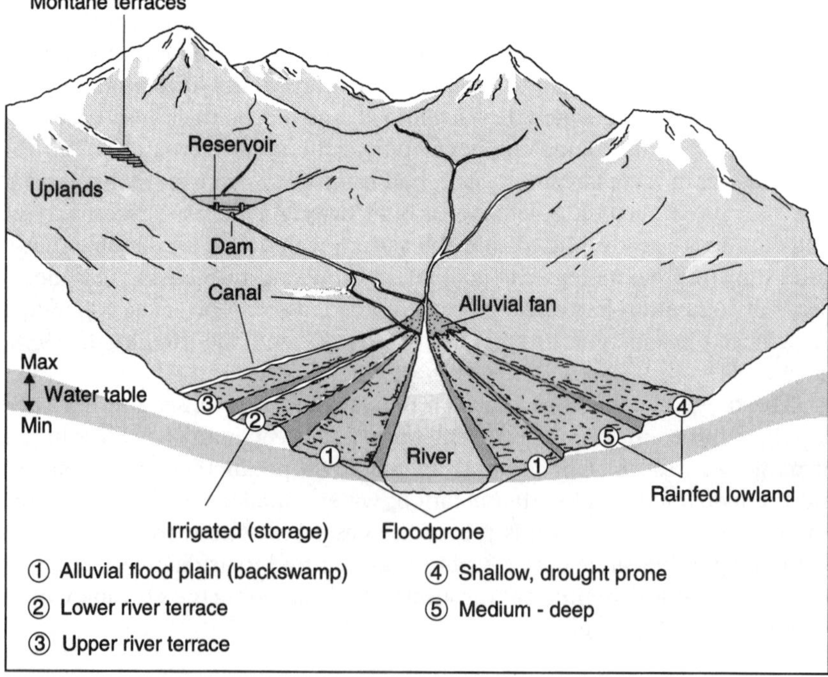

Fig. 3.1. The rice environments and landscape positions.

and the nutrient contents are usually low. The silts derived from basic igneous rocks give rise to the more fertile soils, and the clays, irrespective of origin, usually retain sufficient nutrients to be productive.

As illustrated in Figs 3.1, 3.2 and 3.3 it is normal to find a series of terraces at different levels developed over time. The older and higher terraces are often the most productive, as they can usually be easily puddled to retain water, but allow some through drainage so that they do not become waterlogged and flooded to excessive depths. In the regions of pronounced dry and wet seasons the rate of flow of rivers varies annually. Once the rains (monsoon) have started, the rivers gather water and overflow their banks in the lower reaches so that flooding occurs. Sandy levees form along the river banks, and the finer materials are deposited to form backswamp areas behind the levees (Fig. 3.4). As the river cuts a deeper and deeper channel a series of terraces may be formed at different elevations. When the river approaches the sea, it will slow down further and much of its residual sedimentary load will be deposited. Over the years this will lead to the formation of a barrier at the junction with the ocean and the river course will be diverted. Gradually a deltaic fan will form, marked by the old river courses, and often with several channels open to the ocean. These processes have led to the development of the huge deltas of the Ganges–Brahmaputra, the Huang He (Yellow River) and Yangzi, the Mekong, Irrawaddy and Chao Phraya, and many others of smaller size.

The level of the land surface formed in the delta regions, and hence the type of rice farming that can be practised, depends on several factors. Eustatic

Fig. 3.2. Typical valley development for rainfed lowland rice cultivation.

Fig. 3.3. Terraced hillside, Madagascar.

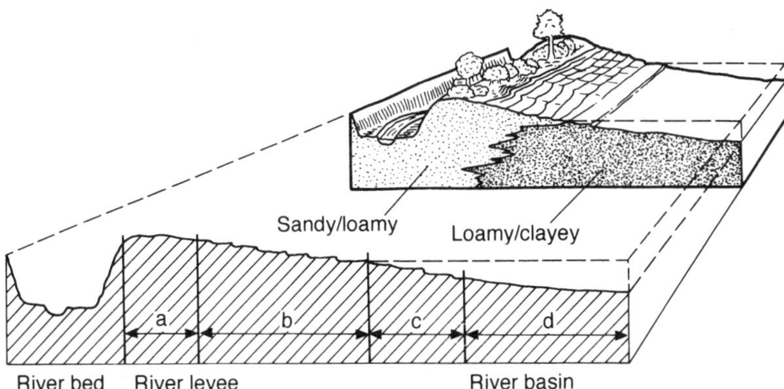

Fig. 3.4. Relationship between type of cultivation and landscape position in a meander flood plain (vertical scale exaggerated).
(a) non-flooded, used for garden and dryland crops.
(b) shallow flooding, used for transplanted rice (semi-dwarf varieties possible).
(c) moderately deep flooding, used for transplanted rice (no semi-dwarf in wet season).
(d) deep flooding, used for direct-seeded rice (deepwater varieties).
(From Moorman and Van Breemen, 1978).

sea-level changes associated with global temperature change have been important over the past 15,000 years. Cooler temperatures in the last 'ice age' from 12,000 to 10,000 BP gave rise to much lower sea levels than at present (Geyh *et al.*, 1979; Yang Huai-jen and Xie Zhi-ren, 1984; Sorensen, 1986; Huang Yu-kun and Chen Jia-jie, 1988). During the subsequent 5000 years sea level has risen by about 100 m, no doubt submerging many of the early sites of rice cultivation. Changes since then have been less dramatic (Table 3.1). The further rise of about 20 m to present mean sea level has taken place more gradually, in accord with temperature changes. Mean sea level has at times been higher than at present – Geyh *et al.* (1979) believe that peat deposits in Malaysia indicate that it may have been as much as 5 m higher – but the more extensive data from East Asia (Table 3.1) indicate that since 2500 BP it has been unusually stable, with changes of little more than a metre above or below present sea level. The predicted rise due to global warming in the next 50 years is still well below rates of change that have occurred in the past.

The level of the land surface is also affected by sedimentation, settlement, earthquakes and other earth movements. The position of coastal boundaries are determined by these earth movements as well as by sea level. Twelve thousand years ago when the sea level may have been as much as 100 m lower than at present there were many land bridges connecting what is now insular Southeast Asia with the mainland. The potential significance of the possible sea-level changes which will arise if global warming continues is considerable, although difficult to predict because of the many different factors which effect the relative level of land and sea (Brammer and Brinkman, 1990).

Table 3.1. Mean sea-level changes in East Asia in the past 15 millennia.

Years BP	Mean sea level, metres above (+) or below (–) present level	Temperature, °C, above (+) or below (–) present
0	0	0
50	+0.4	+0.2
100	–0.2	–0.3
300	–1.0	–1.0
500	+0.8	–0.5
800	–0.5	–0.8
1,100	+1.0	+0.5
1,500	–0.4	0
1,600	+0.2	0
1,800	–1.6	–1.0
2,500	0	0
3,000	–2.4	–0.3
3,500	+0.1	+2.0
4,500	–3.7	–0.2
5,500	+0.2	+1.0
6,000	–7.1	–
7,000	–5.5	+2.0
8,000	–12.1	–2.0
8,300	–9.3	–4?
10,000	–18.5	–
11,000	–30	–
12,000	–40	–
13,000	–80	–8?
15,000	–120	–

Source: Yang Huai-jen and Xie Zhi-ren (1984).

3.2 Rice Production Systems Defined

In the course of the past 10,000 years rice not only spread around the world, but also came to be grown in a wide range of environments, from the hot and humid lowlands to elevations of over 2700 m, and from the equator to as far north as 53 degrees of latitude, and from areas where it is flooded to depths up to 5 m, to uplands where it is never flooded.

The extensive geographical spread of wetland rice cultivation and the long history of rice farming in a wide range of different locations has led to many local terminologies to describe rice production systems (IRRI, 1984a). It has also led to much actual and potential confusion when discussing problems such as sustainability.

Several attempts have been made to classify the environments in which rice is grown, and relate them to the different terminologies to describe rice farming systems (IRRI, 1984a, 1985c, 1993b). Most classifications use water regime as a basis for the classification system. This may be allied to landscape position and temperature regime. The simple classification into irrigated, rainfed lowland, upland and flood-prone now used by IRRI and many others is shown diagrammatically in Fig. 3.5, and related to the dominant farming system practised in each landscape position in Table 3.2. The current IRRI terminology will be used here.

The possible subdivisions within each of the four major categories are many. In the rainfed areas where irrigation water is not available the length of the wet season and the total rainfall, as well as the depth of flooding usually determined by landscape position, are important. In irrigated and other areas temperature regime determines the varieties of rice which can be produced. In upland areas soil fertility as well as rainfall incidence are critically important.

3.2.1 The flood-prone rice environment

The Mekong delta in Vietnam and Cambodia, the Chao Phraya delta in Thailand and the Ganges–Brahmaputra delta in Bangladesh are the three major areas where deepwater and 'floating' rice are still grown. The size of the catchment areas of many of these river systems and the monsoonal rainfall patterns are such that the difference in depth of the river between peak flood

Table 3.2. Characteristics of rice farming systems in the major rice production environments.

A. Storage irrigated rice, with assured year-round water:
  Continuous rice, with one or two crops per year, and occasionally a third rice crop or an upland crop.
B. Diversion irrigated and (favourable) rainfed lowland rice.
  (i) where monsoon season 6 months or more:
    Dry seeded rice, followed by transplanted rice (irrigated when water available), followed by upland crop.
  (ii) where monsoon less than 6 months:
    Transplanted rice (irrigated when water available), followed by upland crop.
C. In rainfed, drought-prone areas (mostly on alluvial terraces):
  One dry or wet seeded rice crop. An upland crop may follow in good season.
D. In flood-prone (deepwater and floating rice) areas:
  One transplanted or wet seeded rice crop. (In some deepwater areas double transplanting is practised.)
E. In upland rice areas.
  (i) under shifting cultivation:
    Dry seeded rice, often interplanted, e.g. with maize, and followed by another upland crop, e.g. cassava.
  (ii) under mechanized cultivation:
    Sole crop dry seeded rice, grown annually for several years, after which a grass pasture may be established.

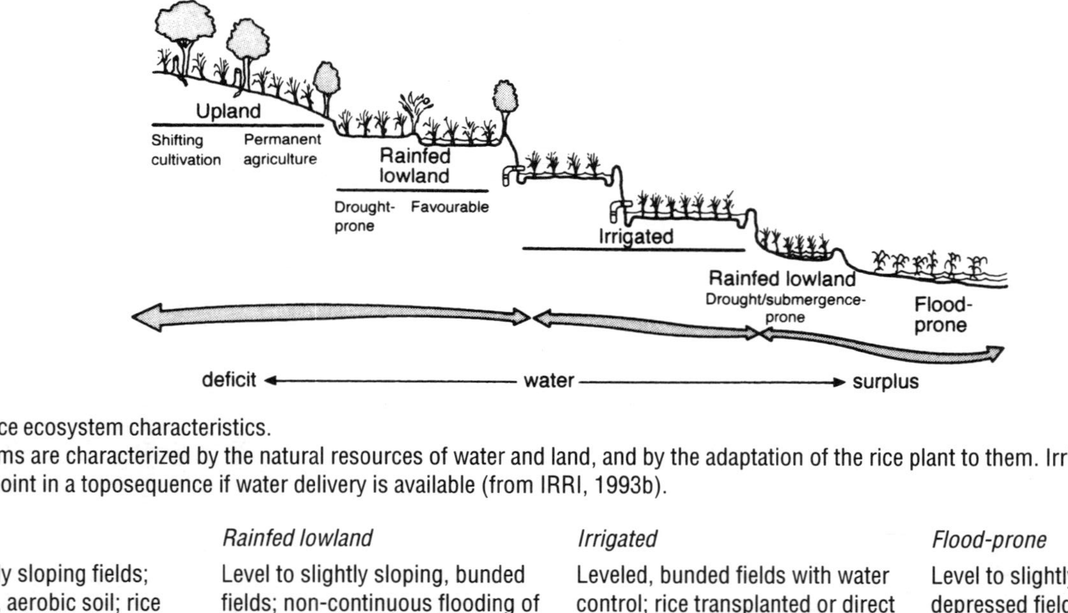

Fig. 3.5. Rice ecosystem characteristics.
Rice ecosystems are characterized by the natural resources of water and land, and by the adaptation of the rice plant to them. Irrigated rice maybe found at any point in a toposequence if water delivery is available (from IRRI, 1993b).

Upland

Level to steeply sloping fields; rarely flooded, aerobic soil; rice direct seeded on ploughed dry soil or dibbled in wet, non-puddled soil.

Rainfed lowland

Level to slightly sloping, bunded fields; non-continuous flooding of variable depth and duration; submergence not exceeding 50 cm for more than 10 consecutive days; rice transplanted in puddled soil or direct seeded on puddled or plowed dry soil; alternating aerobic to anaerobic soil of variable frequency and duration.

Irrigated

Leveled, bunded fields with water control; rice transplanted or direct seeded in puddled soil; shallow flooded with anaerobic soil during crop growth.

Flood-prone

Level to slightly sloping or depressed fields; more than 10 consecutive days of medium to very deep flooding (50 to more than 300 cm) during crop growth; rice transplanted in puddled soil or direct seeded on plowed dry soil; aerobic to anaerobic soil; soil salinity or toxicity in tidal areas.

periods and the time of minimum discharge may be as great as several tens of metres. Thus in spite of the major efforts which have been made to control the flow of these rivers (see e.g. Greer, 1979) the areas behind the river levees are often flooded to depths of several metres. Regularly flooded river basins or backswamp areas are more common in the delta areas, but also occur in other parts of river floodplains. Rice grown in these naturally flooded areas is described as flood prone. Most flood-prone rice is seeded immediately before the arrival of the flood waters, and little more is done until the time arrives for the crop to be harvested (Catling, 1993). In some areas the rice may be transplanted once or twice as the floodwaters advance, in an attempt to save the young rice seedlings from drowning if the floods rise too rapidly for the seedlings to survive. The general relation between rice crop and floodwaters in the flood-prone areas is illustrated in Fig. 3.6.

All of the rainfed lowland rice areas are flooded at some time during the year, but only in those areas where the floodwaters rise rapidly to depths greater than half a metre is the rice crop liable to submergence. Rices used in the areas where rapid water rise occurs have to have more than average elongation ability as well as submergence tolerance. The varieties adapted to the deeply flooded areas are sometimes referred to as floating rice, as the panicles appear to float on the surface of the water as the crop matures.

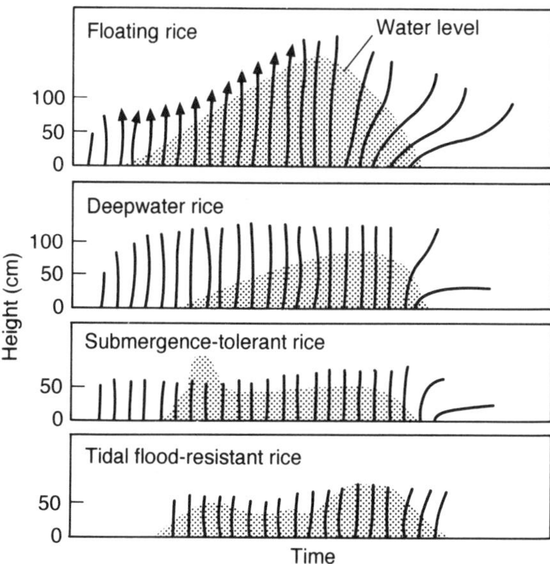

Fig. 3.6. The flood-prone ecosystem includes tall deepwater rices, which tolerate 50–100 cm of flood water; floating rices, which grow rapidly taller to maintain their foliage above floodwaters that can be from 100 to 400 cm deep; submergence-tolerant rices, which can survive flash floods for up to 12 days; and tidal wetland rices, which can survive daily tidal fluctuations and submergence for short periods (IRRI, 1993a).

However the plants retain a rooted foot (or at least toe) hold in the soil (Fig. 3.7). Floating rice varieties also form adventitious roots from the nodes. These are able to absorb nutrients directly from the floodwater.

A special category of deepwater rice is tidal swamp rice (IRRI, 1984b). In the coastal plains of Southeast Asia and a few other areas adjacent to the sea there are considerable areas of swamplands where the natural barriers to the discharge of rivers into the ocean lead to flooding during periods of high discharge. Salinity, the presence of peat in many of the swamps, and the occurrence of acid sulphate soils, introduce special problems for rice cultivation and its sustainability in these areas.

3.2.2 Rainfed lowland rice

Around the more deeply flooded areas are some which are normally only flooded to depths of less than 1 m (IRRI, 1979b, 1986a). The sedimentary soils are deep and soft and lend themselves to easy manipulation to form the bunded paddy fields and rice terraces characteristic of the rainfed lowland rice environment. The bunds serve to retain floodwaters, as well as rainwater which falls during the growing season. The higher terraces may be drought prone. Their usual position in the backswamps also means that they allow little or no drainage. Hence the soil may be strongly reduced, a condition likely to give rise to the formation of organic and other toxins which when formed are not removed by leaching.

The rainfed lowland rice environments are most commonly subdivided according to the normal depth of flooding. An Indian 'rule of thumb' is to subdivide them as shallow, when the depth of flooding does not exceed ankle height, medium deep, when depth of flooding normally reaches the knee, and deep, when the depth of flooding normally reaches the thigh (Khush, 1984). These depths largely determine the variety of rice to be grown. More precise definitions are often recommended but because the seasonal variation in depth of flooding is considerable there is little real advantage in attempting to be more precise. The shallow rainfed category is usefully subdivided further as drought-prone, submergence-prone, drought and submergence-prone, or favourable. There will normally be groundwater at or close to the surface throughout the growing season. Garrity *et al.* (1986) give a more detailed account of the characterization and distribution of the rainfed lowland rice areas.

3.2.3 Upland rice

Upland rice may be defined as rice grown where there are no efforts made to impound water and where there is no natural flooding of the land (IRRI, 1984c; 1986b). Rice is then grown as any other upland crop. It is sometimes

Fig. 3.7. Deepwater rice.

intercropped with maize, and often followed by a legume (Shepherd *et al.*, 1988). The great majority of upland rice varieties tend to be more sensitive to water deficiency than most other crops (IRRI, 1982a; Gupta and O'Toole, 1986).

The term dryland rice rather than upland rice has often been used (e.g. by Huke, 1982). Neither term is entirely satisfactory as upland may be thought to imply land of high elevation, which is not necessarily true, and dryland that the rice is grown in relatively arid areas, which is again not true, although the areas are drier than the wetlands in which most rice is grown. The subdivisions of upland rice environments are as many as the subdivisions of land used for most other crops. Those most used for upland rice have been discussed by Garrity (1984a).

A significant part of the upland rice grown in Africa and South and Southeast Asia is grown by people practising shifting cultivation. This mostly occurs in areas where land is relatively plentiful and population pressure and commercial demands have not yet impinged on an earlier way of life. Nye and Greenland (1960) estimated that there were 200 million people supported by shifting cultivation; more recently Robison and McKean (1992) have stated that there are now more people practising shifting cultivation than ever before. Not all but certainly the majority of those practising shifting cultivation will grow some rice. Nevertheless rice grown under shifting cultivation probably accounts for less than 1% of all rice consumed.

3.2.4 Irrigated rice

The landscape position of most alluvial river terraces (Fig. 3.1) means that they are not normally flooded except when the river is at its peak. By diversion of river water they can however be flooded by leading water through canals and ditches to the points of entry to rice paddies. Hence they are the landform that is most frequently developed to receive water diverted from streams and rivers. The paddy fields are bunded and arranged so that water may flow under gravity from higher to lower fields, and the bunds and outlets constructed so that water depth does not exceed a few centimetres.

They are of course dependent on water being available from the stream or river. This limits the use of many such systems to the monsoon period, and also means that some are subject to periodic drought. They may be described as diversion irrigated.

These systems may be further subdivided according to the location into 'lowland' and 'montane'. The montane terraces of the Ifugao region of the Philippines (Fig. 2.3) and the Himalayan 'hill' regions of Nepal, Assam and Bhutan are spectacular achievements, far less easy to construct than the lowland terraces, and far more vulnerable to damage due to erosion of the mountain areas above the terraces.

A water storage system not only removes the uncertainty of yields being limited by drought but also enables a crop to be taken during the dry season, when radiation is most intense and insect and disease pressure least. Groundwaters if of good quality may also be used for irrigation using pump systems – 'Groundwater' or 'pump irrigated'. Since relatively cheap petrol-driven pumps became available hundreds of thousands have been installed. They are a major feature of the lower reaches of the Ganges–Brahmaputra River systems in eastern India and Bangladesh, in a few areas in Pakistan near the Indus River where the groundwater is not yet saline, and in China wherever there is a water table close to the surface.

3.3 The Present Distribution and Extent of Different Rice Production Systems

As noted previously, rice farming first developed in flood-prone and rainfed lowland areas. Diversion of water from streams and rivers, perhaps using some form of weir to achieve partial control of water flow into the rice paddies, was probably developed 3000–4000 years ago, and tank or pond storage systems perhaps 1000 years later. But until a century ago irrigation of rice paddies from large dam storage systems was of very minor importance compared with diversion irrigation from streams and rivers. Reservoir storage capacity in China was only 30 billion m^3 in 1952, but had increased to 420 billion in 1981 (Runsheng, 1984, quoted by Yu and Buckwell, 1991). Before 1952 most of the 20 Mha of riceland in China were diversion or pump irrigated (Buck, 1937), whereas in 1994 the majority of the 33 Mha grown were storage irrigated.

In Japan it is believed that irrigation was introduced from China about 2400 years ago (Yukawa, 1989). Wood-lined ditches were being used for water distribution by the 2nd century AD, and the first dam was constructed by the end of the 3rd century. The pond system developed slowly but continuously until there are now some 280,000 ponds serving Japan's rice areas. The building of major reservoirs did not commence until about 1960. Many of these are used to service the ponds. All rice in Japan is now storage irrigated.

In South and Southeast Asia the growth of storage systems has been slower, but nevertheless with the considerable development of groundwater systems the total irrigated area is substantial (Table 3.3). Again the major increase in the number of irrigation systems supplied from large reservoirs has taken place since 1950, as has the development of groundwater systems.

Huke (1982) has mapped and quantified the distribution and extent of the four rice farming systems for South, Southeast and East Asia with more detailed studies made of Bangladesh (Huke and Huke, 1983) and other areas (Huke and Huke, 1988, 1990a). The geographic distribution of the different

Table 3.3. Area, yield and production of rice, 1991, from irrigated, rainfed lowland, upland and flood-prone systems.

Region	South Asia	Southeast Asia	East Asia	All Asia	Latin America	Africa	Others	World
System irrigated								
Area	24.1	14.9	34.4	73.4	2.1	1.1	2.1	79.2
Yield	3.59	4.58	5.93	4.9	4.85	5.0	5.8	4.88
Production	86.4	68.3	204.1	358.6	10.2	5.5	12.1	386.6
Rainfed								
Area	20.5	16.1	1.9	38.5	0.4	1.4	0	40.3
Yield	2.41	2.1	3.10	2.3	2.50	2.07	–	2.31
Production	49.5	33.8	5.9	89.2	1.0	2.9	0	93.1
Upland								
Area	7.3	2.4	0.8	10.5	3.7	2.8	0.1	17.1
Yield	0.81	0.96	2.6	1.1	1.57	1.04	1.0	1.2
Production	5.9	2.3	2.1	11.2	5.8	2.9	0.1	20.0
Flood-prone								
Area	5.6	4.4	0	10.0	0.1	1.3	0	11.5
Yield	1.52	1.48	–	1.5	2.0	1.31	–	1.47
Production	8.5	6.5	0	15.3	0.2	1.7	0	16.9
All systems								
Area	57.5	37.8	37.1	133.3	6.3	6.6	2.2	148.4
Yield	2.6	2.9	5.7	3.58	2.73	2.0	5.6	3.50
Production	150.3	111.2	212.1	477.3	17.2	13.1	12.3	519.9

Area, million ha; Yield, t ha^{-1}; Production, million t.
Some smaller countries have been omitted in regional totals, but are included in global totals.
Source: IRRI (1993a)

rice farming systems is shown in Fig. 3.8, together with the relative areas of each in 1991.

Outside of Asia the type of rice farming practiced differs widely between regions. In Europe, the United States, and Australia almost all rice is irrigated, and almost all 'storage irrigated'. The development of rice farming in these areas is of course relatively recent.

In Africa, most rice is produced in Madagascar and Egypt. In Madagascar stream and river diversion systems and 'rainfed lowland' rice, introduced from Indonesia between 1000 and 2000 years ago are dominant (Rabesa Zafera Antoine, 1985). In Egypt the Aswan Dam storage system now feeds the irrigated areas of the Nile delta used for rice production (Momtaz, 1989). Elsewhere in Africa upland rice is widely grown, although it is also produced by a variety of other methods (Buddenhagen and Persley, 1978; IITA, 1984; WARDA, 1984; IRRI, 1985b; Moormann and Juo, 1986). Small amounts of rice have been grown in many indigenous farming systems for probably the

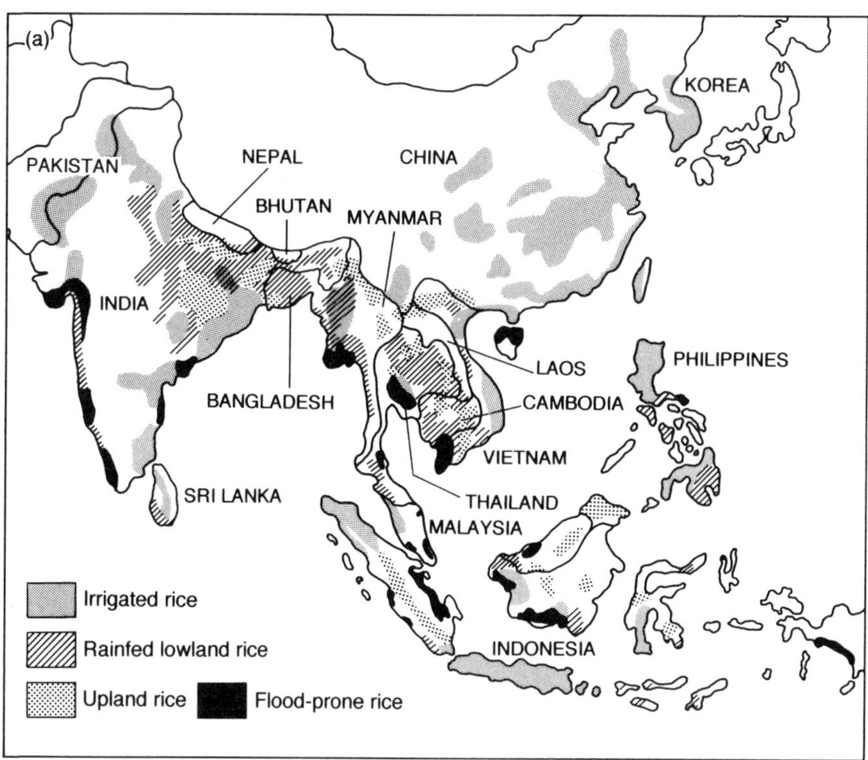

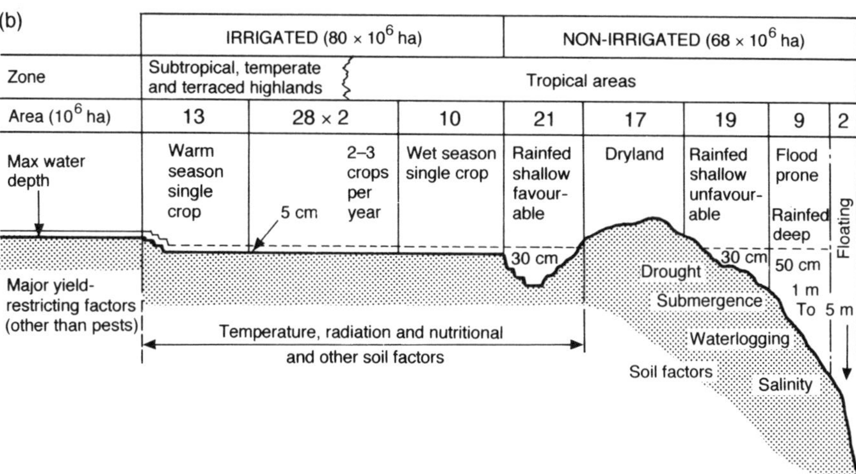

Fig. 3.8. (a) Geographic distribution and (b) relative extent of the major rice production systems (after Garrity, 1984b, and IRRI, 1993b).

last 1500 years (Portères, 1956; Chabrolin, 1977; Carpenter, 1978). The rice may be direct seeded into wet spots, or intercropped with other cereals in shifting cultivation systems. In coastal West Africa rice became a major staple for freed slaves in Sierra Leone and Liberia, possibly introduced from the southern United States. It is mostly grown as rainfed lowland rice although many farmers in the more humid parts of West Africa manage their rice crop in inland valleys by growing rice in a continuum from the lower parts of the valley into areas where the rice has no access to surface water. In the savanna zone of West Africa irrigated rice is produced from relatively small water storage systems, and systems for irrigated rice production are being constructed in a few other areas in Africa. In the traditional *O. glaberrima* areas in the backswamps of the upper Niger River region deepwater rice may still be found, but the amount grown is decreasing as it is replaced by more productive sativas.

In Latin America, upland rice is widely grown as a sole crop in highly mechanized systems (CIAT, 1984; EMBRAPA, 1984) but irrigated rice is becoming increasingly important.

3.4 Rice Yields in Different Production Systems

The productivity of rice can be measured in many ways. Most commonly it is measured in terms of yield per unit area, as land is the most frequent limiting factor. Time, labour, energy, or water consumed can also become limiting, and productivity may then be better considered in terms of these inputs. For rice farming systems other crops produced before or after the rice crop, and animals which may be grazed on the rice lands during the dry season, are often important to the overall productivity of the system (IRRI, 1977, 1982b). The importance of non-rice sources of income in relation to the economic sustainability of rice farming will be discussed in Chapter 7. Here attention is first concentrated on yields per unit area, and then on yield per unit of energy input which reflects a range of factors important to yield.

The yields obtained by farmers are essentially dependent on their choice of variety, management of soil, water and nutrients, and control of pests. Where water supply cannot be controlled yields are essentially dependent on use of a plant type adapted to the conditions. Nutrients supplied by the soil or from the water are less likely to be a limiting factor, although in some favourable sites farmers may be able to obtain an economic return from the use of inputs. Comparison of yields obtained from irrigated, shallow rainfed and upland rice in the Philippines (Fig. 3.9) illustrates typical differences between those obtained from each system. The global and regional production and yields obtained in 1991 from irrigated, rainfed lowland, flood-prone and upland rice are given in Table 3.3.

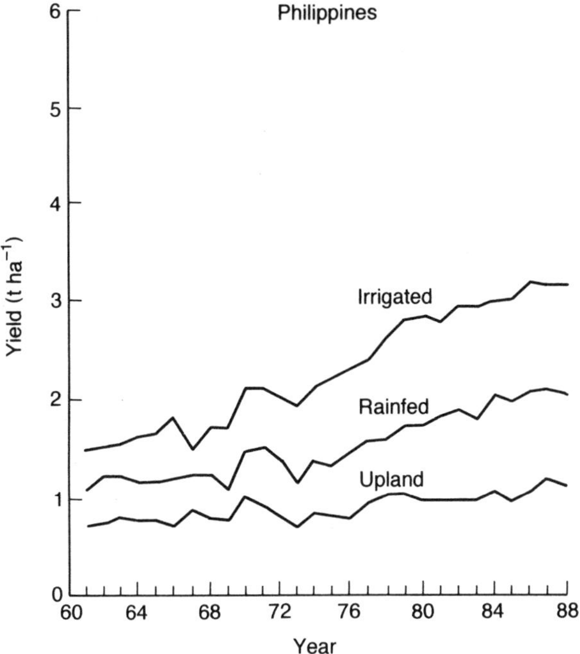

Fig. 3.9. Rice yields in the Philippines by cultural system from 1967/68 to 1988. The irrigated and rainfed data refer only to areas planted with modern varieties (Herdt and Capule, 1983; David, 1991).

3.4.1 The flood-prone system

The global average yield for the flood-prone rice environment is 1.5 t ha^{-1}. This environment includes about 2 Mha of floating rice for which yields are lower than in the remaining 9 Mha of flood-prone rice land, where yields often exceed 2 t ha^{-1} (Catling *et al.*, 1983). In Bangladesh and Thailand (Catling, 1993) and at IRRI yields of flood-prone rice as high as 4 t ha^{-1} have been recorded. Flood-prone rice is rather better in terms of yield per unit area than many shallow rainfed crops, which may be related to the longer period for which the crop is growing. Six months or more is the common duration of flood-prone rice crops, compared with 3–4 months for many shallow rainfed drought-prone rice varieties.

3.4.2 Rainfed lowland rice

Yields from rainfed lowland rice are very variable. The processes of most importance to the understanding of the productivity and sustainability of lowland rainfed rice farming include: the delivery of water to the rice paddy, and its removal from the paddy; the delivery of plant nutrients to the rice crop, from the soil, the water, the sediments deposited in the rice paddy, and any fertilizers or manures which may be used; and the accumulation of salts and the formation of toxins through anaerobic processes in the soil. Where no fertilizers or manure are used the fixation of atmospheric nitrogen by organisms living in or on the paddy water becomes important. In favourable circumstances yields are often comparable with yields obtained from irrigated rice. However, because it is not possible to be certain that water restrictions will not limit yields farmers are often reluctant to use more than low levels of inputs so that full advantage is seldom taken of a favourable season. Selection of the appropriate varieties for these environments is important. Figure 3.10 indicates the yields obtained with several rice varieties across a range of rainfed lowland sites, where the characterization of the site is derived from the mean yield of many rice varieties grown at the site (Seshu, 1986). Varieties such as Mahsuri are available with a wide tolerance to adverse conditions, but which do not yield particularly well when conditions are more favourable. Other varieties such as IR 19431-72-2 do poorly in adverse conditions but perform well in more favourable seasons.

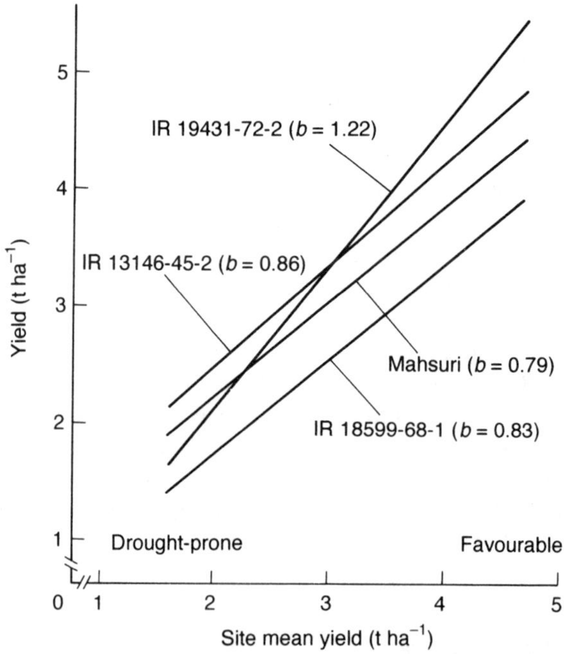

Fig. 3.10. Breeding of rice varieties adapted to rainfed shallow water conditions. Regression of varieties on site means (24 varieties, 13 sites; data from Seshu (1986)). (IR19431-72-2 rice variety selected for favourable sites; IR1316-45-2 variety selected for wide adaptability.)

The global average yield for rainfed conditions is 2.3 t ha^{-1}, but there is a very wide variation, both from year to year and site to site.

3.4.3 Upland rice

Yields of upland rice are subject to the same factors that determine the yields of other upland crops, including the problems of water deficiency to which rice is particularly susceptible, nutrient deficiency and its exhaustion if cropping is continued without adequate replacement of the nutrients removed in the crop, and soil erosion. Yields are generally low, and often less than 1 t ha^{-1}, particularly in the areas of shifting cultivation. In highly favourable circumstances, with adequate fertilization and pest control, yields between 4 and 5 t ha^{-1} have been recorded at some sites in South America (Laing *et al.*, 1984). The average yield of upland rice in Latin America is 1.5 t ha^{-1}, whereas the global average is 1.2 t ha^{-1}. Water deficiency is the most common limiting factor, but weed competition and nutrient deficiencies are also important.

3.4.4 Irrigated rice

The advantage of irrigation is not only that the risk of water shortages is removed. Planting dates can be optimized, and frequently a dry season as well as a wet season crop can be obtained. In the wet season, irrigation water is usually supplementary to that received from rainfall. On most of the lands where rice is grown in Asia more than a metre of rain falls during the life of the wet season crop. Most is retained by the bunds around the paddy fields. Percolation losses of water are usually small because the soils are puddled before planting of the crop. Once water supply is assured other factors may become limiting, but it is normally economic to use manures, pesticides, and weed control practices.

Where sufficient water is available to grow a crop in the dry season yields may be considerably higher than in the wet season (Fig. 1.7). Storage and pump systems with sufficient capacity to provide water through both seasons cover just over one half of the total irrigated rice area (Fig. 3.8). In well controlled irrigated systems at IRRI dry season yields commonly exceed 8 t ha^{-1} whereas wet season yields are often below 5 t ha^{-1}. In California, and Australia where radiation intensities during the rice growing season are very high, yields well in excess of 10 t ha^{-1} are regularly recorded. The average yields for these two regions have exceeded 8 t ha^{-1} for many years. In Egypt where radiation intensity is also high but input levels often lower, national average yields have exceeded 5 t ha^{-1} for several years, and in 1991 reached 7.3 t ha^{-1}.

In addition to high radiation intensity, yields in low rainfall areas, and for dry season as opposed to wet season crops in other areas, are better because pest problems tend to be less important. Both insect and disease pressure are normally low. A further factor contributing to the higher yields obtained from irrigated rice is that considerably more effort has gone into breeding for higher yields from irrigated rice than has gone into breeding for the other rice ecosystems. In the rainfed, flood-prone and upland systems, tolerance to too little or too much water has to be the primary consideration and this severely limits the plant breeders' options.

3.5 Rice Yields and Energy Inputs

The high levels of inputs used to produce irrigated rice enable high yields to be obtained. They also make rice production rather demanding in terms of energy and water use (Huke, 1982; Flinn and Duff, 1985). The efficiency of irrigated systems is dependent on the assumptions used in determining the energy equivalents required for the delivery of irrigation water. If the system is well designed and the efficiency of water use is relatively high, which means better than 50%, then irrigated rice farming compares well with other intensive crop production systems. Evans (1993, p.362) has in fact shown that the data for the wide range of rice production systems assembled by Flinn and Duff (1985) indicates little difference in the efficiency of energy use in rainfed and irrigated systems.

While the highest yields are obtained from dry season irrigated rice, the system is very demanding in its water requirement. Field losses are due to evaporation and transpiration, land soaking and preparation before planting the crop, percolation and seepage losses while the crop is in the field, and surface run-off. If a system could be operated at 100% efficiency the amount of water needed for the rice crop would be the sum of the field losses. In fact distribution losses have mostly been found to exceed these requirements, so that the water requirement for a crop that is in the field for 100 days may be anywhere between about 150% and 300% of the evaporation and drainage requirements. Figure 3.11, based on Levine (1971), illustrates the importance of irrigation system management to actual water use in rice production. Increasing attention is deservedly being given to the relation between sustainability and the availability of water to irrigate rice fields, and water management methods to minimize losses and maximize the efficiency of its use (see e.g. Murty and Koga, 1992; Saleh and Bhuiyan, 1995). As the area of irrigated rice continues to increase the critical importance of water supplies to ensure the maintenance of rice and other food crop production in the face of competition in water demand from urban and industrial users is starting to be recognized even in high rainfall areas where availability of water has in the past been taken for granted (Clarke, 1991; Bhuiyan, 1993; Rosegrant and Svendsen, 1993; IRRI, 1995b). The topic is discussed further in Chapter 6.

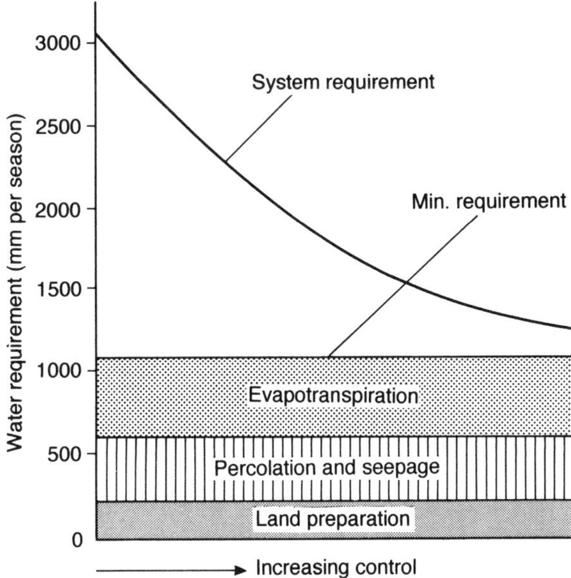

Fig. 3.11. Example of water requirement for component uses in lowland rice as affected by degree of control in the system (modified from Levine, 1971).

The energy and water requirements of rice are both sensitive to the length of time required for the crop to mature. The time required is also important in relation to the opportunity to harvest a second crop in a year or season. Where a dry season water supply is available, the second crop is commonly rice. Otherwise the second crop may be an upland crop grown after rice. Multiple cropping in which a second crop is planted after the rice crop matures has been practised for many years in the ricelands. The importance of the introduction of the champa varieties from Vietnam to China has been mentioned earlier. Much more recently it has been found possible to develop double rice cropping in some favourable rainfed lowland rice areas in the Philippines and elsewhere by direct seeding the first rice crop and following it with a transplanted crop, using short duration varieties (IRRI, 1977, 1982b). More commonly a legume or a short duration upland cereal is grown after an initial rice crop, using the residual moisture in the soil after the rice crop has been harvested. An important example of this type of multiple cropping is the rice–wheat system of the Indian and Pakistani Punjab, the terai of Nepal, and northern Bangladesh and southern China. Rice is the summer crop grown under the monsoon and wheat the winter crop. While these multiple cropping systems are very labour demanding, they are also very productive although, as will be discussed later, doubts have arisen about the sustainability of this system at high production levels.

The obvious advantages to be obtained from multiple cropping have meant that a great deal of effort has been put into breeding of short duration rice varieties. Earlier short duration varieties were low yielding, and pest and photoperiod sensitive. The photoperiod sensitivity had to be bred out of the varieties so that their planting time was not restricted. Plant breeders have now produced several high-yielding, pest resistant, short duration and photoperiod insensitive varieties, which mature in less than 100 days after transplanting, and yield over 60 kg per day (Khush, 1987). Breeders in Japan and China have produced cold-tolerant early maturing varieties adapted to the short summer seasons of latitudes as far north as 54°.

By incorporating pest and disease resistance into rice varieties (Khush, 1977, 1989) and by using biological control methods rather than chemical sprays to control insect pests (Waage and Greathead, 1988) considerable savings in energy use have been achieved.

4 The Biophysical Basis of the Sustainability of Rice Farming

4.1 The Biophysical Components of Sustainability

To sustain the production of any crop requires that water remains available; that the nutrients removed in the crop are replaced in the soil; that the soil is not rendered a less suitable medium for plant growth, by either chemical or physical changes; and that there is no serious build-up of either crop or human pests. Where upland crops are the basis of food production, nutrient depletion, acidification and the build-up of weeds can cause a rapid decline in productivity, particularly in many tropical regions. Unless measures are taken to correct the changes, yields become so low that the land has to be abandoned and a new site found for cultivation. Before the advent of inorganic fertilizers this meant that fields and, at longer intervals, homes, had to be abandoned and a new site developed to enable sufficient food to be grown to maintain the community (Nye and Greenland, 1960; Ruthenberg, 1971; Greenland, 1974). By contrast rice production in the wetlands has proved remarkably stable wherever water supplies are assured.

As shown in Chapter 2, rice farming has been practised in Asia for at least 9000 years. In the early years floods, river migrations and wars often caused movement of people and fields. But water control led to continuing and adequate yields of rice, so that stronger and socially and politically organized communities could develop, resident in stable villages and cities. In wetland sites cultivation appears to have been continuous, much as it is today, although pest problems, using pests in the broad sense to include microbiological, mammalian and weed pests as well as insect pests, and human as well as plant health problems, sometimes persuaded people to leave land fallow for a season, or if it was possible to do so to move to a more favourable site.

Palynological studies in Japan indicate that rice cultivation has probably been practised at the Tsukimino Moor site for at least 1600 years (Fig. 2.2). An interruption appears to have occurred about 900 years ago, for reasons which have not been established, but may relate to the aridity of the climate at that time. The presence of *Oryza* pollen from a depth of 170 cm to the surface with only one small break is strong evidence for cultivation of rice at the site for consecutive periods of the order of 700 and 900 years. At Hemudu in China the archaeological evidence suggests that rice cultivation was practised from about 8500 BP to 3500 BP, at which time it was probably interrupted by rising sea levels. At Khok Phanom Di in Thailand, Maloney *et al.* (1989) found pollen of *Gramineae* which included rice to depths of more than 2500 cm. The lowest depth also yielded charcoal carbon dated to over 4000 BP. Here it appears that rice may have been cultivated at the same site for over 4000 years. At these sites and many like them the biophysical factors of sustainability must have been satisfactorily maintained over a very long period of time.

The sustainability of different rice production systems over very long periods cannot be understood without some appreciation of the chemistry, physics and biology of rice soils. Many advances in the understanding of soil problems have been made in recent years (IRRI, 1982d, 1984d, 1985d, 1987b; IBSRAM, 1991). In this chapter the soil and pest factors determining the biophysical stability of rice production are introduced. In the following two chapters nutrient supply and water requirements are considered in greater depth.

4.2 The Physics, Chemistry and Biology of Paddy Soils in Relation to Biophysical Sustainability

The physics, chemistry and biology of wetland soils differ considerably from those of dryland soils. Not only is oxygen largely excluded from the soil, so changing its chemistry, but terracing, the building of bunds and the covering of the soil with a water layer protects the soil from the most damaging of processes affecting its long-term productivity, soil erosion. The basic properties of wetland soils which have been fundamental to the long-term sustainability of rice farming systems in Asia include:

1. they do not become acid after continuous cultivation for reasons associated with the physical chemistry of flooded soils;
2. because of their position in the landscape nutrients tend to be leached into the soil rather than out of it;
3. phosphorus is usually more readily available to the rice plant because iron is present in the ferrous rather than ferric state, and ferrous phosphates are significantly more soluble than ferric;

4. nutrients adequate for moderate rice yields are replenished by those in the flood and irrigation water, and in silt deposited from floodwaters;
5. the active population of nitrogen fixing organisms in rice paddies helps to maintain a level of organic nitrogen in the soil sufficient to support a modest level of rice production, and
6. as mentioned above, erosion is unlikely to occur because the rice paddies are levelled, surrounded by bunds and covered by water.

These special properties of paddy soils derive in part from the landscape positions in which most rice is grown, the water supply and drainage characteristics of the soil, and intrinsic properties, such as texture and nutritional status. Kawaguchi and Kyuma (1977) reviewed the fertility status of paddy soils in Japan and tropical Asia, and Kyuma (1991) has recently discussed their results in relation to more recent studies. Compared with soils in countries of the Mediterranean region the Japanese soils generally have a higher level of plant-available nutrients, as do the tropical paddy soils (Table 4.1). The Japanese and tropical soils both contain relatively high levels of potassium and other cationic nutrients. The levels of phosphorus and nitrogen in the tropical soils are lower than in the Japanese soils, probably reflecting the higher levels of manuring used in Japan over many years, the derivation of many of them from recent volcanic ash deposits, and the slower rates of mineralization of carbon and nitrogen in the Japanese soils than at the higher temperatures of the tropical soils.

There are also negative factors associated with anaerobism induced by flooding. These include the fact that some heavy metals such as iron and manganese are more soluble in the reduced state and their concentrations may reach toxic levels, and some phytotoxic organic compounds may form during the anaerobic decomposition of fresh organic material. Further, if there is no drainage from the soil the concentration of salt in the soil solution may reach levels at which the growth of the rice plant is affected. Reduction of sulphur to sulphide is often described as an important adverse process in paddy soils, but as will be discussed further below, the problems of sulphide formation are largely negated by its prompt precipitation as iron and other highly insoluble sulphides.

4.2.1 Physical processes in paddy soils important to sustainability

Not directly associated with anaerobism but important to the total productivity of rice soils are the effects of the puddling process. The puddling process used in wet rice cultivation is always important for soils where there would otherwise be a rapid loss of water by seepage and percolation. These are primarily the soils on the alluvial and montane (anthropomorphic) terraces. In upland and alluvial soils where the water table is normally well below the soil surface, puddling

Table 4.1. Chemical characteristics of paddy soils from tropical Asia, Japan and Mediterranean countries (a) analyses of air-dry soils (b) total chemical composition of ignited samples.

(a)

Country	No. of samples	pH	C (%)	N (%)	NH_4-N (mg 100 g^{-1})	Extractable P (mg 100 g^{-1})	Exchangeable (meq 100 g^{-1})		
							Ca	Mg	K
Bangladesh	53	6.1	1.18	0.13	6.1	21.0	7.8	2.7	0.3
Bangladesh*	52		1.52	0.23	2.12	14	8.2	2.5	0.3
Cambodia	16	5.2	1.09	0.10	4.0	1.9	5.4	3.2	0.2
India	73	7.0	0.85	0.08	2.7	21.9	15.0	6.5	0.5
Indonesia	44	6.6	1.39	0.12	14.1	10.0	17.8	6.3	0.4
Malaysia, East	36	4.5	9.66	0.64	30.2	5.0	3.5	3.8	0.4
Malaysia, West	41	4.7	3.36	0.28	14.9	8.2	3.9	5.2	0.4
Myanmar	50	4.8	1.18	0.12	3.5	7.8	7.2	8.4	0.4
Philippines	54	6.4	1.66	0.15	17.2	13.4	14.8	9.3	0.5
Philippines†	52		1.84	0.17		1.6	13	5.2	0.5
Sri Lanka	33	5.9	1.41	0.13	8.4	9.0	5.4	3.5	0.2
Thailand	80	5.2	1.05	0.09	5.2	4.7	7.2	4.3	0.3
Thailand‡	166		1.16	0.10	3.8				
Vietnam	49	4.5	2.49	0.20	7.7	5.1	7.1	5.4	0.4
Japan	84	5.4	3.33	0.29	17.5	46.5	9.3	2.8	0.4
Mediterranean countries	62	6.8	1.82	0.16	7.6	–	15.9	4.6	0.6

(b)

Country	No. of samples	Total (%)			
		P_2O_5	CaO	MgO	K_2O
Bangladesh	53	0.12	1.01	1.18	3.09
Cambodia	16	0.09	0.27	0.40	0.68
India	73	0.12	1.91	1.20	2.55
Indonesia	44	0.18	3.00	1.22	0.72
Malaysia, East	36	0.14	0.26	0.68	1.72
Malaysia, West	41	0.14	0.14	0.51	2.02
Myanmar	50	0.08	0.51	1.84	2.16
Philippines	54	0.17	2.83	1.26	0.99
Philippines†	34	0.2	2.6	2.2	0.8
Sri Lanka	33	0.14	1.00	0.66	1.82
Thailand	80	0.09	0.43	0.52	1.42
Thailand‡	166	0.56			
Vietnam	49	0.06	0.26	1.01	2.70
Vietnam§	14	0.03	2.70	0.20	0.60
Japan	84	0.12	1.86	0.75	2.14

Source: Kyuma (1991), except *Bangladesh from Sattar (1988), †Philippines from Miura *et al.* (1995), ‡Thailand from Cholitkul and Sangtong (1988), and §Vietnam from Uehara *et al.* (1974.)

and flooding the soil for rice leads to reduction of the surface soil to the depth of puddling and a little below, but most of the subsoil is likely to remain oxidized. A 'perched water table' is formed in the puddled layer (Fig. 4.1). These soils with a perched water table are sometimes referred to as 'surface water gleys' or 'pseudogleys'. The soils where the water table is at or above the soil surface may be referred to as 'groundwater gleys' or 'stagnogleys'. More information about the characterisation and classification of wetland soils is given in *Soils and Rice* (IRRI, 1978), *Proceedings of Symposium on Paddy Soil* (Institute of Soil Science, Academia Sinica, 1981) and *Wetland Soils: Characterization, Classification, and Utilization* (IRRI, 1985c).

The stagnogleys are found closer to rivers, in the backswamps, and other low lying positions. All the soil may remain waterlogged and reduced for most or all of the year (Fig. 4.1). These are the soils on which deepwater and flood-prone rice is grown. Generally they are less productive than the surface water gleys.

The pseudogleys on the alluvial terraces, especially those in the higher positions, are where most rice is produced. They show some internal drainage,

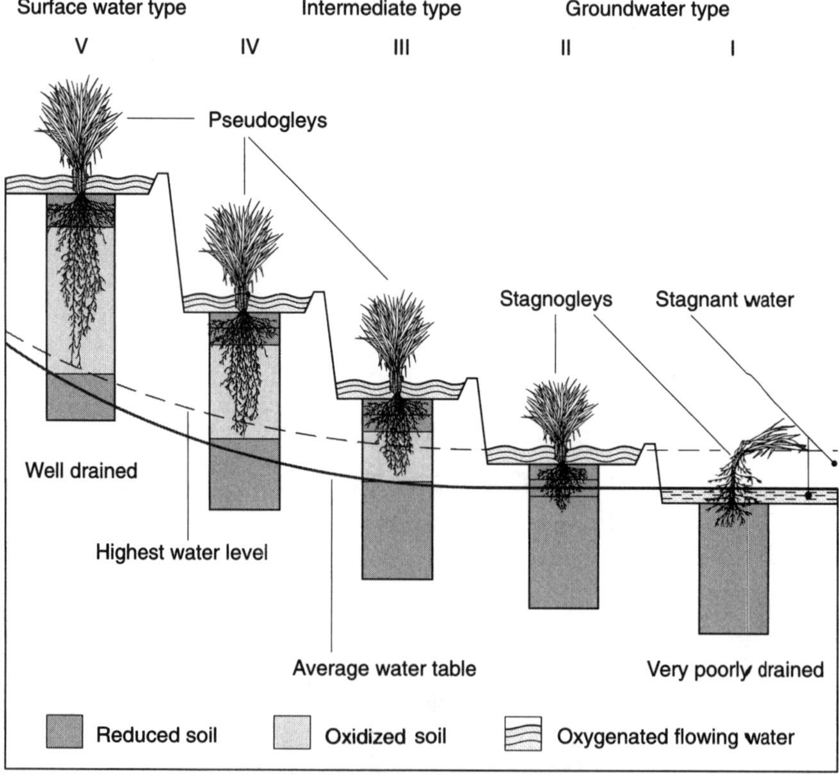

Fig. 4.1. Soil profiles in a wetland landscape (after Kanno, 1962).

whereas in the groundwater or stagnogleys there is no downward water movement unless artificial drainage systems are constructed, usually an expensive operation. Movement of water through the soil helps to remove any toxic compounds which may reduce rice yields. Hence the soils in the higher landscape positions are those usually associated with higher rice productivity. To maintain the high productivity of the pseudogleys it is essential that the surface drains are kept open, to allow water to drain through and across rice paddies, a lesson learnt many years ago by the Chinese and others as they developed their water control systems. Figure 4.2 is an example from China (Fu Ming-hua, 1981) of the relation between the 'high-yield paddy soils' in slightly elevated landscape positions and those in lower positions. To increase the productivity of the stagnogleys subsurface drainage has long been used in Japan (Kohno, 1992) to supplement the surface drains.

The puddling process – cultivation of the soil when it is saturated with water – destroys the soil aggregates and disperses the fine clay particles (Greenland, 1981; Sharma and De Datta, 1985a, b, 1986). The puddled layer becomes soft so that it is easy to plant rice seedlings. The dispersed clay particles tend to settle to the base of the cultivated layer, making it more densely packed and less permeable (Koenigs, 1963; Sanchez, 1973a, b; Greenland, 1985; Adachi and Inoue, 1991; Pagliai and Painuli, 1991). The compaction of the layer at the base of the cultivated soil due to animal and human trampling and the passage of cultivation implements enhances the effects of puddling on

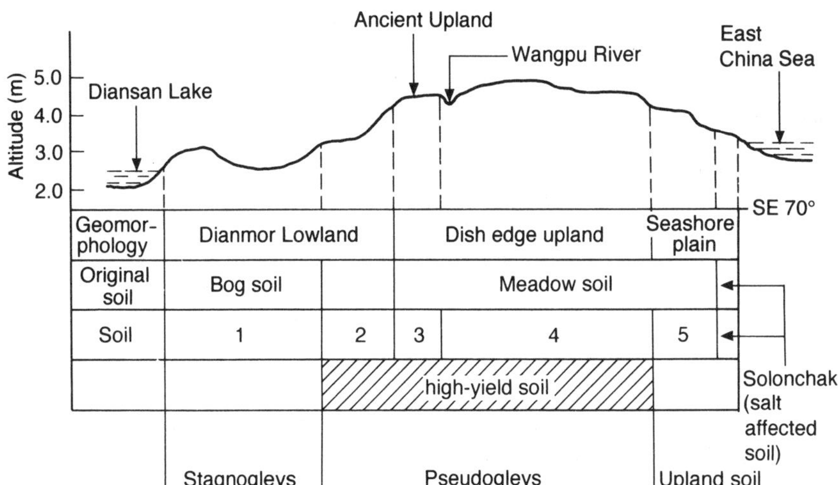

Fig. 4.2. The profile of geomorphology and soil distribution in the suburbs of Shanghai. 1. Greyish-blue lowland paddy soil; 2. Greyish-yellow lowland paddy soil; 3. Greyish-brown upland paddy soil; 4. Brownish-yellow upland paddy soil; 5. Sandy loam paddy soil (after Fu Ming-hua, 1981).

permeability by creating a 'traffic pan', which is a common feature of many paddy soils (Dexter and Woodhead, 1985).

The puddling process is aimed to reduce percolation sufficiently so that water demands are not excessive, but not so much that drainage is totally prevented. In addition puddling should level the surface of the paddy field so that the applied water covers the soil uniformly. This allows a depth of 2–10 cm of water to be maintained above the surface of the soil. Such shallow flooding is optimal for the rice crop (Sanchez, 1973a; De Datta, 1981). The water depth and movement across the paddy fields can be conveniently controlled by opening and closing small gaps in the bunds around the paddy field.

In addition to reducing water losses from the soil there are other desirable effects of puddling, such as destruction of weeds, and lowering of soil strength in the puddled layer above the pan (Salokhe and Shirin, 1992). Reduction of soil strength makes cultivation, transplanting and root growth in the softened puddled layer easier. These effects in softening the topsoil and controlling weeds are of course important whether or not drainage of the soil can occur, and so puddling is and has been a normal practice on the stagnogleys of the backswamps and other locations which are naturally flooded, as well as on the pseudogleys where it is important to reduce the rate of drainage.

When an upland crop is grown after rice, dispersal of the soil by puddling usually has the undesirable effect of producing dense clods when the soil dries (Fig. 4.3). This makes it difficult to prepare a seedbed in which to establish

Fig. 4.3. The physical state of soils after flooding and drying.

upland crops. The traffic pans which are a common feature of paddy soils will also have the undesirable effect of restricting root development not only of the rice crop but also of an upland crop that may be grown after rice (Prihar *et al.*, 1985; Pagliai and Painuli, 1991). The water retained in the subsoil on which the post-monsoon season crop depends cannot then be fully exploited.

4.2.2 Chemical processes in paddy soils in relation to sustainability

By retaining water in the soil, air is excluded. In the absence of oxygen soil organisms seek their oxidizing source (terminal electron acceptors) from other soil constituents, and the soil is thereby reduced to a less oxidized condition (Ponnamperuma, 1984a). The most visible change that occurs when soils are reduced is that the red and brown compounds of ferric iron are reduced to blue-grey compounds of ferrous iron. Before this change occurs organisms will have reduced nitrate ions to nitrogen, and manganic manganese to manganous. Where soils contain much ferric iron or manganic manganese they may be buffered at a redox potential above which deleterious changes do not occur. If all the ferric and manganic ions can be reduced oxygen will be taken from sulphates, which are reduced to sulphide. In more strongly reduced conditions carbon metabolism does not lead to the formation of carbon dioxide, but to the production initially of organic acids and then of methane and other hydrocarbons. It is theoretically possible that phosphates could also be reduced, with the formation of phosphine. The ease of reduction of different materials can be determined quantitatively by electrochemical methods and expressed in terms of a redox potential (Fig. 4.4).

These reductions are all brought about by microorganisms, and require an energy supply if they are to occur. Thus in a flooded sterile soil, little change will take place, and in a non-sterile soil with a only a limited supply of organic material as a source of energy for the organisms the reduction process soon comes to a halt. If only thermodynamic considerations operated the reductions would occur sequentially with each beginning only when the previous stage has been completed. However, microsite variability allows several reactions to occur simultaneously. Thus much decomposable organic matter occurs in the soil as rather large particles. Around these particles reduction is more intense, so that except in a well-stirred soil the redox potential is lower around the decomposing organic particles than in the bulk soil. This accounts for the fact that apparently anomalous reductions have sometimes been reported. For example, sulphate has been reported to be reduced to sulphide at redox potentials as high as +400 mV (Howarth and Jorgensen, 1984, quoted by Lefroy *et al.*, 1992).

The greater solubility of ferrous than ferric compounds means that there is a tendency for the concentration of cations in solution to be higher than in aerobic soils (Yu Tian-ren, 1985). Both ferrous and other cations are found at

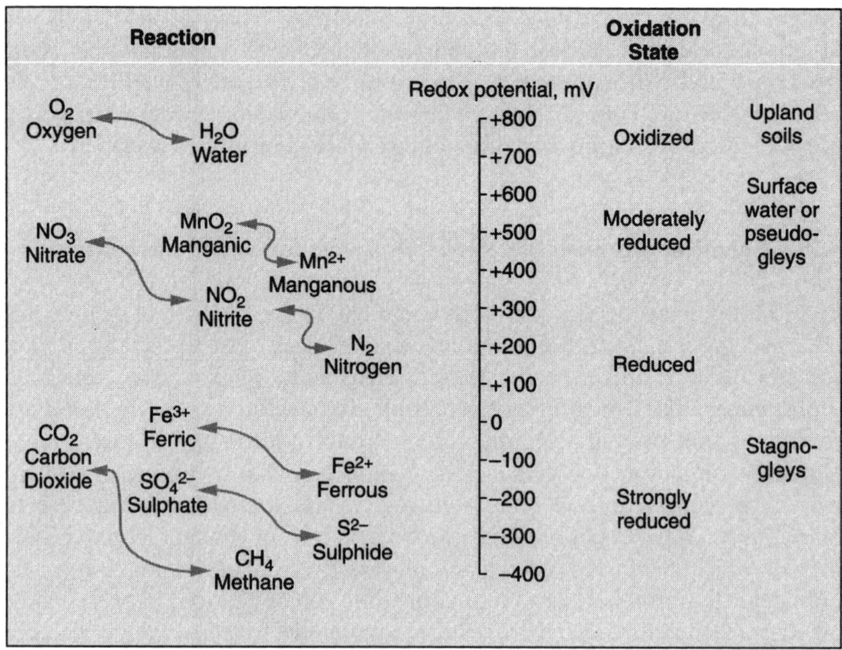

Fig. 4.4. Oxidation–reduction processes in flooded soils and the redox potentials at which changes occur (after Rowell, 1981).

higher concentrations because other cations are displaced from the soil exchange complex by ferrous ions.

The details of the chemistry of iron compounds in flooded soils are quite complex, because their solubility is affected by factors such as the acidity (pH), the redox potential and the presence of organic compounds which can form complexes with the ferrous ions (Patrick and Reddy, 1977; Yu Tian-ren, 1985). Similarly the actual solubility of phosphorus depends on several factors, although in many paddy soils pH and redox potential often dominate. The more strongly reduced and the more acid the conditions, the more phosphorus is found in solution (Kirk and Saleque, 1995).

Adjacent to the rice roots there will normally be an oxidized zone about 3–5 mm thick (Trolldenier, 1981; Liu Zhiyu *et al.*, 1990). This arises from the transport of oxygen through the aerenchyma of the rice plant to the root surface. Ferrous iron is oxidized at the root surface, where it is precipitated as ferric hydroxide and much acidity is released (Kirk, 1990; Kirk *et al.*, 1990; Begg *et al.*, 1994). The ferric hydroxide stains the rice roots brown. This is usually an indication of a healthily growing root system. The precipitation of ferric hydroxide prevents excess ferrous ions entering the root, and is important to the prevention of iron toxicity (Ottow *et al.*, 1983). In strongly reduced soils of high iron or manganese content insufficient oxygen may reach the root surface to oxidize all the ferrous iron and manganous manganese reaching the root surface. A black stain of ferrous sulphide may then form on the surface of the root, a sure indication of excessively reduced conditions, and a normal accompaniment of iron toxicity.

The susceptibility of rice to iron and manganese toxicity differs considerably between rice varieties, and with other soil conditions such as the supply of potassium (Ottow *et al.*, 1983; Trolldenier, 1988). The levels of iron and manganese which can be tolerated by rice plants are relatively high. For instance Cheng and Ouellette (1971) give 7000 ppm as the critical tissue concentration for manganese toxicity and Yoshida (1981) 300 ppm for iron toxicity. The actual levels at which yield loss occurs depend very much on the supply of other nutrients.

The solubilization of iron and manganese under anaerobic conditions means that they move downwards in the soil profile when percolation occurs. When they meet an oxidized soil layer, as they will normally do in the better drained paddy soils, the iron and manganese will precipitate. If the soil dries to this depth the pan may become partially cemented by hydrous oxides, increasing the strength of the pan and its resistance to root penetration (Koenigs, 1950). Kanno *et al.* (1964) and Motomura *et al.* (1970) have discussed the presence of this type of feature in Japanese soils. They suggest that formation of the cemented pans can occur in 200 years. Irreversible cementation of this sort is rare, although traffic pans are a common feature of paddy soils. The rarity is presumably because the usually high and changing level of the water table from season to season seldom allows the subsoil to become sufficiently dry. The presence of organic agents which form stable complexes with the iron and manganese compounds, and the change in the level of the soil surface due to silt deposition, and other factors will also affect the translocation and cementation process.

As noted earlier the reduction processes are all associated with the activities of soil microorganisms. Organic substrates are essential if microorganisms are to grow actively. In most soils there is normally at least some organic material available to them. Following the harvest of a crop there are always root and other residues available, and during the growth of the crop root exudates as well as dead root material become available. Further, algal and other organisms grow on the surface of the paddy water and take their carbon supply from the air; as they die they provide a further substrate for other organisms. Roger and Watanabe (1984) quote dry weight productivity of some aquatic macrophytes falling mostly in the range 1–5 t ha^{-1} per season, and also note some much higher figures. The average amount of organic material accumulating in paddy soils from sources other than the rice crop has been estimated (Neue, 1985) to be from 1–2 t ha^{-1} per cropping season. Additions from roots, stubble and straw will be equal to this or rather greater. Thus there will normally be sufficient organic substrate available for reduced conditions to prevail when a paddy soil is flooded, although the intensity of reduction will differ between soils and at different times in any one soil. When irrigation water is kept flowing across a rice paddy it usually carries dissolved oxygen, and so maintains an oxidized layer at the soil–water interface. If there are high concentrations of ferric and other material present in the oxidized state the soil will be buffered against too strongly reduced conditions. However when large quantities of fresh organic material are added the majority of soils become strongly reduced, and copious amounts of methane and other hydrocarbons are evolved. Hence the old name of 'bog gas' for methane. Methane and other hydrocarbons are more active greenhouse gases than carbon dioxide, and so their evolution from rice paddies is a topic of environmental concern.

In addition to the effects of oxidation and reduction flooded soils differ from others in the way that acidity and alkalinity are controlled. The pH of the floodwaters is determined by the concentration of carbon dioxide in the water. If the water is in equilibrium with the atmosphere this means that the pH is close to 6.0, or nearly neutral. At this pH most plant nutrients are able to maintain solution concentrations well suited to plant growth. Soils which are initially acid when flooded tend to equilibrate with the floodwater and become less acidic in a matter of days or weeks. The neutralization is aided by the reduction of iron, which releases hydroxyls to neutralize the acidity. Soils which are initially alkaline move towards neutrality, usually a little more slowly (Fig. 4.5). The partial pressure of carbon dioxide in the floodwater buffers the carbonate and reduces the pH (Ponnamperuma, 1972), sometimes aided by the organic acids which are formed during the initial stages of the decomposition of organic matter.

In the paddy water above the soil other important changes occur. Photosynthetic algae are almost ubiquitous in paddy water. During the day when they are photosynthesizing they take carbon dioxide from the water, with the consequence that the pH of the water rises, usually sufficiently for the water to become quite alkaline (Fig. 4.6). At night the pH falls as the algae are no longer photosynthesizing (Fillery *et al.*, 1984; Bouldin, 1986).

Around pH 6 the concentration of phosphate in the soil solution is largely dependent on the concentration of iron, with which it forms sparingly soluble iron phosphates. As mentioned above, ferrous phosphates are slightly more soluble than ferric, so that reduction usually releases extra phosphate to the soil solution (Fig. 4.7). While this means that growth of the rice crop is enhanced, it also means that phosphate may be more liable to loss from paddy soils by seepage and percolation. An example of the kinetics of ferrous iron on flooding of a paddy soil is given in Fig. 4.8.

4.2.3 Organic matter in paddy soils in relation to sustainability

Much has been written about the importance of organic matter in tropical soils, and the difficulty of maintaining adequate levels because of the rapid rate at

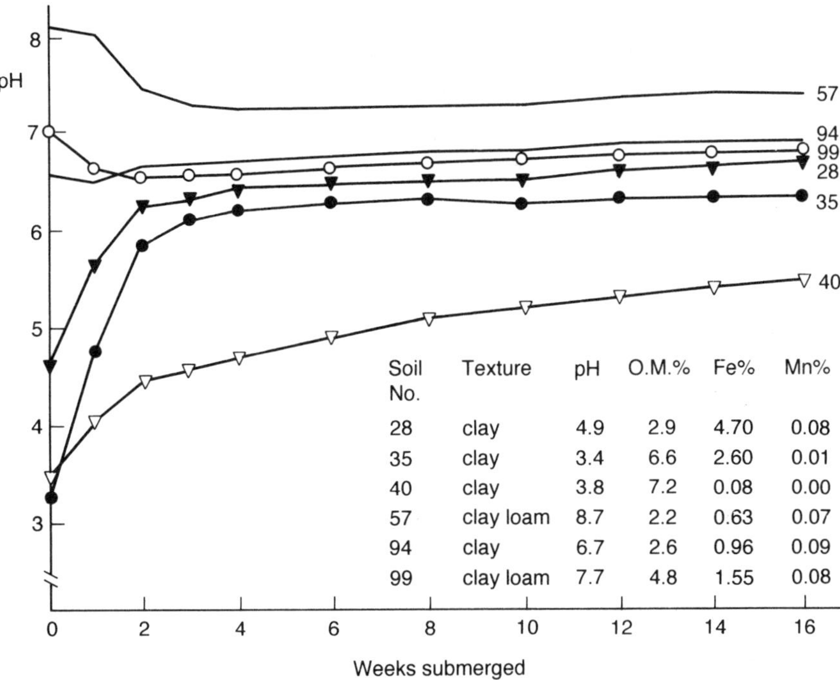

Soil No.	Texture	pH	O.M.%	Fe%	Mn%
28	clay	4.9	2.9	4.70	0.08
35	clay	3.4	6.6	2.60	0.01
40	clay	3.8	7.2	0.08	0.00
57	clay loam	8.7	2.2	0.63	0.07
94	clay	6.7	2.6	0.96	0.09
99	clay loam	7.7	4.8	1.55	0.08

Fig. 4.5. Trends in the pH values of some submerged soils (Ponnamperuma, 1972).

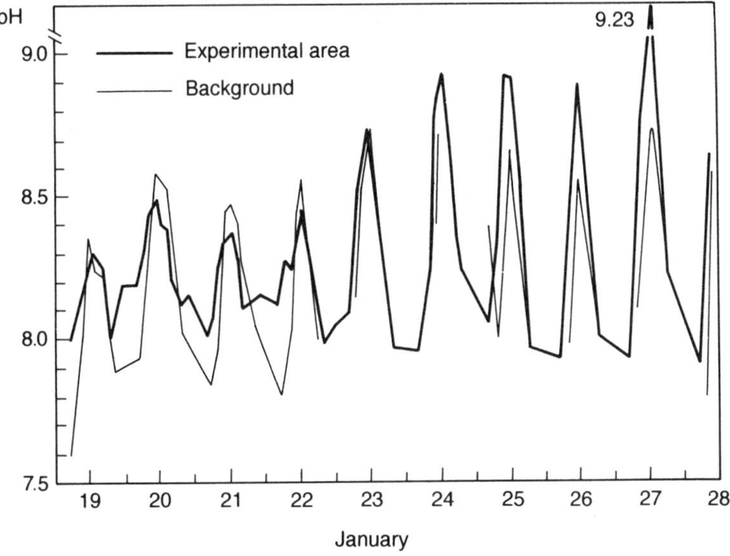

Fig. 4.6. Diurnal changes in the pH value of paddy water (IRRI, 1982e).

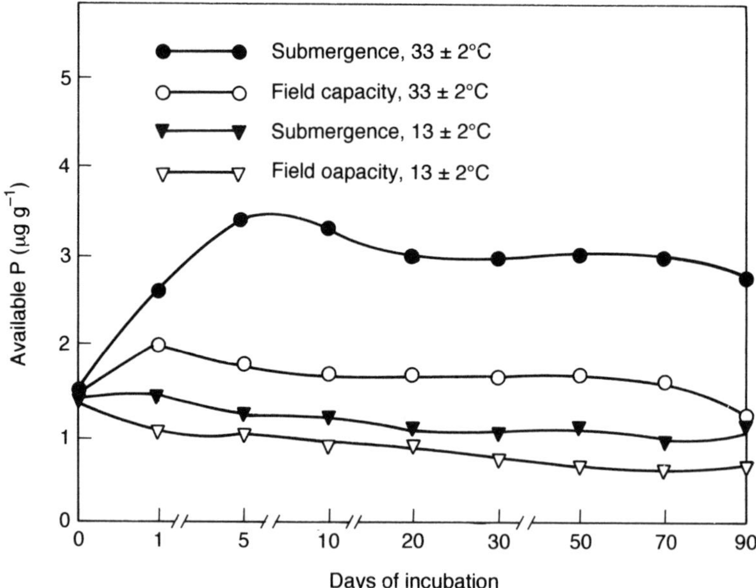

Fig. 4.7. Effect of soil submergence on the concentration of phosphate in the soil solution (Morris and Meelu, 1985).

which it is oxidized (Greenland *et al.*, 1992). Although the difficulty of maintaining a satisfactory level of organic matter in tropical upland soils has often been exaggerated, the higher temperatures do produce a more rapid rate of decomposition. Slower oxidation of organic material in wetland soils of the temperate zone usually leads to accumulation of organic matter and often to the formation of peats. Direct extrapolation of this difference between upland and wetland soils to the tropics would lead to the expectation that organic matter levels in paddy soils would exhibit higher levels of organic matter than tropical upland soils. The mean levels of organic carbon in the topsoils of many paddy soils from tropical Asia are given in Table 4.1. The overall mean is close to 2%. The soils from East and West Malaysia and some from Vietnam however appear to be atypical, with pH values 4.5 and 4.7, and much higher carbon contents than those from other countries. Most of these samples came from acid peaty swamps which had been reclaimed for rice production, and a few came from acid sulphate soils (Kawaguchi and Kyuma, 1977). If these soils are excluded the mean is close to 1%. This figure may be compared with the range of 1.27–1.81% given for 510 upland samples (oxisols and ultisols) from the Cerrado region of Brazil (Sanchez, 1981). The level of organic matter in the soil is affected not only by aeration and temperature but also by clay content and the quantities of organic matter which are returned to the soil each year. The mean clay content of the paddy soils other than the Malaysian and Vietnamese

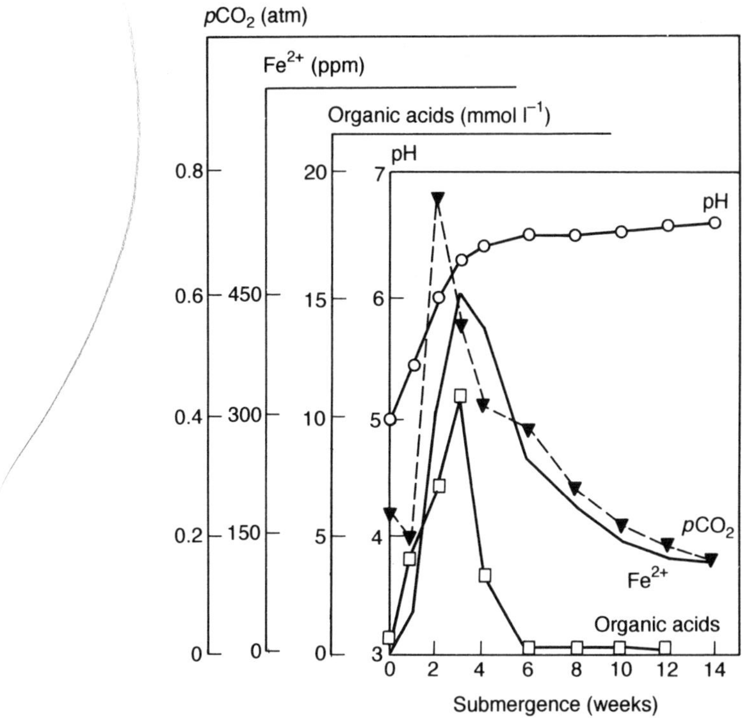

Fig. 4.8. Changes in pH, pCO_2, water-soluble Fe^{2+}, and organic acids in flooded stagnogley (Luisiana) soil at 30°C during 12 weeks following flooding (Ponnamperuma, 1985).

soils was 38.4%; that of the Cerrado soils was not given but they would be expected to be lower. It is also probable that the return of organic matter to the soil from the sparse Cerrado vegetation would be less than from the rice in the paddy soils. Nevertheless it is clear that many paddy soils, in spite of the fact that they are flooded for much of the year, do not contain significantly more organic matter than many upland soils in the tropics. It is also clear that there are some acid paddy soils which do accumulate organic matter; it may be that the difference is partly due to a difference between pseudogleys and stagnogleys, and partly to the difference in soil acidity; only in the acid stagnogleys is there a strong tendency for the mineralization of organic matter to be seriously hindered.

Neue (1991) has made a comparison of the decomposition rates of ^{14}C-labelled rice straw in comparable upland and wetland sites in the Philippines and Thailand (Fig. 4.9). There was little difference at either site between the rates of decomposition under flooded or aerated conditions.

The occurrence of tropical peats is relatively uncommon. They account for less than 10% of the world's peatland (Andriesse, 1988). The only extensive occurrences in the tropics are in Sumatra and East Malaysia (Moncharoen,

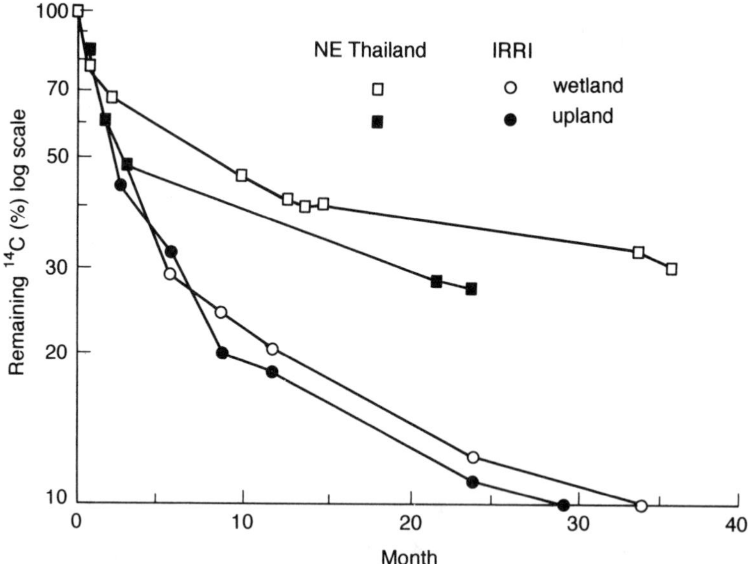

Fig. 4.9. Decomposition of ^{14}C-labelled rice straw in upland and wetland soils from Thailand and the Philippines (IRRI) (Neue, 1991).

1985). This again indicates that mineralization of organic matter in flooded conditions in the tropics is less hindered than under poorly aerated conditions in the temperate zone, accumulation only occurring in the swampy areas under very acid conditions – Andriesse (1988) states that the pH of the great majority of tropical peats falls in the range from 3.0 to 4.5.

A rigorous study of the dynamics of organic matter decomposition under paddy soil conditions in the tropics is needed, but it does appear that decomposition, at least of freshly added material, is rather rapid. The processes that lead to the production of methane and other hydrocarbons probably increase more rapidly with temperature increases than do the processes of aerobic decomposition.

When paddy soils dry each year at the end of the wet season, and again when the dry soil is first moistened at the start of a new monsoon, rapid oxidation of organic matter occurs for a short time. Thus oxidation as well as reduction processes will contribute to the losses of carbon from paddy soils, although it has mostly been assumed that the principal loss is in the form of methane and other hydrocarbons rather than as carbon dioxide (Neue and Scharpenseel, 1984; Watanabe, 1984; Greenland and De Datta, 1985; Neue, 1985, 1991).

The anaerobic decomposition of organic materials in paddy soils has been much studied, chiefly because of the production of phytotoxic materials such as aliphatic and phenolic acids during the decomposition process (Takajima,

1964; Tsutsuki, 1984; Watanabe, 1984). Many of the potentially phytotoxic products tend to be transitory, so that the extent of damage is difficult to assess (Cannell and Lynch, 1984). Ponnamperuma (1985) studied the dynamics of release of organic acids on flooding of a stagnogley (Fig. 4.8). The release of the acids reached a peak within about 2 weeks of flooding but soon thereafter fell to a low level. Sharma and De Datta (1992) found no evidence that addition of rice straw caused any increase in the phytotoxic effects of flooding in pseudogleys under tropical conditions. They also found little effect of percolation rate unless the soil contained more than 5% organic matter, or large quantities of decomposable organic matter had been added. They concluded that only in soils of higher organic matter content is it necessary to maintain a measurable percolation rate to avoid accumulation of phytotoxins. When organic manures are used it is generally advisable to allow time for initial decomposition to be complete before transplanting rice. This may well be an important reason why the traditional practice in most paddy lands is to allow about 4–6 weeks for 'land soaking' before transplanting the rice crop, and why the tradition became established in Japan and China, where large quantities of organic manures were used, to ensure a percolation rate of at least 10 mm per day, and to practise mid-season drainage. Both practices remove phytotoxic organic compounds from the root zone. Although common in China and Japan, mid-season drainage is seldom practised in warmer climates. This is presumably because the higher temperatures enable the toxins to be decomposed sufficiently rapidly that plant growth is little affected, even though the soils remain flooded (Sharma *et al.*, 1989).

The case for the use of organic manures rather than inorganic fertilizers is often made on the grounds that improving soil conditions by addition of organic matter is essential to maintain soil fertility and to sequester carbon so that it does not contribute to the greenhouse effect. As far as paddy soils are concerned, adding organic matter may increase the soil fertility, but is highly disadvantageous in terms of the greenhouse effect, because decomposition of the organic material under flooded conditions produces methane rather than carbon dioxide, and the greenhouse effect of methane is several times greater than that of carbon dioxide. Adding organic materials to upland soils is also generally advantageous because they stabilize aggregates and assist air and water movement. These effects are of course disadvantageous in paddy soils, where the purpose of puddling is to break down any aggregates and disperse the soil particles. Reaggregation of a paddy soil on drying after flooding is advantageous where an upland crop is grown after rice, but much remains to be learnt about the effects of organic matter in this regard.

As far as plant nutrition is concerned, organic matter can be critically important to the supply of nitrogen, sulphur and micronutrients to the rice plant. These are released in plant-available form as the organic matter mineralizes. Neue's results (Fig. 4.9) show that freshly added organic materials can break down rapidly under flooded conditions. Much less has been established about the rate of breakdown of humified soil organic matter. Cassman *et al.* (1995) believe that after rice has been grown under continuously flooded conditions for several years the soil organic matter becomes increasingly resistant to decomposition. Hence less nitrogen and other nutrients are released to plants. To maintain yields requires that larger amounts of manures and fertilizer must be added, even though the total organic matter content of the soil may be increasing. Detailed characterization of the soil organic matter from long-term experiments has indicated that there is indeed a change in the chemical composition of the extractable humic fraction (Olk *et al.*, 1995). Much further information is needed to establish the importance of these changes under a range of conditions. It is particularly important that more information be collected on the dynamics of organic matter breakdown.

4.2.4 The biology of paddy soils in relation to sustainability

Knowledge and understanding of the biology of flooded soils is still inadequate (Roger and Kurihara, 1991). Thus while it has been known for many years that algal growth in rice paddies is greater than on the surface of wet but not waterlogged soils, the ecology of the algal predators has until recently received almost no attention. As will be discussed below, the ecology of the predators is important in assessing how much nitrogen is fixed by indigenous populations of blue-green algae, as well as how effective inoculation of the soil with strongly nitrogen-fixing blue-green algae can be. Aquatic oligochaete worms are known to be important to the breakdown of organic materials in wetland conditions, but only recently have they received any attention (Simpson *et al.*, 1993, 1994). The ecology of other soil organisms is also important to nitrogen fixation, as well as oxidation and reduction processes and nitrogen transformations, and the presence and persistence of pest and disease organisms, both those affecting the rice plant (Mew *et al.*, 1994) and those which affect humans (IRRI/PEEM, 1988; Roger and Bhuiyan, 1990; Simpson *et al.*, 1994).

Within a flooded rice field there are several locations in which there are different redox states, in addition to those due to microsites around particulate organic matter. This is illustrated in Fig. 4.10 for an irrigated paddy field at the flowering stage of the rice plant. Provided the water is not allowed to become stagnant it will be oxygenated and the first centimetre or so of soil at the soil–water interface will remain oxidized. As noted above, there will also be an oxygenated zone around the rice roots. Most of the soil layer below the immediate surface will be reduced, the intensity of reduction depending on the availability of organic substrates for the organisms conducting the reduction, and the presence of oxidized material such as manganic oxides, which can buffer the soil at a rather high redox potential of about +0.45 volts (Rowell, 1981). Where there is no strong buffering and adequate organic substrates, reduction will proceed to the stage where sulphide and methane may be evolved.

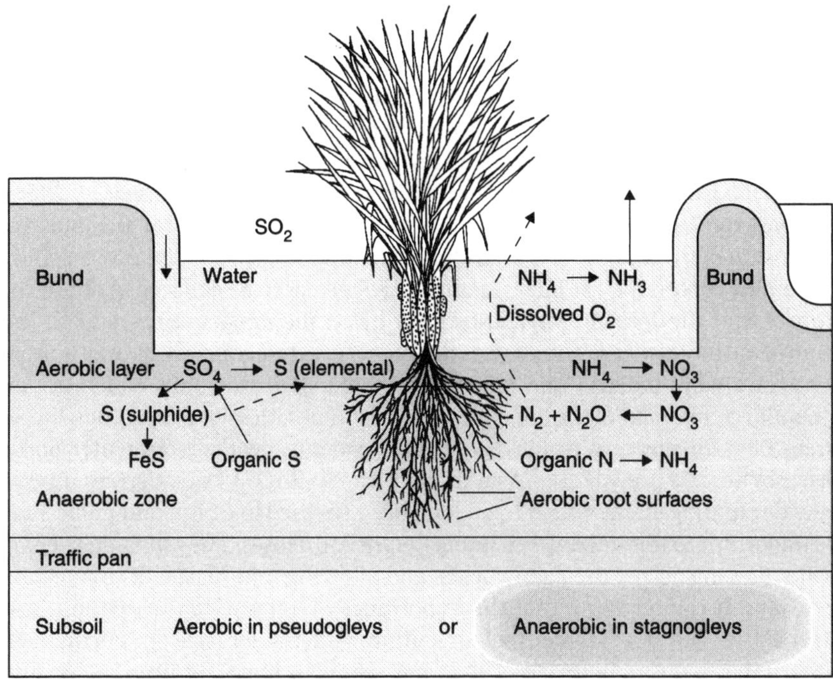

Fig. 4.10. Aerobic and anaerobic sites in a flooded rice field, and processes affecting transformations of nitrogen and sulphur.

Sulphide is highly toxic to almost all organisms, but because of the very low solubility of ferrous and some other sulphides it is precipitated in the soil and does less damage to plant growth than might be expected (Ponnamperuma, 1972; Yoshida, 1981). Phosphates have generally been assumed not to be reduced (Burford and Bremner, 1972), although a recent report has indicated that in laboratory experiments phosphine was lost from brackish and saline marsh soils at rates up to 6.5 ng m^{-2} h^{-1} (Devai and Delaune, 1995).

Landscape and seasonal changes in rainfall interact in determining the length of time for which a soil is flooded and the depth to which it is flooded. Both have important effects on the ecology of the ricelands. Perhaps most importantly in terms of sustainability, the marked seasonality in most rice production systems from flooded to dry conditions has prevented the persistence of many pest organisms. Only as irrigation of the rice crop has extended into the dry season enabling two or three crops to be taken in one year has pest persistence become more widespread. As noted earlier (Chapter 2), double rice cropping developed in the warmer parts of China when the short duration rice varieties were introduced from Vietnam at least 1000 years ago. It has increased rapidly in recent years as irrigation systems have been developed.

IRRI economists believe that double cropping of rice is now practised on more than 28 of the 120 Mha of ricelands in the world. The labour demands and the fact that there is now little yield advantage in growing three lower yielding crops than two higher yielding crops has meant that the very intensive systems in which three or more crops are grown in a year are rather rare.

Concern is often expressed about the relation between the spread of irrigation systems and the increased habitat for water-borne disease vectors. Goonasekere and Amerasinghe (1988) for instance say that the massive expansion of irrigation in Asia has 'resulted in vector-borne diseases reaching epidemic proportions'. While it is obviously correct to be concerned about the impact that the spread of irrigation may have, the impact appears to be less than some have feared. The vectors that are of greatest concern are mosquitoes which transmit malaria and Japanese encephalitis, and snails which are an essential component of the life cycle of schistosomes. In spite of the much larger areas of rice fields and water distribution systems, deaths from water-borne diseases are less prevalent in Asia than in Africa (FAO, 1987). Certain aspects of water management which have long been stressed in China and Japan may be important to minimizing the impact of these diseases. They include keeping water flowing across the paddy fields, and allowing mid-season drainage for a few days. It is interesting that the importance of 'intermittent irrigation' was stressed in the development of irrigation systems for rice in Portugal in 1941 (Hill and Cambournac, 1941) but seems to have not been adopted in many areas subsequently. Although as has been mentioned good drainage is beneficial to rice production, it may well have had important effects in limiting the populations of the vector species in rice paddies. What is certainly true is that a great deal more needs to be learnt about the factors which determine the populations of the important disease vectors, not only in the paddy waters but also in the reservoirs and canals by which the water is distributed (IRRI/PEEM, 1988).

4.2.5 The importance of green manures and Azolla *to the biophysical sustainability of rice farming*

Many leguminous plants have been used in rice-based farming systems in Asia (Table 4.2) to provide human and animal food and to enrich the soil. Those which have most commonly been grown are sunn hemp (*Crotalaria juncea*), dhaincha (*Sesbania aculeata*), pillipesera (*Phaseolus trilobus*), and berseem (*Trifolium alexandrinum*) among those used for animal feed, and mungbean (*Vigna mungo*), soybean (*Glycine max*) and cowpea (*Vigna unguiculata*) among those which have been used to provide grain for human consumption.

Joffe (1955) reported that green manuring, the practice of growing a crop primarily to improve the soil and benefit a subsequent crop, was used in China almost 3000 years ago. Min Zong-dian, quoted by Liu Chung Chu (1988),

Table 4.2. Some plants and trees used to provide green manure for rice-based cropping systems.

Common name	Botanical name
Green manure crops	
Sunn hemp	*Crotolaria juncea*
Dhaincha	*Sesbania aculeata*
Pillipesera	*Phaseolus trilobus*
Milk vetch	*Astragalus sinicus*
Hairy vetch	*Vicia villosa*
Cow vetch	*Vicia craccia*
Common vetch	*Vicia sativa*
Joint vetch	*Vicia americana*
Burr clover	*Medicago hispida*
Pith plant	*Aeschynomene aspera*
(Several)	*Sesbania* spp.
Khesari	*Lathyrus sativus*
Lespedeza	*Lespedeza thunbergii*
Berseem	*Trifolium alexandrinum*
Grain legumes used as green manures (beans harvested, residues incorporated)	
Mung bean/green gram	*Vigna radiata*
Mung bean/black gram	*Vigna mungo*
Cowpea	*Vigna unguiculata*
Soybean	*Glycine max*
Hyacinth bean	*Lablab purpureus*
Chickpea	*Cicer arietinum*
Pigeon pea	*Cajanus cajan*
Woody species grown on bunds or around paddy fields (leaves and branches used as green manure)	
Agati, Baculo	*Sesbania grandiflora*
Sesban	*Sesbania sesban*
Ipil-ipil	*Leucaena leucocephala*
Mother of cacao, quickstick	*Glyricidium sepium*
Pith tree	*Aeschynomene elaphroxylon*

(Allen and Allen, 1981, list many more leguminous species used as green manures, and Brewbaker, 1985, many more leguminous trees and shrubs with potential as green manure providers.)
Source: IRRI (1988).

noted that the use of milk vetch as a green manure had been practised in the Yangzi valley for more than 1000 years; Qi-xiao Wen (1984) says that the practice dates from the 3rd century AD, and Chang (1976b) that it has been practised from at least the 6th century AD. The total area where green manuring was practised in China is reported by Chen Li-zhi (1988) as 1.3 Mha in 1949 and over 10 Mha in the late 1970s. Gu Rong-shen and Wen Qi-xiao (1981) give 8 Mha in 1981, a figure which includes the area in which *Azolla*, the water fern which harbours active nitrogen-fixing blue-green algae, was being used as a green manure.

Liu Chung Chu (1979) and the Soil and Fertilizer Institute at Hangzhou (SFI, 1985) give the area on which *Azolla* was used as a green manure in China in the 1970s as 1.3 Mha, and the Azolla Research Centre, Hanoi, (ARC, 1985) indicates that the corresponding area in Vietnam was 100,000 ha. These figures may be compared with the total wetland rice areas at that time of about 34 Mha in China and 5 Mha in Vietnam.

Egawa (1974) and Ishikawa (1988) report that wild grasses, tree leaves and weeds were traditionally used as green manures in Japan, but that Chinese milk vetch was brought to Japan about AD 800 and used as green manure. Subsequently the use of legumes increased through the Edo era (1603–1867) until in 1909 there were 360,000 ha manured in this way. The area then increased to over 500,000 ha, or about 20% of the total, by 1934, since when it has diminished until the area on which green manures are now used in Japan is negligible.

Other leguminous plants which have been used as green manures for rice include woody perennials, and annual grain legumes from which a harvest of beans may be taken before incorporation of the crop residues in the soil. Most leguminous material used as green manure in South and Southeast Asia appears to have been grown as much for the value of the seeds harvested for human consumption as for the advantages derived by the succeeding rice crop.

The amount of nitrogen fixed depends on how well the legume or *Azolla* grows. There are ample data to show that the amount of nitrogen provided to the succeeding rice crop is most commonly equivalent to the 30–50 kg supplied to the crop by 50–100 kg N ha^{-1} of inorganic fertilizer (IRRI, 1988). The total fixed will be greater than this because some fixed nitrogen is removed in those parts of the plant consumed by man or animals, part is lost to the atmosphere and some remains in the soil.

The stem nodulated legumes, particularly *Sesbania rostrata* and *Aeschynomene aspera* (Fig. 4.11), are fast growing and also active nitrogen fixers, even when growing in standing water (Rinaudo *et al.*, 1988; Becker *et al.*, 1990). Paul (1945) appears to have been the first to suggest the use of a stem nodulated legume as a green manure for rice. He noticed the stem nodules on *Aeschynomene aspera*, the pith plant, but interest in its potential seems to have died with the interest in pith helmets. Interest was revived when Dreyfus and Dommergues (1981) recognized the potential of the stem nodulated *Sesbania rostrata* in West Africa, and Alazard (1985) and Alazard and Becker (1987) described the stem and root nodulation of *Aeschynomene* spp.

Several other species are now known to form stem nodules (Ladha *et al.*, 1992b). The rhizobia associated with them are active nitrogen fixers, are seed borne, so that inoculation is unnecessary, and have their own photosynthetic

Fig. 4.11. Stem nodulation of *Sesbania rostrata.*

system using light of a different wavelength from that used by plants, so that they are not necessarily competitive with the host plant (Eaglesham *et al.*, 1990). Although there is still much to be learnt to optimize methods for their agronomic management (Becker *at al.*, 1995), the stem nodulating legumes open new possibilities for sustaining nitrogen supplies for the rice crop (Ladha and Garrity, 1994).

Other green manures such as *Sesbania aculeata* may be grown on the bunds around the rice paddy. Others are grown before or after the rice crop with the consequent chance that much of the nitrogen fixed may be lost before the rice crop is planted. Much recent attention has been focused on the advantages to be obtained by planting leguminous trees on upland sites. The foliage can be used for composting or as a green manure, and the trunks and branches cut for fuel or building materials (Brewbaker and Hutton, 1979; Burley, 1985; Brewbaker and Glover, 1988). *Leucaena* spp. grow wild in south China, and the use of their foliage for composting has long been advocated (Thorp, 1936). The extent of their use again depends on economic factors, principally on their value for supplying animal feed and building poles.

Azolla has been used as a green manure in China for over 1000 years (Liu Chung Chu and Zheng Weiwen, 1989) and probably for over 2000. *Er Ya*, a book written in China 2000 years ago, mentions 'ping' but it is not certain that this was a reference to *Azolla* or to other water weeds, as 'ping' is now commonly translated as duckweed although correctly it should be translated as water plant. The current Chinese words for *Azolla* are 'man jiang hong', but Liu

Chung Chu and Zheng Weiwen (1989) state that it is also called 'hong ping', which means red ping and probably refers to *Azolla pinnata*, and 'lu ping' or green ping which may refer to another species, or to another strain of *A. pinnata*, as several varieties are known which remain green as long as they have adequate supplies of phosphorus. Liu Chung Chu and Zheng Weiwen (1989) quote various early Chinese texts referring to ping and its uses, but perhaps the most convincing about the importance of ping to rice cultivation is a legend first recorded 700 years ago (see Box 4.1).

Perhaps a more authentic record of the potential uses of *Azolla* is contained in a book published in China about AD 540 *The Art of Feeding the People – Essential Techniques for People's Welfare* by Jia Si-sie (or Jia Ssu Hsieh). A modern Chinese translation has been published recently (Miao, Q. Y., 1982). About the end of the 19th century Hong Binwen published an article in *Long Xue Bao*, the earliest agricultural journal in China, in which he describes the practice of growing 'ping' as fertilizer. He gives considerable detail:

> In Spring, the ping plants grow floating on the water surface, which can control weeds; in mid-Summer, ping plants die and degrade, resulting in colour change of water in the fields, best for the young plants; later on the degraded plants go into soil and become fertilizer. The young plants fed with this fertilizer, grow very fast. At first ping plants weigh 50 to 100 kg mu^{-1} (0.75 to 1.5 t ha^{-1}), and reach more than 1000 kg mu^{-1} (15 t ha^{-1}) at the end of the growing season. One ping plant can produce nine baby plants in a night. That is really miracle. Finally ping plants all die out. But in Winter, farmers in mountainous areas look for ping plants and

Box 4.1. The legend of the discovery of *Azolla*.

Once upon a time there was a landlord who hired a farm labourer, Mr Yang, and a housemaid, Miss Chen. Mr Yang and Miss Chen liked each other and finally fell in love. But the cruel and merciless landlord was not happy at all. He told them they could not get married unless they could grow two crops of rice with early crop harvest of 2 dan mu^{-1} (1.5 t ha^{-1}) and a late crop of 6 dan mu^{-1} (4.5 t ha^{-1}). During that time it was very difficult to grow two crops with half of the harvests demanded by the landlord. It was obviously impossible. The two lovers could not do anything but wished for God to help them.

One night, both Mr Yang and Miss Chen had the same dream. They saw a myriad of sun rays shining from the sky, and a fairy came out of the clouds. The fairy threw some glowing things to the paddy fields. Suddenly the rice plants grew rapidly. Next morning, Mr Yang and Miss Chen coincidentally rushed to the paddy fields at the same time without knowing the other had the same dream. They found there were many small plants floating on the paddy fields, that is, 'lu ping'. That year their two crops of rice harvested more than what was demanded by the landlord. So they got married. Since then this loving couple started business of growing 'lu ping' and supplying seed plants to other farmers. To express their gratitude to the fairy, they made a clay sculpture and built a temple called The Goddess Ping.

take them to warm places to grow. In Spring farmers from other places go there to buy seed plants. It is a secret where they get their seed plants. So nobody can follow them there. The seed plants initially weighing four liang (200 g) can grow rapidly and eventually cover 100 mu (6.7 ha). Ping, the green manure, a top beneficial plant.

(This account and the legend above are quoted from the book by Liu Chung Chu and Zheng Weiwen, 1989, and have been translated by Bin-cheng Zhang of CAB INTERNATIONAL.)

Azolla was given little attention outside of China and the neighbouring countries until the 1970s when it made an important contribution to the ability of North Vietnam to maintain its rice production during the Vietnam war. Subsequently it has received considerable attention (Liu Chung Chu, 1979; Lumpkin and Plucknett, 1980, 1982; Gu Rong-shen and Wen Qi-xiao, 1981; Li Zhuo-xin, 1982; Shi-ye Li, 1984; IRRI, 1987a; Liu Chung Chu and Zheng Weiwen, 1989).

Azolla has undoubtedly contributed considerably to rice productivity in China, and probably for more than a millennium. Its greatest contribution is likely to have been to the maintenance of high yields in the more favoured areas (Fig. 4.12). Although use of *Azolla* as a green manure is still practised in China and Vietnam, as labour costs are increasing so the economics of the use of *Azolla*, as well as other green manures, has become less and less attractive. The extent of green manuring is decreasing, while *Azolla* is finding its greatest use in aquaculture (MacKay, 1995).

Fig. 4.12. *Azolla* in paddy fields in Fujien Province, China.

4.3 The Pests of Rice and their Relation to Sustainability

There are many pests of the rice crop. They include insect pests, diseases, weeds, rats and birds. Weeds, rats and birds are essentially ever present as part of the rice production system. Insect pests and diseases have coevolved with the rice crop. Their much shorter generation time means that they have always been a threat to sustainability, in that they can adapt rapidly to changing ecological conditions. If some other factor does not change to control their numbers the rice crop may be largely consumed by them. As the food of the pest disappears it too will decline. Recovery then becomes possible. The surviving rice varieties are those best adapted to the changed circumstances, including perhaps tolerance to the pest. Thus in history pest problems are somewhat self-regulatory, although serious difficulties often arise while a new ecological balance is established.

Plagues of rice pests are not a new phenomenon. In the *Book of Lord Shang*, written over 2000 years ago it is stated:

> Indeed, if farmers are few, and those who live idly on others are many, then the state will be poor and in a dangerous condition. Now, for example, if various kinds of caterpillars, which are born in spring and die in autumn, appear only once, the result is that the people have no food for many years. Now, if one man tills and a hundred live on him, it means that they are like a great visitation of caterpillars. (McNeill and Sedlar, 1970)

It is difficult to guess to what species the caterpillars that inspired this quotation belong. Armyworms or cutworms are the most likely. The quotation does make clear that plagues of damaging pests were well known at that time. The value of application of white arsenic to the roots of rice plants is reported from Chinese writings of the 4th century AD, and of the value of pouring oil on the paddy water to control insect pests from the 12th century (Konishi and Ito, 1973). A Chinese report, made in the 17th century states:

> two disasters strike; in mid-Autumn rice panicles are blasted by the cold, and turn black and mottled . . . the other strikes in mid-Summer . . . leaf hoppers appear and munching noisily, devour all the leaves completely. The peasants say these two disasters fall from the heavens. (Bray, 1984, p.550)

It is not clear whether the cause of the blast problem was disease or the effects of cold, but the peasants were close to the truth regarding the leafhoppers, as it has now been established that the leafhoppers are transported in air currents from the tropics to southern China, and literally fall from the sky onto the rice crops (Heong and Sogawa, 1994).

The first reliable reports of blast disease were made in Japan, at the beginning of the 18th century (IRRI, 1963). Outbreaks of planthoppers were also first recorded from Japan, but much earlier, in AD 701 and 815, and 16 outbreaks were reported from southern Japan in the 18th century (Konishi and

Ito, 1973). Several measures used to control insect pests in rice fields in Japan in the 19th century are illustrated in Fig. 4.13.

These methods were probably no more or less effective than early pest control systems in Europe, which included burning ivy fumes to deter bats, treating seed with weasel ashes to deter mice, steeping seed in wine to protect them from seedling diseases, and using salt to control weeds (Graham-Bryce, 1983). Some traditional farmer practices still in use in the Philippines to protect rice crops against pests involve various rituals, and the recitation of prayers and incantations, as well as such practices as burning the bark of the baluno tree to control stemborers, hanging rags soaked in shark oil to control rice bugs and placing branches of the patulong tree in the corners and at the centre of the field to deter all pests (Fujisaka *et al.*, 1989).

Some 100 insect species, 74 diseases and 1800 weed species have been reported to afflict rice, of which 30 insects, 16 diseases and 15 weeds are considered to be economically important (Norton and Way, 1990).

Fig. 4.13. Control of planthoppers on rice by a bonfire (extreme bottom right), whale oil pouring (two peasants with wooden pails), and beating with a cluster of bamboo. A pan above fire at the top right indicated melting the whale oil (after Ôkura-Nagatune's *Zyokô-roku-Kôhen*, 1844) (Konishi and Ito, 1973).

4.3.1 Weed control

Removal by hand, and by cultivation, water management and crop management have been the traditional control methods. Only recently have they been supplemented by herbicides (De Datta, 1981; IRRI, 1983a; Grayson *et al.*, 1990). Wet cultivation and puddling are in fact effective in the control of many weeds, although aquatic weeds can still cause significant yield losses. When rice is transplanted into well prepared soils the loss of potential yield may still be as high as 30%. A greater weed problem occurs when rice is direct seeded, an increasingly common practice where labour costs are high, or when early establishment is required in rainfed systems. Yield losses as high as 80% have then been recorded (IRRI, 1983a). A study in Indonesia showed yield losses due to weeds averaged 50%. In this instance the yield losses were reported to be independent of production system or season (Zoschke, 1990). With transplanted rice the C_4 plants *Echinochloa crus-galli* and *Cyperus rotundus* are much the most serious weeds, whereas in dry seeded rice these species as well as other members of the *Poaceae* and *Cyperaceae* are joined by many other C_3 plants (IRRI, 1983a).

The most common method of weed control in the tropics is still hand weeding, although small hand pushed rotary weeders are popular wherever row planting is used. Herbicides are used only where labour costs are relatively high, as in many of the direct-seeded areas. Where herbicides are used both a pre-emergence treatment with, for example, butachlor, and a post-emergence treatment with, for example, propanil, are usually necessary for good control. The use of 2,4-D has also proved surprisingly effective for the control of both sedges and grassy weeds of rice (De Datta, 1981).

4.3.2 Diseases

Disease control has primarily depended on the development of resistance or tolerance of the rice plant to the pest, although fungicides are also used. Stable resistance to fungal and bacterial diseases is difficult to develop, as the causal organisms have a very short life cycle and new races develop rather rapidly, able to attack the formerly resistant variety. Ensuring that genes from several different sources are present in a variety can help to make resistance longer-lived. Varietal resistance needs to be allied to cultivation methods which minimize the opportunity of the pests to pass from one crop to the next, and where it is economic, allied also to chemical control methods. It has recently been shown that biological control, by inoculation of rice seed with antagonistic bacteria (Mew *et al.*, 1994) also promises to be an effective technique for controlling some diseases of rice.

4.3.3 Insect pests

There are many insect pests of rice (Table 4.3) but those commonly causing major yield losses are relatively few. They include the stem borers (*Chilo* spp., *Tryporyza* spp., *Sesamia* spp.), leafhoppers (*Nephotettix* spp.), planthoppers (in Asia the brown planthopper, *Nilaparvata lugens*; in Latin America *Sogatodes* spp.), the gall midge (*Orseolia oryzae*), whorl maggot (*Hydrellia* spp.), leaf folder (*Cnaphalocrocis medinalis*) and rice bugs (*Leptocrisa* spp.).

Table 4.3. Economically important pests of tropical rice.

Insect pests	**Rice diseases**
Vegetative stage	Bacterial blight
Armyworms and cutworms	Bacterial leaf streak
Grasshoppers, katydids, and field crickets	Bakanae
Mealybug	Brown spot
Rice black bugs	False smut
Rice caseworm	Grassy stunt virus
Rice gall midge	Narrow brown leaf spot
Rice green hairy caterpillar	Rice blast
Rice green semilooper	Rice ragged stunt
Rice hispa	Sheath blight
Rice leaf folders	Sheath rot
Rice stem borers	Stem nematode
Dark-headed stem borers	Stem rot
Pink stem borer	Tungro virus
Striped stem borer	White tip
White stem borer	Yellow dwarf disease
Yellow stem borer	
Rice thrips	**Weed pests of rice**
Rice whorl maggots	*Commelina benghalenis*
Seedling maggots	*Cyperus difformis*
Reproductive stage	*Cyperus iria*
Rice brown planthopper	*Cyperus rotundus*
Rice green leafhopper	*Dactyloctenium aegyptium*
Rice greenhorned caterpillar	*Digitaria ciliaris*
Rice skippers	*Echinochloa colona*
Rice white leafhopper	*Echinochloa crus-galli*
Rice whitebacked planthopper	*Eleusine indica*
Rice zigzag leafhopper	*Fimbristylis miliacea*
Smaller brown planthopper	*Monochloria vaginilis*
Ripening stage	*Paspalum distichum*
Rice panicle mite	*Portulaca oleracea*
Rice seed bugs	*Scirpus martimus*
	Spenoclea zeylanica

Source: Norton and Way (1990).

Of these pests stem borers probably have caused most extensive damage in the past and may well continue to do so, as they are hard to control. Other insect pests such as the green leaf hopper which transmits tungro disease are more important as vectors of virus diseases than because of the direct damage they do to the rice plant. The most spectacular damage is caused by sudden outbreaks of plagues of planthoppers. The brown leafhopper can devastate large areas, where the rice is completely killed by hopperburn.

The extent of losses to insect pests has grown as the intensity and frequency of planting and the areas planted to rice and to a single variety of rice have increased. Where a single variety has been widely grown it has provided an opportunity for a pest to multiply to epidemic proportions if the variety is a suitable medium for its rapid increase. The early rice varieties with high yield potential and responsiveness to nitrogen fertilizers released at the start of the green revolution in the 1960s provide an example. They had no resistance to the brown planthopper, then regarded as a minor pest. Consequently when the variety IR8 which lacked resistance to the hopper was widely planted in Luzon, the 'rice bowl' of the Philippines, there was a rapid build up of the pest. The average rice yields in Central Luzon, the area most affected, had increased significantly from 1966 to 1970. They then remained static for the next 4 years while control methods were developed (Fig. 4.14).

The most important step towards control was through development of rice varieties with resistance to the pest (IRRI, 1979c, 1985e). When the pest first appeared as a threat the immediate response was to encourage the use of pesticides, not only against the brown planthopper but against a range of insect pests. As understanding of the ecology of the pests developed it was recognized that natural control agents were often present and also killed by the insecticide (Heinrichs *et al.*, 1982). Therefore spraying with insecticides was best done judiciously if at all (IRRI, 1984e).

The many insects which were not pests but fed on the brown planthopper and other damaging insects were 'friends of the rice farmer' (Kenmore *et al.*, 1984; IRRI, 1987c). After spraying with wide spectrum insecticides the 'friends' were often killed. As the pest survivors often multiplied more quickly than the surviving natural enemies, use of the insecticide led to a resurgence of the pest and a worse situation than before the pesticide was used (Heinrichs *et al.*, 1982).

Proper pest management practices are now recognized to require an integration of control methods, including appropriate agronomic practice, host plant resistance, biological control, and judicious use of pesticides (Reissig *et al.*, 1986; Kenmore *et al.*, 1987; Norton and Way, 1990). Kenmore (1991) has shown in Indonesia how effective such integrated pest management strategies (IPM) can be.

Rice varieties bred at IRRI and many other research stations are now rigorously screened for resistance to most of the known major pests of rice (Table 4.4). While IR8, released in 1966, had only limited pest resistance,

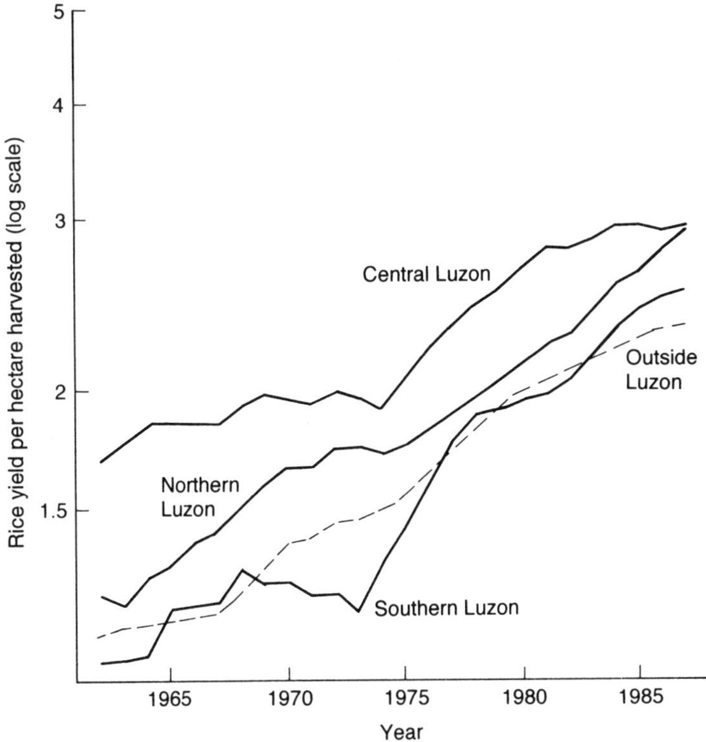

Fig. 4.14. Average rice yield trends in the Philippines by region, 5-year moving averages (Hayami and Otsuka, 1994).

IR36 and IR42 released several years later incorporated genes for resistance to the major insect pests and diseases prevalent in the Philippines. Figure 4.15 shows the pedigree of IR36 and 42, both of which were selected from the same cross. The effects of their resistance on the stability of the yields obtained from them is illustrated in Fig. 4.16. Trials in the Philippines conducted between 1973 and 1986 show that while the yields of IR36 and IR42 were consistently in the range 6–8 t ha^{-1}, that of IR8, in the same trials, ranged from 2.7 to 7.9 t ha^{-1}. IR36 is probably the most popular and widely grown variety of any crop, although rivalled by two or three wheat varieties. In 1982 more than 11 Mha were planted to IR36 (IRRI, 1985a). In spite of its wide distribution IR36 has remained remarkably resistant to pests. In more recently bred and widely grown varieties an even wider range of resistance genes have been incorporated: in IR72 no less than 22 landraces were used as sources of desired genes.

Sustaining yields at relatively low levels, and with few areas planted to more than one crop each year, was possible with simple cultivation methods

Table 4.4. Pest resistance at IRRI of new varieties released since 1966.

	Disease or insect reactions							
Variety	Blast	Bacterial blight	Tungro	Grassy stunt	Green leafhopper	Brown planthopper	Stem borer	Gall midge
IR5	MR	S	S	S	R	S	MR	S
IR8	S	S	S	S	R	S	S	S
IR20	MR	R	S	S	R	S	MR	S
IR22	S	R	S	S	S	S	S	S
IR24	S	S	S	S	R	S	S	S
IR28	R	R	R	R	R	R	MR	S
IR32	MR	R	R	R	R	R	MR	R
IR36	MR	R	R	R	R	R	MR	R
IR38	MR	R	R	R	R	R	MR	R
IR42	MR	R	R	R	R	R	MR	R
IR46	MR	R	R	R	R	R	MR	R
IR50	S	R	R	R	R	R	S	–
IR54	MR	R	R	R	R	R	MR	–
IR58	MR	R	R	R	R	R	S	–
IR60	MR	R	R	R	R	R	MR	–
IR62	MR	R	R	R	R	R	MR	–
IR64	MR	R	R	R	R	R	MR	–
IR66	MR	R	R	R	R	R	MR	–
IR68	MR	R	R	R	R	R	MR	–
IR72	MR	R	R	R	R	R	MR	–
IR74	MR	MS	R	R	R	R	MR	–

R = resistant; MR = moderately resistant; MS, moderately susceptible;
S = susceptible; – not known.
Source: Khush (1995).

and farmer selection for varietal resistance. The rapid growth of population and the consequent expansion of rice production has been achieved largely as a result of scientific advances but has introduced a potentially dangerous vulnerability to insect pests. Future disasters are avoidable, but only if continuing scientific attention is given to the occurrence of pest damage, and its control.

4.3.4 Mammalian and bird pests

Rats are sometimes reported to cause more damage to wetland rice crops than all other pests combined (Quick, 1990). They are omnipresent and usually responsible for losses of the order of 10% of yield. Their control is difficult. They are also serious carriers of disease. Birds are often a problem, and again the damage they cause is difficult to control.

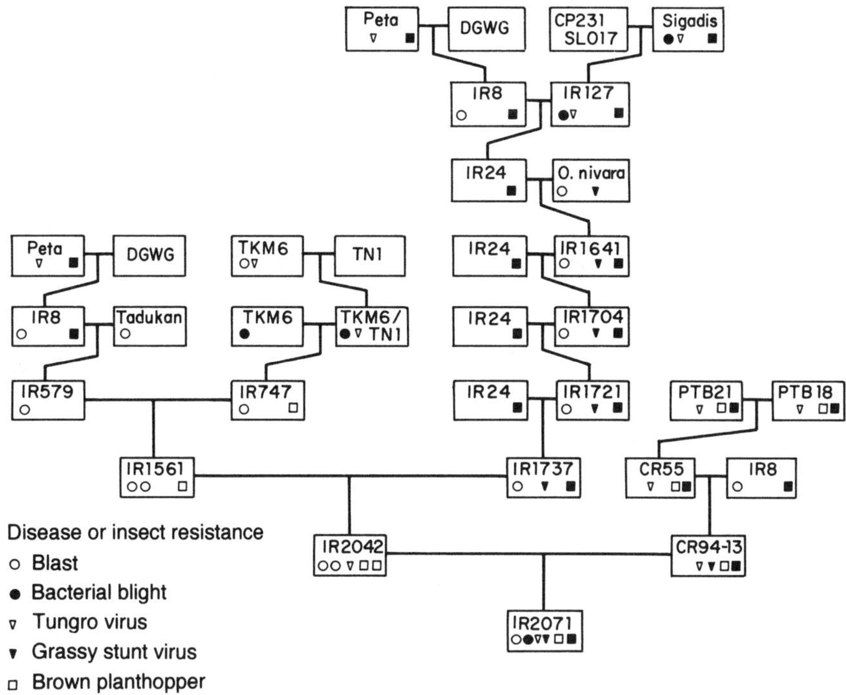

Fig. 4.15. The pedigree of IR36 and IR42, which were selected from IR2071. Thirteen varieties were involved in the ancestry of these varieties (Khush, 1978).

4.4 Biodiversity and its Significance to the Biophysical Sustainability of Rice Farming

Biodiversity within the tribe *Oryzae*, the genus *Oryza* and the species *O. sativa* and *O. glaberrima*, as well as among the general biological population of rice fields is important to the biophysical sustainability of rice farming (Chang and Vaughan, 1991; Jackson, 1995).

4.4.1 Biodiversity of rice and related species

The reason for the importance of diversity within the tribe, genus and species of rice is that it provides the gene pool from which resistance to new pests and adverse soil and other conditions can be sought, and new and improved varieties can be developed. Rice and the wide range of pests which affect it have coevolved. If evolution of the rice species is seriously slowed the more rapid changes in pest species could have serious consequences. Potentially the loss of

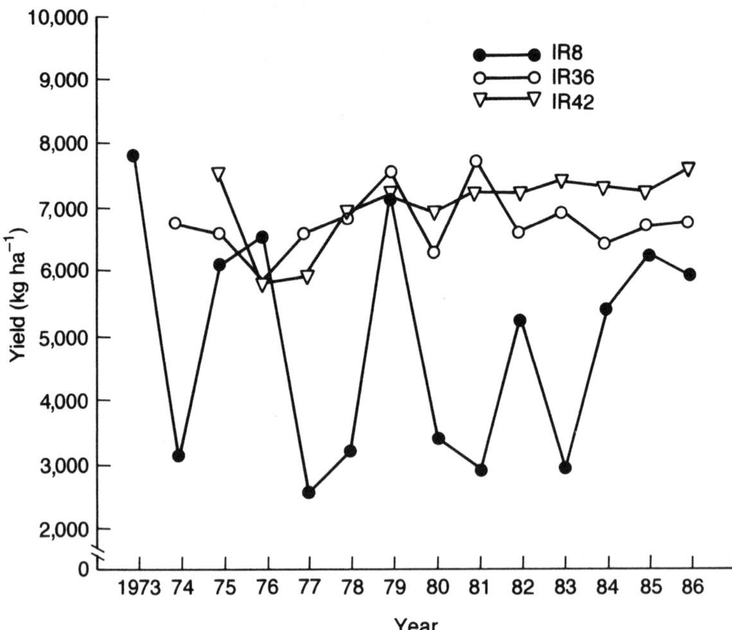

Fig. 4.16. Yield stability of IR36 and IR42 compared with that of IR8 (Khush, 1990). Data from replicated yield trials at IRRI.

biodiversity in rice could lead to the disappearance of the possibility to improve or maintain rice yields. The increasing use of a limited number of high yielding varieties does indeed threaten the natural evolution of the crop.

The number of species within the *Oryzae* tribe is not particularly large (Table 4.5). Hence genetic erosion is a real threat. However an active programme of collection, conservation and maintenance of the germplasm of *O. sativa* and related species has been conducted since the 1960s at the International Rice Research Institute, where more than 80,000 accessions are currently held. There are also considerable collections at many national genetic resource centres. Together these collections offer some protection against the threat posed by genetic erosion. Advances in rice biotechnology (Khush and Toenniessen, 1991) and in rice genetics (IRRI, 1986c) are providing new tools to enable the rice breeder to make fuller use of the potential offered by the genetic diversity which exists. Recent progress in genetic engineering (Gatehouse *et al.*, 1992) presents the additional opportunity to add to the gene pool by modification of genes of rice, and to transfer genes from non-rice plants.

Within the *O. sativa* species there has been extensive adaptation to different environments. Traditionally a separation into three subspecies, *indica, japonica* and *javanica,* has been made (Grist, 1975). The difficulties that breeders have sometimes experienced in making fertile crosses within as well as between

Table 4.5. (a) Taxa and (b) genera, in the genus *Oryza*, and the Tribe *Oryzeae*, showing species complexes, number of species and genome groups.

(a)

Species complex	Taxa	Genome group	Distribution
O. sativa complex			
	O. glaberrima	A^gA^g	Africa (mainly West)
	O. barthii	A^gA^g	Africa
	O. longistaminata	A^lA^l	Africa
	O. sativa	AA	Worldwide
	O. nivara	AA	Tropical and subtropical Asia
	O. rufipogon	AA	Tropical and subtropical Asia
	O. meridionalis	A^mA^m	Tropical Australia
	O. glumaepatula	$A^{gl}A^{gl}$	South America
O. ridleyi complex			
	O. longiglumis	Tetraploid	Irian Jaya, Indonesia
	O. ridleyi	Tetraploid	SE Asia
O. meyeriana complex			
	O. granulata	Diploid	S and SE Asia
	O. meyeriana	Diploid	SE Asia
O. officinalis complex			
	O. officinalis	CC	Tropical and subtropical Asia
	O. minuta	BBCC	Philippines
	O. eichingeri	CC	Sri Lanka, Africa
	O. rhizomatis	CC	Sri Lanka
	O. punctata	BBCC, BB	Africa
	O. latifolia	CCDD	Latin America
	O. alta	CCDD	Latin America
	O. grandiglumis	CCDD	South America
	O. australiensis	EE	Australia
	O. schlechteri	Diploid	Papua New Guinea
	O. brachyantha	FF	Africa

(b)

Genera	No. of species	Distribution	Tropical (T)/temperate (t)
Oryza	22	Pan-tropical	T
Leersia	17	Worldwide	t + T
Chikusiochloa	3	China, Japan	t
Hygroryza	1	Asia	t + T
Porteresia	1	South Asia	T
Zizania	3	Europe, Asia, N. America	t + T
Luziola	11	N. and S. America	t + T
Zizaniopsis	5	N. and S. America	t + T
Rhynchoryza	1	S. America	t
Maltebrunia	5	Tropical and S. Africa	T
Prosphytochloa	1	S. Africa	t
Potamophila	1	Australia	t + T

Source: Jackson (1995)

these subspecies has shown that a better characterization of the large differences known to exist is needed. A recent proposal by Glaszmann (1986) based on the allelic pattern of isozyme loci has shown that a grouping into six classes can be of greater help in understanding the evolution of the genus, and so to geneticists and breeders in the development of new varieties.

4.4.2 Biodiversity of microorganisms and invertebrates in rice fields

The importance of biodiversity among microorganisms and invertebrates has seldom been given adequate attention (Hawksworth, 1991). Invertebrates and microorganisms are important to the maintenance of soil fertility, and to the limitation of damage due to pests. It is generally assumed that diversity in populations equates to stability, and so to maintain soil fertility and minimize pest damage it is essential, or at least desirable, to maintain maximum diversity.

The natural diversity of meso- and microbiological populations has almost certainly contributed to the long-term sustainability of rice farming. As cultural practices have become more uniform and chemicals have been more widely used to eliminate pest species, it has been widely assumed that population diversity will be reduced. In fact there are few irrefutable data to support this hypothesis (Roger *et al.*, 1991). Table 4.6 summarizes the data that was available in 1990. More recent records of species numbers tend to be lower than the older ones, which provides some support for the hypothesis that diversity in ricefields is declining, but the data come from different environments, and different techniques were used to estimate the diversity. Thus the measures of diversity are not really comparable. Comparison of arthropod diversity in farmers' fields in the Philippines and at the International Rice Research Institute, using the same methods at each location, have recently been made. The greatest diversity was found on the experimental fields at the Institute, where heavy use had been made of fertilizers and pesticides, and the least in the rice fields of the Ifugao rice terraces at Banaue, where only traditional varieties of rice had been grown, without fertilizer or pesticides, for many years (Fig. 4.17). This throws some doubt on the hypothesis and highlights the need for more experimental data to be obtained.

In terms of soil fertility the greatest importance of biological diversity is to nitrogen fixation and mineralization of organic matter. In terms of pest management it is most important in connection with the biological control of pests, using pests in the widest connotation. The use of nitrogen fertilizers certainly tends to inhibit nitrogen fixation, but inorganic phosphorus fertilizers encourage it. These effects however relate to specific nitrogen fixing organisms rather than the population diversity. The influence of pesticides on the mineralization of soil organic matter is similarly varied. Part of the population of mineralizers may initially be killed by the pesticide, but because of the

Table 4.6. Quantitative records of species/taxa in wetland ricefields.

1. Number of species recorded by Heckman in 1975 in a 1-year study of a single field in north-eastern Thailand (six samplings)

Sarcodina	31	Cyanobacteria	11
Ciliata	83	Algae	166
Rotifers	50	Pteridophyta	3
Platyhelminths	7	Monocotyledonae	25
Nematoda	7	Dicotyledonae	10
Annelida	11	Pisces	18
Mollusca	12	Amphibia/Reptilia	10
Arthropoda	146	Total	590

2. Number of species/taxa of aquatic invertebrates, excluding protozoa, recorded by different authors

• Heckman (1979) (species), one traditional field, 1-year study (Thailand)	183
• Lim (1980) (taxa), 2-year study of pesticide application (Malaysia)	39
• Takahashi *et al.* (1982) (taxa), four fields, single samplings (California)	10–21
• International Rice Research Institute (1985g) and Roger *et al.* (1987) (species), single samplings in 18 fields with pesticide applied (Philippines and India)	2–26

3. Records of arthropod species in ricefields over one crop cycle

• Kobayashi *et al.* (1973): study in 1954/55 of several fields by net sweeping (Shikoku, Japan)		450
• Heong *et al.* (unpubl. data): study in 1989 of five ricefields by suction (Philippines):	Fields considered separately:	146, 125, 116, 92, 87
	Five fields combined	240

Source: Roger *et al.* (1991) who give references to the original works.

simultaneous removal of competitors the mineralizers may quickly recover to function more effectively than before (Roger *et al.*, 1991). Several insecticides reduce the numbers of ostracods which graze on the nitrogen fixing algal population, so that nitrogen fixation is enhanced. Insecticides also tend to reduce the numbers of zooplankton in paddy water and so limit the growth of the fish population wherever rice–fish farming is practised. At the present time there is too little information to enable firm conclusions to be drawn about the balance of detrimental and favourable effects of intensification of agricultural practices on soil fertility. What is needed is more long-term experimentation to enable crude generalizations based on limited data to be replaced by established fact.

Rather more is known of the relation between diversity and pest management. The importance of biological control to management of rice pests has already been discussed. Using insecticides disrupts the insect population and pest outbreaks arise which can cause serious damage until the previous population balance is restored. The practice of growing a crop itself produces

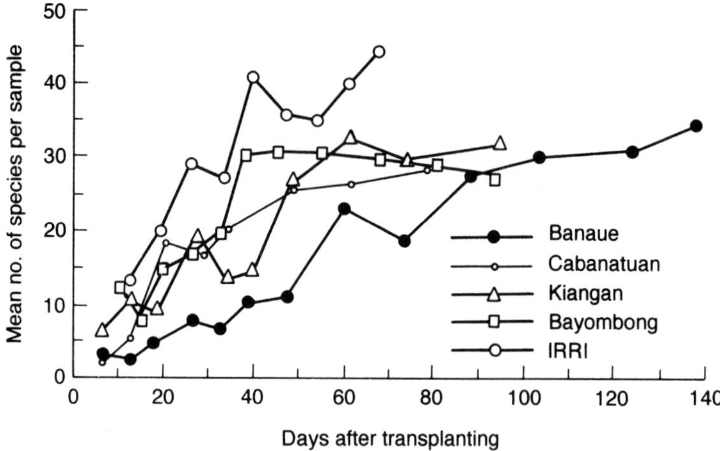

Fig. 4.17. Dynamics of arthropod diversity during a crop cycle in five ricefields in the Philippines (Banaue: rice terrace where rice has grown without agrochemicals for centuries; Cabanatuan, Kiangan, Bayombong: farmers' fields with agrochemical use; IRRI, experimental plot on the IRRI farm). (Roger *et al.*, 1991).

an environment which is less diverse than a natural environment, and encourages the pest species which live on the crop. Where large areas are planted to a single variety, and if there is no break in the annual cultivation cycle, that pest species may multiply to achieve epidemic proportions. The economic success of green revolution technology has meant that most farmers have sought to grow high yielding varieties, so that exceptionally uniform environments have been created over large areas. Where adequate attention has not been given to pest control pest outbreaks have tended to become more common. In the past the diversity in the biological populations of rice fields and the surrounding vegetation has contributed significantly to the maintenance of populations of the predators of the pests, and so to the sustainability of rice farming. As cropping intensifies and less and less land is left uncultivated so the environment becomes more and more uniform (Waage, 1991), and pest outbreaks more likely to occur.

Maintaining the Nutrient Requirements of Rice

5

The first essential of biophysical sustainability is to ensure that sufficient nutrients remain available in the soil for succeeding rice crops, in spite of withdrawals in earlier rice crops and other losses. Changes in the soil nutrient supply reflect the balance between inputs and outputs. If the outputs exceed the inputs by more than can be supplied from the reserves in the soil, the soil degrades and yields decline.

It is possible to quantify the soil change, or balance, B, of inputs, I, and outputs, O, from what is known of the inputs and outputs from different rice production systems, as:

$$B = I - O$$

Inputs come from rainfall, R, irrigation and floodwater, F, sediments, S, nitrogen fixation, N, and manures and fertilizers, M. Outputs are due to crop removals, C, seepage and percolation, P, and volatilization, V. Thus:

$$B = R + F + S + N + M - C - P - V$$

In the following sections values for each of the components in the equation will be considered for different rice farming systems. Attention will be focused on wetland conditions. The components of the nutrient balance for rice produced in upland conditions does not differ from that of the cereals normally grown under upland conditions, and upland rice production is no more or less sustainable than that of the other cereals. In contrast wetland conditions present several special features in relation to the sustainability of nutrient supplies for crops. Most importantly, nutrient removals by leaching and erosion tend to be major factors in the balance of upland soils, whereas in the wetlands inputs from irrigation and floodwater, including the sediment deposited by floods, tend to be larger than leaching losses, and erosion does not occur.

5.1 Inputs of Nutrients to Rice Production Systems

There are significant inputs to rice-based cropping systems from rain, irrigation and floodwaters, and sediments. In addition biological nitrogen fixation in wetland systems is quantitatively different from that which occurs in upland soils, and in the history of rice farming organic manures of various origins have probably played a greater role than in other farming systems.

5.1.1 Additions from rainfall, and flood and irrigation water

Data on nutrients falling onto land in rainfall and as dust in a range of tropical locations are collected in Table 5.1. There is a considerable volume of literature relating to the nutrients deposited in rain and as particulate airborne deposits. A bibliography is given by Wetselaar *et al.* (1981). Although concentrations of nutrients tend to be lower than in rainfall in more industrialized regions, the amounts deposited are at least comparable because of the larger volume of rain falling in the Asian rice areas. The quantities are quite variable in both time and space, amount of rainfall, and distance from urban and industrial centres and from the ocean having important effects. The data in Table 5.1 do not reflect the large contributions that may be made by volcanic dust. Such contributions tend to be localized and occur infrequently. The contribution of volcanic dust is probably negligible in the delta areas where most rice is grown, but should not be neglected when nutrient balances are calculated for the volcanic regions of such areas as Java and parts of the Philippines.

Some of the nutrients in the rainfall may not benefit the land where the rain falls initially, because in heavy storms some runs off to lower landscape positions. The ultimate beneficiary is then most likely to be the deepwater rice areas. In constructing the hypothetical nutrient balances discussed below for diversion irrigated and rainfed lowland rice systems average nutrient deliveries in 2000 mm of rain have been taken as 12 kg ha^{-1} of N, 0.2 of P and 12 of K.

Nutrients supplied in flood and irrigation water can also be significant. Typically the amount of water used by a rice crop is of the order of 500–1000 mm in addition to the rainfall (cf. Figs 3.8 and 6.2). Concentrations

Table 5.1. Nutrients in rainfall (kg ha^{-1} year^{-1}).

Country	Rainfall mm year^{-1}	N	P	K	Ca	Mg	S	Reference
Japan	1700	5	0.2	4				Yatazawa, 1977
Malaysia	2500	20	0.2	12	38	3		Shorrocks, 1972
Bangladesh	2390	13	0.3	12	36	8	3.5	Abedin Mian *et al.*, 1991
Brazil (Amazonia)	2130	6	0.2	–	1	2		Klinge, 1977
Ghana (Kade)	1460	15	0.04	18	12	11		Nye, 1961

of nutrients in the floodwater cover a wide range, differing between sources, and within a single source during the year, as well as from year to year. Takahashi (1965) collected nutrient concentration data for rivers in Japan and Thailand, Vacharotayan and Takai (1983) for specific sites in Thailand and Abedin Mian *et al.* (1991) for a specific site in Bangladesh. Yatazawa (1977) gives an estimate of the mean nutrient delivery to rice fields in Japan from irrigation water. The quantities of nutrients supplied to a rice paddy by 1000 mm of irrigation water containing the mean concentrations reported by Yatazawa are given in Table 5.2, together with actual figures for one year for the sites in Thailand and Bangladesh, and the estimate for Japan. The amount of nitrogen supplied is comparable with that in rainfall, the amount of phosphorus is less and generally negligible, and potassium much greater. A summary of the data on sulphur contributions (Lefroy *et al.*, 1992) indicates a considerable range of values in both rainfall and irrigation water. The mean deliveries of nutrients by irrigation and floodwaters depend on the source of the water and the amounts reaching the rice field, adjusted where necessary for the overflow from the ricefield where this occurs.

5.1.2 Additions from sediments

Nutrient deposition in sediments may be expected to vary even more widely than the contributions from rainfall and irrigation and floodwater. The amount of sediment deposited at any one location will depend on position in the landscape, and on local differences in the level of the soil surface, as well as on the amount of sediment carried by the river flooding the area. This in turn will depend on the landscape through which the river flows and the soil and rainfall characteristics in the river catchment. The nutrient content of the sediment will depend on soil and geological characteristics of the areas from which the sediments are derived. Major seasonal differences in the amounts of silt carried by the rivers, and the depths to which the rice paddies are flooded, also occur. Whitton and Rother (1988) for instance report a difference in the silt deposited at a site at Manikganj bil in Bangladesh 20 times higher in 1987 than in 1986. Unusually high floods were experienced in Bangladesh in 1987. Greater deposition may be expected to occur in periods of heavy flooding, when not only is the amount of floodwater depositing sediment at any one site likely to be greater, but the extent of erosion of the soils where the sediments are derived is also likely to be greater.

Determinations of mean sediment deposition rates derived from measurements made for only a few years may give a misleading impression of the long-term mean deposition rate, if they miss major floods. The best average rates are those obtained by dating sediments by depth, using time scales of centuries or millennia. This may be done by using the data from archaeological sites where it has been shown that rice was grown over a long period, and ^{14}C

Table 5.2. Nutrients in irrigation water from rivers and wells.

	Bangladesh	Bangladesh		Thailand Surin		Thailand Chum Pai		Thailand	Japan	Japan	Philippines IRRI
	Mixed canal and well water, 713 mm	Floodwater 4000 mm		'Irrigation water' 1000 mm		'Irrigation water' 1000 mm		All major rivers	All irrigation water for rice paddies	All major rivers	Deep wells
	kg ha^{-1}	ppm	kg ha^{-1}	ppm	kg ha^{-1}	ppm	kg ha^{-1}	ppm	kg ha^{-1}	ppm	kg ha^{-1}
N	1.3	0.04	1.6	0.5	5	0.8	8	0.14	16.5	0.31	3
P	–	0.09	3.6	–	–	–	–	0.003	0.4	0.007	0.6
K	19	1.9	16	19	190	5.8	58	2.5	27	1.2	170
Ca	149	9.8	392	18	180	8.5	85	19.8		8.8	240
Mg	89	3.0	120	1.4	14	3.3	33	3.7		1.9	180
S	62	–	–	4.9	49	8.6	86	1.1		3.7	35
Fe	0.4	0.7	28	0.9	9	1.3	13				
Mn	0.2			0	0	0.05	5				
Zn	0.02										
Cu	0.008										
Reference	Abedin Mian *et al.*, 1991	Whitton and Rother, 1988		Vacharotayan and Takai, 1983				Takahashi, 1965	Yatazawa, 1977	Takahashi, 1965	Dobermann *et al.*, 1996

dated material has been recovered from depth. At Tsukumino Moor in Japan for instance material dated to 3150 ± 85 BP has been recovered at 3000 mm (Yamanaka, 1979). This corresponds to a mean rate of deposition close to 1 mm per year. At Kok Phanom Di in Thailand Maloney *et al.* (1989) report that pollen probably from rice was found to a depth of 2 m. Material recovered from that depth was dated to 4000 BP, corresponding to a mean sedimentation rate of about 0.5 mm $year^{-1}$. It is possible that some movement of the dated material may have occurred in the soil profile. Such movement (pedoturbation) is associated with the activities of burrowing members of the soil fauna, and displacement of material in cracks in the soil when it is subject to extreme wetting and drying, or freezing and thawing. None of these are common in flooded paddy soils. The consistency in the rate of deposition in the Tsukumino Moor soil profile, as indicated by the carbon dates at different soil depths, also indicates that little pedoturbation has occurred.

Direct measurements of sediment load have been reported which correspond to similar rates of deposition. Patnaik (1978) for instance reports data from Cuttack in eastern India indicating that 2–5 mm of sediment were deposited annually. Collection of sediment in traps for 1–5 years in Bangladesh gave rates from 179 to 2384 g m^{-2} $year^{-1}$ (Catling, 1993, p.48). One millimetre of soil per hectare with a bulk density of 1.0 corresponds to 10 t ha^{-1}, or 1000 g m^{-2}. Thus these measurements indicate depositional rates between 0.179 and 2.384 mm $year^{-1}$. Uehara *et al.* (1974) estimated the mean rate of deposition of sediment in areas flooded by the lower reaches of the Mekong River to be 1 mm $year^{-1}$. Hence the range of mean annual sedimentation rates in the rice areas of southern Asia is probably between 0.1 and 5 mm of soil per year, although in individual years there will be large departures from this range.

The nutrients added to the soil by sediments can be calculated from the depositional rates if the nutrient concentrations in the sedimented material are known. These concentrations differ considerably depending on the source of the sediments. A further complication in calculating what they will contribute to rice crops is that not all of the nutrients in the sediments will be available to crops. In constructing the nutrient balances discussed below this problem has been avoided by the simple expedient of assuming that most of the nutrients contained will be released by weathering over the course of centuries, so that total contents can be used.

The wide range in amount deposited and the nutrient concentration contained make it difficult to arrive at a specific figure for nutrient additions to any one soil. To obtain an idea of the quantities of nutrients that may be added in sedimented material Table 5.3(a) has been constructed, using: (i) a low sedimentation rate of 0.1 mm $year^{-1}$ and the lowest national average nutrient content for paddy soils given by Kyuma (1991); (ii) a higher sedimentation rate of 1.0 mm $year^{-1}$ and the highest national average nutrient content given by

Kyuma; and (iii) a sedimentation rate of 1.0 mm year^{-1} and the means for tropical Asian paddy soils taken from Kyuma's data given in Table 4.1.

Actual data for sites in China, Bangladesh and the Mekong delta are in Table 5.3(b). They are at the lower end of the the ranges in Table 5.3(a) except for two potassium contents which are below the addition derived from the low concentration and low deposition level. Some data for actual river sediment

Table 5.3. Addition of nutrients to paddy soils from sediments.

	N	P	K	Ca	Mg
(a) Hypothetical: Concentrations correspond to highest, lowest, and average total nutrient concentrations in tropical soils in Asia as given by Kyuma, 1991; deposition rates correspond to 0.1 mm or 1 t ha^{-1} as a low rate; or 1.0 mm or 10 t ha^{-1} year^{-1} as an average rate.					
(i)					
Low concentration (%)	0.08	0.04	0.60	0.12	0.30
Low deposition rate (kg ha^{-1})	0.8	0.4	6.0	1.2	3.0
(ii)					
High concentration (%)	0.64	0.07	2.70	2.0	0.9
High deposition rate (kg ha^{-1})	64	7.0	270	200	90
(iii)					
Average concentration (%)	0.17	0.05	1.20	0.17	0.76
Average deposition rate (kg ha^{-1})	8.5	2.5	60	8.5	3.8
(b) Measured annual nutrient additions from sediments					
(i) Guangdon, China (kg ha^{-1}) (FAO, 1977)	1.8	0.6	0.45		
(ii) Bangladesh, 5 deepwater sites (kg ha^{-1}) (Whitton *et al.*, 1988)	9	2	8		
(iii) Mekong delta, 3 sites (kg ha^{-1}) (Uehara *et al.*, 1974)	–	1	3.2	50	4
(c) Actual sediment composition, and additions from deposit of 10 t ha^{-1} (1 mm)					
(i) Yellow River (Huanghe) silt					
On 28-7-90 (%)	–	0.07	2.23	5.01	2.26
On 23-7-93 (%)	–	0.09	2.20	8.04	3.33
Average (%)	–	0.08	2.22	6.53	2.80
For full deposit of 1.0 mm (kg ha^{-1}) (Fullen *et al.*, 1995)	–	8.0	222	653	280
(ii) Mekong River sediment, 3 sites (%)	–	0.04	0.44	3.24	1.05
Full deposit of 1.0 mm (kg ha^{-1}) (Uehara *et al.*, 1974)	–	4.0	44	324	105
(iii) Jemmu and Meghna River sediments, 4 sites (%)	0.5	0.11	–	0.63	0.55
For deposit of 1.0 mm (kg ha^{-1}) (Whitton *et al.*, 1988)	50	11	–	63	55

compositions are in Table 5.3(c). In all cases the concentrations of phosphorus and most other elements are higher in the sediments than in the soils where they were deposited. Thus use of soil rather than sediment concentrations in calculating the rates of deposition given in Table 5.3(a) are more likely to lead to an underestimate of the contribution that sediments make to the nutrient balance.

5.1.3 Additions from biological nitrogen fixation

Biological nitrogen fixation, particularly by blue-green algae living on the surface of paddy water, has been widely assumed to make an important contribution to the sustainability of rice-based farming systems. There are in fact several sources of biologically fixed nitrogen contributing to the nitrogen nutrition of rice plants. The different sites are illustrated in Fig. 5.1. Some free-living nitrogen-fixing bacteria live in the soil, but most are found on the surface of rice roots, and on the surfaces of rice and weed stems and leaves. Nitrogen-fixing blue-green algae live at the air–water interface, the soil–water interface, float in the paddy water, and are epiphytic on the surfaces of rice plants and weeds. A particularly favourable site is below the leaf surfaces of the water fern *Azolla*, as discussed in Chapter 4. All known species of *Azolla* contain the nitrogen-fixing blue-green alga *Anabaena azollae* (Ladha and Watanabe, 1987) and the alga rather than *Azolla* is the nitrogen fixing agent, although it is customary to refer to nitrogen fixation by *Azolla*.

Many leguminous plants have been used as green manures for rice (Table 4.2), including both annuals sown before the rice crop, and perennials, mostly woody species, not necessarily grown in the rice paddies but on the bunds around them or on adjacent land (IRRI, 1988). The nitrogen fixing bacteria associated with legumes normally live symbiotically in nodules on the roots of the legumes, although stem nodulating species which grow in waterlogged soils and fix nitrogen efficiently, appear to have greater potential (Becker *et al.*, 1990; Ladha *et al.*, 1992b).

The quantities of nitrogen added to the soil and to the rice crop by these various nitrogen fixing organisms have been the subject of much study. The amounts fixed depend on several factors, including soil type, water quality, temperature and management of the rice crop and production system. As far as nitrogen fixing bacteria and blue-green algae are concerned the organisms seem to be almost always present. Their activity is determined by how favourable the conditions are for their growth. Although claims have often been made of the advantage to be obtained by inoculating soils with blue-green algae, the evidence is at least equivocal. Normally the adapted species of nitrogen fixing algae are present in the paddy soil (Roger and Kulasooriya, 1980; Roger, 1991). Their numbers are controlled by factors such as predation by snails. Their activity depends very largely on the availability of phosphorus in

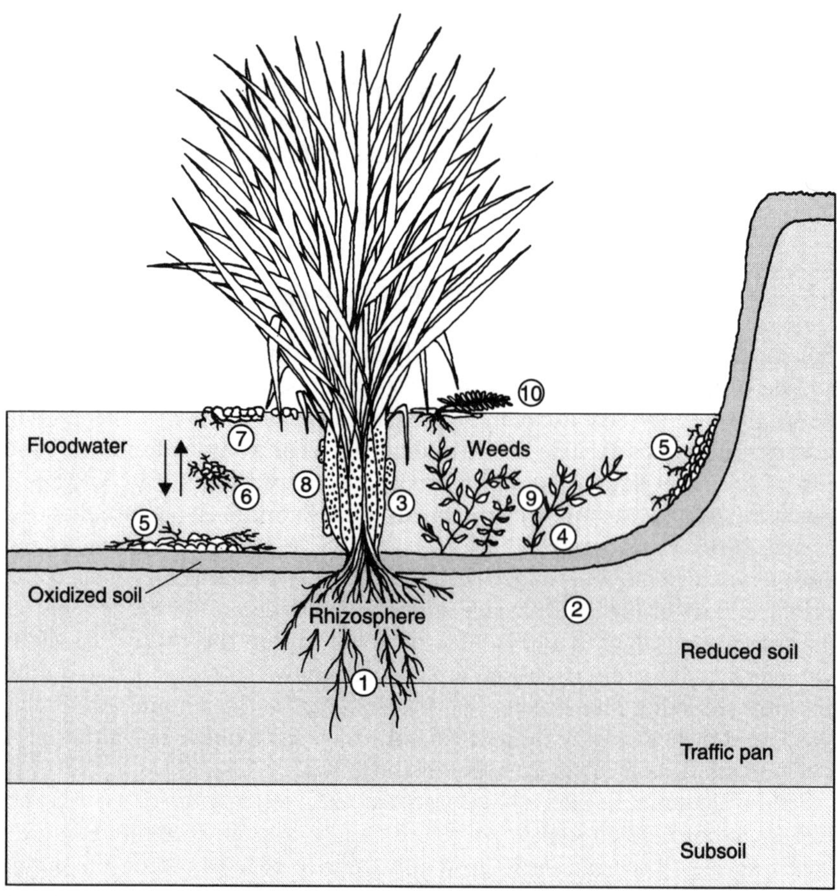

Fig. 5.1. The microenvironments within a flooded rice field (Roger and Watanabe, 1986) showing locations for nitrogen fixing organisms.
Nitrogen fixing bacteria: (1) associated with rice roots, (2) in the soil, (3) epiphytic on rice, (4) epiphytic on weeds. Blue-green algae: (5) at soil–water interface, (6) free floating, (7) at air–water interface, (8) epiphytic on rice, (9) epiphytic on weeds, (10) symbiotic on *Azolla*.

the paddy water, active fixation requiring rather high levels of phosphate to be available to the algae. High levels of inorganic nitrogen in the soil or paddy water tend to inhibit fixation.

Roger and Kulasooriya (1980) reported 21 studies in which the amount of nitrogen fixed by free living blue-green algae was measured. The results showed a range from zero to 80 kg ha^{-1} per crop, with a mean of 27 kg. The total that might be fixed per year in multiple cropping systems could be twice as large, although it would be influenced by seasonal factors, most importantly by aridity in the dry season of monsoonal areas, and cold in the winter in more temperate areas. Subsequent studies (Roger and Ladha, 1992) have given values from 0.2 to 50 kg ha^{-1} in the period required to produce a rice crop.

Estimates of nitrogen fixation by bacteria growing in the rice paddies have been made by several techniques (Roger and Ladha, 1992). Pot experiments, with and without rice plants included, and with and without light excluded from the water standing on the soil surface, have established that nitrogen fixing bacteria associated with rice roots can make a significant contribution (App *et al.*, 1980, 1984), and that the magnitude of the contribution differs between rice varieties (App *et al.*, 1986). Extrapolation on an area basis from pot to field is unrealistic, but gives an addition rate of about 7 kg ha^{-1} per crop. This is in accordance with the information derived from field studies using other techniques (Roger and Ladha, 1992). Incorporation of straw into rice paddies provides additional sites and energy sources for heterotrophic fixation by bacteria, which Roger and Ladha suggest may contribute as much as 2–4 kg N t^{-1} of straw incorporated.

Much greater rates of fixation have been recorded for the *Azolla–Anabaena* symbiosis. *Azolla* may be grown in the rice paddy with the rice crop, or in special ponds, or before the crop and incorporated as a green manure. *Azolla* also requires an adequate supply of phosphorus to flourish, and insect predators often attack the *Azolla*. Nevertheless in favourable circumstances *Azolla* can fix nitrogen at high rates. Roger and Ladha (1992) quote measured rates of 20–150 kg ha^{-1} in experimental plots, and 10–50 kg ha^{-1} when grown with the rice crop in field trials. Watanabe (1987) reports results of a 4-year field trial in which experiments were conducted at 37 sites in ten countries. A crop of *Azolla* was allowed to develop for 30 days in the paddy water in the land soaking period, and incorporated in the soil before transplanting of the rice crop. After the rice had been transplanted a second *Azolla* crop was established in the rice field. The total fresh weight of the two *Azolla* crops averaged 30 t ha^{-1}, and provided nitrogen equivalent to that obtained when 50 kg N ha^{-1} were applied as urea fertilizer. While it is clear that the environment beneath the fronds of the *Azolla* plants is a very favourable site for the nitrogen fixing algae that live with the *Azolla*, there is still much scope for optimizing the conditions for rapid growth and development of the symbiosis (IRRI, 1987a), and for optimizing the methods for utilization of *Azolla*.

Total quantities of nitrogen which may be fixed biologically in rice production systems are listed in Table 5.4. In compiling the table it has been recognized that the major factor limiting biological nitrogen fixation is usually the amount of phosphorus available to support the activity of the nitrogen fixing organism. Hence the fixation rates will usually be greater in manured or fertilized soils than in the systems where yields are poorer. While phosphate stimulates nitrogen fixation, inorganic nitrogen is often inhibitory. Nitrogen balance data collected from long-term trials in Japan and elsewhere (Table 5.5) show negative balances when nitrogen fertilizers are used, and positive balances when they are not. Interactions between the nitrogen fixing agents

Table 5.4. Nitrogen fixed biologically in rice-based cropping systems.

	Irrigated paddies with internal drainage manured and/or fertilized			Rainfed lowland rice in naturally flooded paddies; no internal drainage			Deepwater rice, no manures or fertilizers			Upland rice, small addition animal manure		
	N fixed (kg ha^{-1} year^{-1})		Contributed to rice crop (kg ha^{-1} year^{-1})	N fixed (kg ha^{-1} year^{-1})		Contributed to rice crop (kg ha^{-1} year^{-1})	N fixed (kg ha^{-1} year^{-1})		Contributed to rice crop (kg ha^{-1} year^{-1})	N fixed (kg ha^{-1} year^{-1})		Contributed to rice crop (kg ha^{-1} year^{-1})
	Range	Median		Total	Median		Range	Median		Range	Median	
By free-living and root associated bacteria	5–15	8	5	0–10	5	3	0–4	2	1	0–4	2	1
By free-living blue-green algae	10–140	54	30	2–30	20	10	2–20	10	5	1–10	5	2
By grain legumes grown before or after the rice crop	20–120	60	30	10–40	20	10	0	0	0	10–40	20	10
By legume green manures	50–150	100	60	10–80	40	20	0	0	0	20–80	50	20
By *Azolla*/*Anabaena azollae*	20–140	60	40	5–50	20	5	0	0	0	0	0	0
In loppings of woody species												
(i) of *Leucaena leucocephala*	80–260	150	75	40–100	70	35	0	0	0	80–260	150	75
(ii) of other species	10–150	80	40	5–50	30	15	0	0	0	10–150	80	40

Data compiled from several sources, including Ladha *et al.*, 1992a; IRRI, 1988; App *et al.*, 1980, 1986; Whitton *et al.*, 1988; Giller and Wilson, 1991.

Table 5.5. Effect of nitrogen fertilizers on the nitrogen balance in some rice production trials (after Greenland and Watanabe, 1982).

Site	Annual cropping pattern	Duration (years)	Treatment	N fert. added (F)	Soil change ((N2 – N1)/Y)	Crop uptake (C)	Balance* ((N2 – N1)/Y) + C – F	Ref.
Japan								
Aomori	Rice	21	PK	0	–20	45	+25	(a)
Lat. 41N			NPK	57	–35	66	–26	
Kagawa	Rice/barley	21	PK	0	–42	80 (55)	+38	(b)
Lat. 34N			NPK	152	–18	154 (96)	–21	
Shiga	Rice/wheat	40	No fert.	0	–1.7	41 (30)	+39	(c)
Lat. 35N			PK	0	–13.1	67 (51)	+34	
			NPK	152	+2.2	112 (74)	–38	
Philippines								
Los Banos	Rice	12	No fert.	0	+30	116	+146	(a)
Lat. 14N								
Maligaya	Rice	8.5	No fert.	0	+30	91	+121	(a)
Lat. 14N								
Thailand								
Chainat	Rice	2	–N	0	+47	58	+105	(d)
Lat. 15N	Rice/fallow	2	+N	240	+39	139	–62	
			–N	0	+28	36	+64	

*N1 = initial soil N; N2 = final soil N; Y = number of years experiment conducted; C = N removed in crops; F = N added as fertilizer.
Balance, (N2 – N1)/Y + C – F, is the nitrogen fixed per year, less any losses due to leaching or volatilization.
All data in kg ha^{-1} year^{-1}.
Figures in parentheses are the nitrogen uptake by the rice crop.
References: (a) Koyama and App, 1979; (b) Ando, 1975; (c) Takahashi, personal communication; (d) Firth *et al.*, 1973.

are still little understood. As Roger and Ladha (1990) conclude a method is badly needed to determine *in situ* the contribution of biologically fixed nitrogen to nutrition of the rice crop.

5.1.4 Additions from manures and fertilizers

There are written records to establish that the practice of composting weeds, human and animal excreta, and using bones and fishmeal as soil amendments was well established in China and South Asia during the first millennium BC (Qi-xiao Wen, 1984; Abrol, 1990). Chinese texts written in AD 960 describe organic manures in detail (Chang, 1976a). Bray (1984) includes a recipe for the pelleting of human faeces with other organic wastes and potash ascribed to

Chu Hsi published in AD 1170, and Ho Ping-ti (1959) states that briquettes of this kind were sold in city markets for many years. The approximate nutrient composition of the range of organic materials now used as manures is given in Table 5.6. The amounts of nutrients applied to the soil have been calculated for an addition of 2 t ha^{-1} of manure. It is difficult to know what rates were in fact used before about 1900. Kanazawa (1984) gives a maximum rate for Japan of 6.5 t ha^{-1} reached in 1965, and a range for different regions in Japan from 3 to 10 t ha^{-1}. Qi-xaio Wen (1984) estimated that in China it averaged 2.7 t ha^{-1} in 1979. At that time organic manures were used in conjunction with inorganic fertilizers, and it may safely be assumed that rates at which organic manures were applied in the past would have been somewhat higher. The difficulty of transporting manually more than the fresh weight equivalent of 4 t ha^{-1} to the rice paddy and incorporating it with a simple plough drawn by a water buffalo means that rates greater than 10 t ha^{-1} of fresh material, which would be approximately equivalent to 4 t ha^{-1} dry, would have been rather uncommon.

Tang and Stone (1980) give a detailed breakdown of the total amounts of organic manures used in China between 1952 and 1977 (Table 5.7). Many assumptions had to be made in compiling these figures from a range of sources. They include an estimate of river and pond sediment transported to the fields. If it is assumed that one-third of the total was applied to the rice crop then approximately 10 t ha^{-1} wet weight were used. Yu and Buckwell (1991) give

Table 5.6. Macronutrient composition of some organic wastes, and amounts of nutrient added to soils when 2 t ha^{-1} of organic wastes are applied.

	Composition (%)			Nutrient content of 2 t ha^{-1}		
Manure	N	P	K	N	P	K
Human faeces[a]	1.0	0.2	0.3	20	4	6
Cattle faeces[a]	0.3	0.1	0.1	6	2	2
Pig faeces[a]	0.5	0.2	0.4	10	4	8
Green manures[a]	2.7	0.3	1.6	55	6	32
'Organic manure'[b]				7	3	5
Soybean cake[b]						
Milk vetch[c]				7	1	4
Pig manure[c]				9	2	10
Cotton seed meal[d]	6.6	1.1	1.2	132	22	24
Soybean meal[d]	7.0	0.5	1.3	140	10	26
Farmyard manure[d]	0.6	0.1	0.5	12	2	10
Pig manure[de]	1.0	0.3	0.7	20	6	14
Poultry manure[d]	1.6	0.5	0.8	32	10	16
Dairy manure[d]	0.7	0.1	0.5	14	2	10

(a) From Qi-xiao Wen, 1984; (b) from Kanazawa, 1984; (c) from Lu Ru-kun, 1981; (d) from Tisdale and Nelson, 1966; (e) Rerkasem and Rerkasem (1991) give ranges 0.49–1.25, 0.27–2.44 and 0.65–2.13 t ha^{-1} for quantities applied.

Table 5.7. Use of organic manures in China (all crops). Sources and estimated quantities applied.

	Million tonnes, wet weight							
Year	Night soil	Pig manure	Draft animal manure	Green manure	River and pond sediment	Compost	Oil cake	Total
1952	186	130	422	11	114	74	5	942
1957	237	184	514	39	152	79	5	1209
1967	324	358	599	96	152	81	3	1613
1977	398	492	767	168	152	104	5	2081

(Detailed notes of the assumptions used in preparing these estimates are given by Tang and Stone (1980). The values should be treated with appropriate reserve.)
Adapted from Tang and Stone (1980).

the area cultivated for rice in China in 1952 as 28 Mha out of a total cultivated area of 140 Mha. On the reasonable assumption that rice would receive more than its fare share of the manures applied the rate of 10 t ha^{-1} appears to be reasonable. Outside of China, Korea and Japan the rates used would probably have been rather lower. In some of the intensive rice farming areas of China, several forms of organic manure, silt carried from canals and reservoirs, and in recent years inorganic fertilizers, have been applied to each rice crop. A detailed study of the practices used in Jiangsu Province in China has been given by Wiens (1984). Two rice crops and an upland crop were grown sequentially each year. The nutrient balances estimated by Wiens provide information on the quantities of manures and fertilizer used for each crop (Table 5.8). Of the rice straw, about half was returned to the soil, and half used for animal bedding, fuel and fodder.

The tradition of using manures for rice has a long history in India, but the increasing demand for animal manure as fuel has meant that the quantities used to maintain soil fertility have fallen. In Southeast Asia the use of animal or human excrement as a manure is less common than in East Asia. Inclusion of a grain legume as a crop following rice is however widespread, as is the practice of grazing water buffaloes on the weeds in the paddy areas during the dry season. The quantities of legume residues returned to the soil will often be less than 1 t ha^{-1} as the yields of the legume tend to be low. Nevertheless nitrogen additions can be quite large. In the balances presented below 48 kg ha^{-1} per crop of 0.5 t ha^{-1} has been used for a grain legume grown after a crop of irrigated rice, and used as a green manure as well as a source of food, 26 kg ha^{-1} for a similar crop grown after rainfed rice when soil moisture conditions are likely to be less favourable, and 76 kg ha^{-1} for a crop yielding 1 t ha^{-1}, and grown under favourable conditions in a rice–rice–grain legume system.

The use of inorganic fertilizers for rice is a relatively new phenomenon in rice production. Commercial manufacture of fertilizers began in Japan about

1880, but for the next 70 years fishmeal, and cotton and rapeseed meals and soybean cake were the main offf-arm sources of plant nutrients for rice production (Barker *et al.*, 1985). In the late 1950s approximately one-third of the nutrients added to rice paddies in Japan came from organic sources. By the 1970s this fell to less than one-fifth, or about 50 kg of nutrients per hectare. At that time the use of inorganic fertilizers exceeded 300 kg ha^{-1}. This

Table 5.8. NPK nutrient balance for a rice–rice–vetch (GM) cropping system, Jiangsu Province, China.

	Rice 1	Rice 2	GM	Total (kg ha^{-1} $year^{-1}$)
Inputs (kg ha^{-1} $crop^{-1}$)				
N				
Compost	162	–	–	
Stable manure	26	9	–	
Straw	–	12	–	
NH_4HCO_3	79	69	11	
Total	267	90	11	N 368
P				
Compost	17	–	–	
Stable manure	1	1	–	
Straw	–	0.5	–	
Superphosphate	9	6	3	
Total	27	7.5	3	P 37.5
K				
Irrigation water	15	15	–	
Fertilizer	80	8	–	
Compost	90	–	–	
Stable manure	8	6	–	
Straw	–	2	–	
Total	193	31	–	K 224
Crop removal (kg ha^{-1} $crop^{-1}$)				
N	101	83	0	N 184
P	17	14	(16)	P 31
K	130	100	(112)	K 230

Balance kg ha^{-1} $crop^{-1}$	N	P	K
Total input from fertilizer and manures per annum	368	37.5	224
Total crop removal per annum	184	31	230
Crude annual balance	+184	+6.5	–6

(Assumes no percolation, volatilization or other losses.)
Source: Wiens (1984).

represents a large increase from the rates applied earlier. For instance, Kanazawa (1984) states that in southwest Japan in 1931 about 20 kg ha^{-1} of nutrients were used from organic manure and soybean cake, and a similar amount from ammonium sulphate, superphosphate and potassium sulphate.

The total amount of inorganic fertilizer used in China in 1952 was only 78,000 t. If all had been applied to rice this would have amounted to little more than 3 kg ha^{-1}. By 1984 the amount of inorganic fertilizer used had increased 200-fold (Yu and Buckwell, 1991). He Kang (1985) stated that in 1985 about half of all inorganic fertilizer used in China was applied to rice. Thus the amount of inorganic fertilizers used on rice may have reached approximately 300 kg ha^{-1} year^{-1} by about 1985, or something of the order of 200 kg ha^{-1} crop^{-1}, after allowing for the fact that many rice farms in China grow more than one crop of rice per year. This represents a large addition per crop as substantial amounts of organic manures continue to be used with the inorganic fertilizers (Qi-xaio Wen, 1984; Zhao Qi-guo, 1990).

In other parts of Asia there was also little use of inorganic fertilizers on rice before 1970, except in Korea and Taiwan, where Japanese influence was strong. Part of the reason was the widely held belief that tropical rice would not respond to higher levels of nitrogen. This was generally true of the tropical rice varieties then grown, which were tall and leafy, so that higher rates of nitrogen tended to increase vegetative growth and cause them to lodge. The panicles then dropped into the paddy water and rotted. The semi-dwarf varieties popularized by the 'green revolution' were short, stiff-strawed (Fig. 1.5) and responded well to nitrogen fertilizers (Fig. 5.2). Since 1970 use of inorganic fertilizers has increased considerably in most South and Southeast Asian countries. This is illustrated for Indonesia, India and the Philippines in Fig. 5.3. Most countries do not publish statistics for fertilizer use by crop, and for those countries who do the amounts used for irrigated and rainfed rice are quite different. The uncertainties of water supply tend to inhibit fertilizer use on rainfed crops. The data presented in Fig. 5.3 have therefore been calculated assuming that fertilizer was used on irrigated rice only. This will not affect the data for Japan and South Korea where almost all rice is irrigated, but may slightly exaggerate the rates per hectare shown for Indonesia, India and the Philippines.

5.2 Losses of Nutrients from the Soil under Rice Cultivation

The processes by which nutrients are lost from paddy soils include crop removals, drainage losses and losses by volatilization. Losses by soil erosion which are often important in other systems do not normally affect wetland systems, although they do of course affect upland rice.

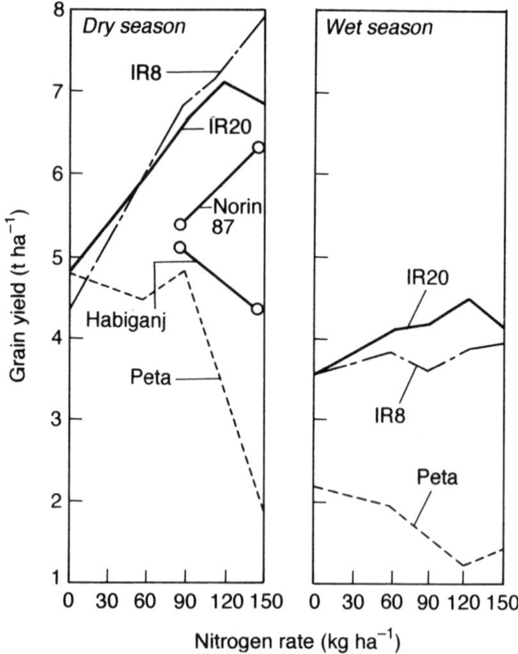

Fig. 5.2. Response of short, stiff-strawed rice varieties (IR8, IR20 and Norin 87) and traditional tall rice varieties (Peta and Habiganj) to applied nitrogen (IR8, IR20 and Peta data from De Datta and Malabuyoc, 1976, and Norin 87 and Habiganj data from Hayami and Otsuka, 1994).

5.2.1 Nutrient removal in the rice crop

The nutrients contained in a crop of rice are dependent on the variety, the yield and the conditions in which the crop is grown. Data for the quantities of nutrients contained per tonne of grain and straw are given in Table 5.9 for a number of varieties, locations and yields. If it is assumed that in the early years of rice cultivation harvests were collected by simple removal of the rice panicle, and that straw was left to rot in the field, then the only nutrients removed from the rice field would have been those contained in the grain. The quantities of nutrients removed in grain yields of 1 t ha^{-1} are between 10 and 15 kg ha^{-1} of nitrogen (N), 1 and 5 kg ha^{-1} of phosphorus (P), and 1.5 and 7 kg ha^{-1} of potassium (K). For all other nutrients the amounts removed are less than 1 kg ha^{-1} t^{-1}. Higher yields and more productive varieties, and straw taken for various uses or burnt lead to much larger removals. Other nutrients may be lost when the straw is burnt, as some ash is often blown away. There are no reliable data about the quantitative importance of such losses.

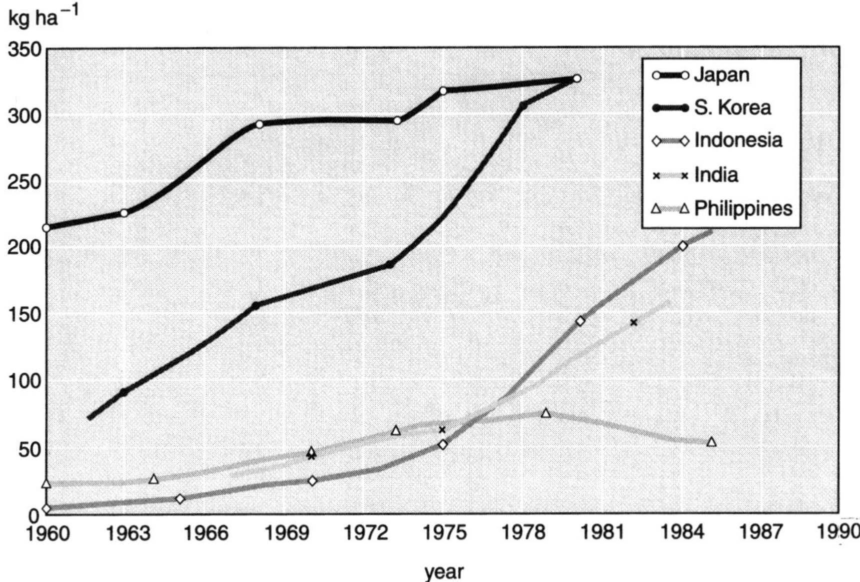

Fig. 5.3. Fertilizer use on rice ($N + P_2O_5 + K_2O$) in Japan and South Korea, 1960–1978 and in India, Indonesia and the Philippines, 1960–1985 (Alexandratos, 1995).

5.2.2 Seepage and percolation losses

Typical amounts of percolation from paddy soils are between 500 and 1500 mm in the period between transplanting and harvest (see e.g. Miranda and Levine, 1978). Concentrations of nutrients in the drainage water will of course differ depending on the soil and soil management. Yatazawa (1977) has collected data from a number of lysimeter experiments in Japan, Gong Zi-tong (1985) has given data for nutrient concentrations in percolation water from a paddy soil in China, Abedin Mian *et al.* (1991) for one in Bangladesh, and Sharma and De Datta (1985b) for two sites at IRRI in the Philippines (Table 5.10). Of the sites at IRRI one was a poorly drained stagnogley and the other a well-drained pseudogley, and data were provided for each when puddled and not puddled.

Use of manures and other inputs in Japan and China was higher than in most other areas, the soil in Bangladesh was developed in fertile sediments from the Brahmaputra, and the soils on the IRRI farm had been well fertilized in the past and received irrigation water from deep wells which is particularly rich in potassium. Thus these concentrations should approximate to the upper limits of those likely to have occurred in the period when yields did not exceed 2 t ha^{-1}. The phosphorus concentrations are in accord with those which Ponnamperuma (1972) found for submerged soils of clay or clay loam texture.

Table 5.9. Nutrient contents of rice grain and straw at harvest.

	kg t^{-1}						
Variety	N	P	K	Ca	Mg	S	Ref.
Grain (rough rice)							
Many	9.6–13	1.7–3.9	1.5–3.7	0.1–0.8	0.6–1.5	0.4–1.6	1
Peta	11.2	2.1	2.8	0.5	1.3	0.8	2
IR8 (1)	13.3	4.4	7.1	0.4	1.5	0.8	2
IR8 (2)	10.9	2.0	3.1	0.5	1.1	1.0	3
IR36	14.6	2.6	2.7	0.1	1.0	0.6	4
Upland	23	4.8	6.7	0.9	0.2		5
Straw							
Peta	12.1	3.6	45	–	–	–	2
IR8 (1)	5.4	0.9	29	–	–	–	2
IR8 (2)	5.3	0.8	14	3.9	2.6	0.7	5
IR36	9.0	0.6	28	3.2	1.6	0.4	3
Upland	4.6	0.7	12	1.7	1.4	–	4

	ppm					
	Fe	Mn	Zn	Cu	B	Ref.
Grain (rough rice)						
Many	14–60	17–94	1.7–31	2–11	–	1
Peta	166	43	16	2	25	2
IR8 (1)	116	52	18	3	25	2
IR8 (2)	38	48	12	5	4.1	3
IR36	200	60	20	2	16	4
Straw						
IR8 (2)	200	560	30	3	8.9	5
IR36	180	370	20	2	19	3

References: 1, Juliano and Bechtel, 1985; 2, Yoshida, 1981; 3, De Datta, 1981; 4, De Datta, 1989; Sanchez, 1976.
IR8(1) Sample from IRRI farm, yield 8.7 t ha^{-1}.
IR8(2) Sample from Maligaya Research Station, yield 7.9 t ha^{-1}.

The nutrient concentrations given in Table 5.10 can be converted to losses in kilogrammes per hectare per 1000 mm of drainage water percolating through the root zone by multiplying by ten. The percolation from the Bangladesh soil was measured as 1850 mm in the year. The losses given for the Japanese soils are the annual means for many paddy soil lysimeter experiments conducted in Japan.

The nutrient losses from the paddy soils may be compared with losses from upland red earths in China and northeast India. The largest losses from upland

Table 5.10. Nutrient concentrations in waters percolating from some wetland and upland rice-based cropping systems.

Country	System	N	P	K	Ca	Mg	Reference
China	Paddy rice	–	0.47	3.1	5.9	1.9	Gong Zi-tong, 1985
Japan	Paddy rice	0.7–3.1	0.07–0.3	2.87–3.2	–	–	Yatazawa, 1977
Bangladesh	Paddy rice	0.92	0.09	2.1	26.7	22.7	A. Mian *et al.* 1991
Philippines	Paddy rice						
	poorly drained						
	puddled	1.0	0.17	14.3			Sharma and De
	not puddled	0.91	0.15	12.5			Datta, 1985b
	well drained						
	puddled	0.96	0.11	7.0			Sharma and De
	not puddled	1.04	0.12	10.2			Datta, 1985b
China	Upland rice	–	0.25	0.59	4.4	1.2	Gong Zi-tong, 1985
India	Upland rice	0.10	0.02	–	–	–	Ramakrishnan, 1992
Japan	Upland rice	1.9	0.0	1.9	–	–	Yatazawa, 1977
		S	**Fe**	**Mn**	**Zn**	**Cu**	
Bangladesh	Paddy rice	0.16	1.2	0.03	0.01	0.005	A. Mian *et al.*, 1991
Philippines	poorly drained						
	puddled		0.53	1.15	0.11		Sharma and De
	not puddled		0.30	0.52	0.07		Datta, 1985b
	well drained						
	puddled		0.54	0.92	0.06		Sharma and De
	not puddled		1.70	1.50	0.05		Datta, 1985b

Concentration in mg l^{-1}. Losses in kg ha^{-1} for leachate of 1000 mm are $10 \times$ mg l^{-1}.

soils are almost always of nitrogen in the form of nitrate. Nitrate seldom forms or persists in paddy soils because of the reduced conditions and so losses of nitrogen by leaching from paddy soils are smaller, and in the form of ammonium. The losses of phosphorus from the paddy soils are greater than from the upland soils, in accordance with its greater solubility in reduced soil conditions. The losses of phosphorus by leaching may be comparable with the amounts removed in grain. Losses of cationic nutrients may exceed the crop removals considerably.

The majority of paddy fields are on alluvial fans, and river terraces. They belong to the pseudogleys, the well drained, high yielding paddy soils described by Chen Jia-fang and Li Shi-ye (1981) and Gong Zi-tong (1985) among others (Figs 4.1 and 4.2). In the valley bottoms and the lower parts of the delta areas where the permanent water table is at or above the soil surface for most of the year the soils are stagnogleys. There is no vertical percolation of water through the soil. There may however be lateral flow or seepage. Except in boundary positions the amounts of nutrients brought into the paddy will equate closely to what moves out, so that the net loss is nil. The same applies to nutrient transfers by surface flow. The losses by this type of transfer will therefore not be considered in subsequent discussion. Nutrients leached into groundwaters from higher landscape positions may become available to rice grown in lower positions, although amounts transferred in this way are likely to be small.

5.2.3 Losses by volatilization

The only nutrient elements affected by volatilization are nitrogen and sulphur. Losses of nitrogen by volatilization can be considerable. Conversion of nitrate to nitrous oxide and nitrogen occurs rapidly after the flooding of most soils. The mineralization of organic nitrogen produces ammonium. Only in aerobic conditions is nitrate produced from the ammonium. Thus losses by denitrification, the reduction of nitrate to gaseous forms of nitrogen, are only significant when there is an alternation of wet and dry conditions. This occurs most frequently in drought-prone rainfed lowland rice soils. The magnitude of the loss depends on the amount of nitrate formed, which in turn depends on the amount of available organic nitrogen in the soil. Addition of nitrate to a flooded paddy soil can lead to volatilization of 70% of the added nitrogen in a few days (Reddy and Patrick, 1986). Thus there is a potential for substantial losses of nitrogen by denitrification. More relevant to conditions where nitrate is not added to flooded soils is the fact that in laboratory studies as much as 30% of the total soil organic nitrogen can be lost from a soil subject to continued wetting and drying (Reddy and Patrick, 1975). In the field losses of nitrate at the end of the dry season may be reduced if a dry season crop is taken (Buresh *et al.*, 1989; Buresh and De Datta, 1991; George *et al.*, 1994), or the weeds allowed to develop and returned to the soil (George *et al.*, 1992, 1993, 1995).

In addition to losses of nitrogen following wetting and drying, some losses occur while the soil remains flooded. This is because there is an oxidized layer at the soil–water interface and around the rice root (Fig. 5.1). Ammonium diffusing into these zones may be nitrified, and the nitrate formed diffuse into the anaerobic zone where it is promptly denitrified. The denitrification process can produce both nitrous oxide and nitrogen gas. Most studies have found a much greater amount of nitrogen gas (dinitrogen) than nitrous oxide, possibly because the nitrous oxide can itself be reduced to dinitrogen (Reddy and Patrick, 1986). In very strongly reducing conditions, which arise when there is a plentiful supply of readily decomposed organic matter present, nitrate reduction can proceed by a different pathway to form ammonium rather than dinitrogen, a process referred to as nitrate dissimilation.

Most ammonium in rice soils is adsorbed on the cation exchange sites on the clay and humic material in the soil. Hence only a low concentration is found in the soil solution and in the paddy water in equilibrium with the soil solution. What is found in the paddy water is subject to volatilization as ammonia gas whenever the pH of the floodwater rises above 7.5.

Volatilization of carbon dioxide as the temperature of the paddy water increases during the day can cause an increase in pH, but algal growth is the primary cause of the paddy water becoming alkaline. When ammoniacal fertilizers or urea are added to the paddy water, losses as ammonia can be large and rapid. Measurements of ammonia in the air above rice paddies in the Philippines (Fig. 5.4) have shown that 20–40 kg ha^{-1} can be lost in 10 days. The rate of loss is determined by atmospheric conditions at the air–water interface, high temperatures and rapid air movement leading to large losses (Fillery *et al.*, 1986). Once the crop canopy has closed little air movement occurs at the air–water interface so that late additions of urea do not suffer significant losses.

The nutrient balance study in Bangladesh (Abedin Mian *et al.*, 1991) showed a deficit of 50 kg ha^{-1} of nitrogen in a year, ascribed to volatilization processes, and other studies using ^{15}N labelled materials (De Datta and Buresh, 1989) showed losses presumed due to volatilization of 30–40 kg ha^{-1} when the labelled material was added to the paddy water, and of 10–15 kg ha^{-1} when it was incorporated in the soil. By combining bulk aerodynamic determinations of loss as ammonia with ^{15}N measurements De Datta *et al.* (1989) were able to show that up to 60% of applied fertilizer nitrogen could be lost by volatilization, of which about three-quarters was lost as ammonium and the balance presumably as dinitrogen and nitrous oxide. Further substantial losses of nitrogen by volatilization may occur if the straw is burnt. Rice straw usually contains between 5 and 10 kg N t^{-1}, and most or all of this will be converted to oxides of nitrogen in a burn.

Most sulphur in straw is also converted to volatile oxides in a burn. It is also possible for sulphur to be lost by volatilization as hydrogen sulphide, but sulphide, if formed, is promptly precipitated in the soil as ferrous and other sulphides. If hydrogen sulphide does escape, the atmospheric residence time of hydrogen sulphide is extremely short, so that it is unlikely that measurable sulphur losses occur in this way. Oxides of sulphur formed when straw is burnt also have a short atmospheric residence time and will be returned to the soil in rainfall, although not necessarily at the same site (Fox and Hue, 1986).

5.3 The Macronutrient Balance of the Soil under Rice Cultivation

5.3.1 Before 1960

From the above assessments of nutrient gains and losses from paddy soils it is possible to assess how the macronutrient balance of rice production systems was sustained. There are many factors which affect the inputs and outputs at

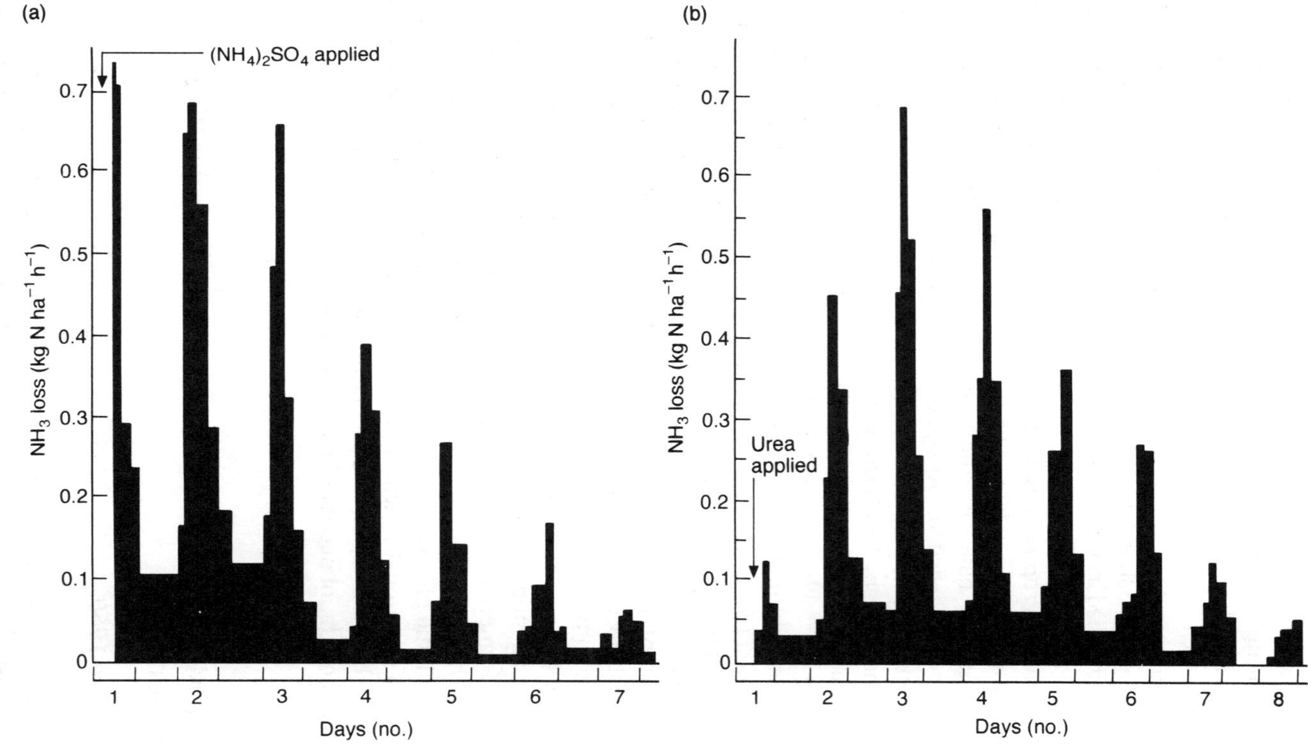

Fig. 5.4. Ammonia evolution from a flooded rice field after addition of (a) ammonium sulphate and (b) urea (from IRRI, 1983b).

any one location, but by making reasonable assumptions balances for typical irrigated, rainfed lowland, deepwater, and upland rice may be calculated. The results of such calculations are presented in Table 5.11. They are intended to represent typical balances in the period before the use of inorganic fertilizers, in monsoonal areas with mean annual rainfall of the order of 2000 mm. The balance for irrigated rice represents an area where water is obtained by stream diversion, so that only one rice crop can be grown each year, but a grain legume yielding 0.5 t ha^{-1} beans is grown after the rice crop. The calculations are based on yields of rice (grain) of 2 t ha^{-1} when 2 t ha^{-1} of animal manure are used. A second calculation has been done for a yield of 1 t ha^{-1} of rice and 0.25 t ha^{-1} of mungbeans. The grain to straw ratio has been taken as 2 : 3 (harvest index of 40%) except for deepwater rice where a grain to straw ratio of 1 : 4 has been assumed. Annual additions of sediment are average levels for both amount deposited (1 mm $year^{-1}$) and nutrient content. Total rather than available nutrient content in the sediment has been used. This exaggerates the short-term advantage to be derived from the sediments, because much of the nutrient content will not be immediately available to crops. However, as discussed previously, the minerals in the sediments will weather over time, and a high proportion of the nutrients they contain will be released in the long term.

The balance calculated for rainfed lowland rice is based on the assumption that there is no loss by drainage as the site is in a lower landscape position (the soil is a stagnogley). The nutrients derived from the natural flooding have been calculated assuming 250 mm of floodwater, and again 2000 mm of rainfall. Percolation losses would be negligible, so that any nutrients in rain and floodwaters would add to the nutrients in the soil. It has also been assumed that the application of 2 t ha^{-1} of manure, and the nitrogen derived from a grain legume grown on the residual water in the soil after the rice crop has been harvested, enabled a rice yield of 2 t ha^{-1} grain and 3 t ha^{-1} straw to be obtained, as well as 0.5 t ha^{-1} of mungbeans.

For deepwater rice it has been assumed that the water supply from which nutrients are derived is 2000 mm of floodwater, and the same 2000 mm of rainfall. There is no percolation, and no manure is used. Some surface runoff may occur, but as nutrients may be absorbed directly from the rainfall and floodwater by the nodal roots of the rice plants, it is reasonable to assume that all the nutrients in the water are utilized, in spite of their low concentration.

For upland rice it is also assumed that no manure is used, that the rice is interplanted or relay planted with a grain legume such as cowpeas, and the grain yields are 1 t ha^{-1} of rice grain and 0.25 t ha^{-1} of cowpeas.

The net balances (Table 5.12) show that for diversion irrigated and rainfed lowland rice, and the conditions assumed, yields of 1 t ha^{-1} or less would have been sustainable, provided that most of the straw was left in the field. For yields of 2 t ha^{-1} phosphorus would have had to be drawn from the soil reserves. For upland rice nitrogen and potassium as well as phosphorus would have had to be drawn from soil reserves. Few soils are able to support such withdrawals

Table 5.11. Typical annual nutrient balances for rice soils pre-1960 calculated from probable inputs and outputs, in kg ha^{-1}.

	N	P	K	Notes
(a) Diversion irrigated; cropping system: rice–mungbean.				
Inputs				
Rainfall	12	0.2	12	2000 mm
Irrigation water	8	0.1	25	1000 mm
Sediments	17	5	120	Deposit of 1 mm
Nitrogen fixation	83	–	–	BGA 30, legume 48, bacteria 5
Manures	12	2	10	2 t ha^{-1} FYM added
Total	132	7.3	167	
Outputs				
Grain	24	5	6	2 t ha^{-1} harvested
Straw	18	2	75	3 t ha^{-1} removed
Bean crop	18	4	8	0.5 t ha^{-1} harvested
Percolation	10	2	30	1000 mm
Volatilization	4	0	0	
Total	74	13	113	
Net balance	+58	–5.7	+48	Straw removed
	+76	–3.7	+123	Straw retained
For yield 1 t ha^{-1} rice and 0.25 t ha^{-1} beans	+73	+0.8	+130	Straw retained
(b) Rainfed lowland rice; cropping system: rice–mungbean.				
Inputs				
Rainfall	12	0.2	12	2000 mm
Floodwater	2	0	6	250 mm
Sediments	17	5	120	Deposit of 1 mm
Nitrogen fixation	59	–	–	BGA 30, legume 26, bacteria 3
Manures	12	2	10	2 t ha^{-1} FYM added
Total	102	7.2	148	
Outputs				
Grain	24	5	6	2 t ha^{-1} harvested
Straw	18	2	75	3 t ha^{-1} removed
Bean crop	18	4	8	0.5 t ha^{-1} harvested
Percolation	0	0	0	
Volatilization	4	0	0	
Total	64	11	89	
Net balance	+38	–3.8	+59	Straw removed
	+56	–1.8	+134	Straw retained
For yield 1 t ha^{-1} rice 0.5 t ha^{-1} beans	+68	+0.7	+137	Straw retained

Table 5.11. *Continued*

	N	P	K	Notes
(c) Deepwater/flood-prone rice; cropping system: one rice crop.				
Inputs				
Rainfall	12	0.2	12	2000 mm
Floodwater	4	0.5	12	2000 mm
Sediments	17	5	120	Deposit of 1 mm
Nitrogen fixation	6	–	–	BGA 5, bacteria 1
Manures	0	0	0	
Total	39	5.7	167	
Outputs				
Grain	24	5	6	2 t ha^{-1} harvested
Straw	48	6	200	8 t ha^{-1} removed
Percolation	0	0	0	
Volatilization	0	0	0	
Total	72	11	206	
Net balance	–33	–5.3	–62	Straw removed
	+15	–0.7	+138	Straw retained
(d) Upland rice; cropping system: rice relay cropped with cowpea.				
Inputs				
Rainfall	12	0.2	12	2000 mm
Nitrogen fixation	32	–	–	Legume 30, bacteria 2
Total	44	0.2	12	
Outputs				
Grain	18	4	5	1 t ha^{-1} harvested with higher nutrient content than others
Straw	12	1	50	2 t ha^{-1} removed
Peas	10	2	4	0.25 t ha^{-1} harvested
Percolation	50	2	10	1000 mm
Volatilization	2	0	0	
Total	92	9	69	
Net balance	–48	–8.8	–59	

A positive balance indicates addition to soil reserves, a negative balance depletion of soil reserves. BGA, blue-green algae; FYM, farmyard manure

for an extended period, so that yields would have continued to fall unless additional sources of nutrients were found and used.

For wetland soils, only where sediments were heavier and richer in phosphorus content than the levels assumed in constructing the balances would yields of 2 t ha^{-1} or more have been sustainable. In favourable sites yields larger than 2 t ha^{-1} may have been harvested, but the size of the phosphorus deficits make it unlikely that yields could have been sustained at

Table 5.12. Collected net annual NPK balances (kg ha^{-1}) before 1960.

	N	P	K	Notes
Diversion irrigated rice				System: rice, beans
Grain yield, rice 1 t ha^{-1}, beans 0.25 t ha^{-1}	+73	+0.8	+130	BNF 59 kg ha^{-1}, +straw
Grain yield, rice 2 t ha^{-1}, beans 0.5 t ha^{-1}	+76	−3.7	+123	BNF 83 kg ha^{-1}, +straw
Grain yield, rice 5 t ha^{-1}, beans 0.5 t ha^{-1}	+52	−9.2	+124	BNF 83 kg ha^{-1}, +straw, +4 t ha^{-1}manure
Rainfed lowland rice				System: rice, beans
Grain yield, rice 1 t ha^{-1}, beans 0.5 t ha^{-1}	+68	+0.7	+137	BNF 59 kg ha^{-1}, +straw
Grain yield, rice 2 t ha^{-1}, beans 0.5 t ha^{-1}	+56	−1.8	+134	BNF 59 kg ha^{-1}, +straw
Deepwater rice				System: rice
Grain yield, rice 2 t ha^{-1}	+15	+0.7	+138	BNF 6 kg ha^{-1}, +straw
Grain yield, rice 2 t ha^{-1}	−33	−5.3	−62	BNF 6 kg ha^{-1}, −straw
Upland rice				System: rice
Grain yield, rice 1 t ha^{-1}, peas 0.25 t ha^{-1}	−48	−8.8	−57	BNF 22 kg ha^{-1}, −straw

BNF, biological nitrogen fixation.

this level. Potassium would have become limiting as well as phosphorus in most situations, particularly where much straw was removed from the site.

The importance of returning at least part of the rice straw to maintain the potassium level is quite clear. In most instances only part of the rice straw is left on the paddy fields after harvest. The straw and stubble may be grazed, so that the nutrients are returned to the soil through the animal. Most actual balances would probably fall somewhere between the estimates for straw removed and straw retained.

Although phosphorus supplies would not have been adequate for a 2 t crop, sufficient nitrogen should be provided by 2 t ha^{-1} of animal manure plus nitrogen fixation to sustain yields at that level. If *Azolla* was used in place of the grain legume more nitrogen would have been returned, and more importantly less phosphate would have been removed, so that the high yields demanded by the cruel landlord (p.88) might have been sustainable. But it could only have happened in an area where the phosphorus and potassium supplied from sediments and manures were also high.

For deepwater rice systems a yield of 2 t ha^{-1} is sustainable provided that little or no straw is removed. A rather high grain to straw ratio has been used, characteristic of varieties tolerant to water depths up to 2 m and more (Catling, 1993). Catling considers the importance usually assigned to the nutrients supplied to deepwater rice by sediments as a myth. However he was considering the short-term supply of nutrients to the crop from the annual sediment deposit. As can be seen from Table 5.11c the major source of nutrients for deepwater rice is the sediment deposited each year. The phosphorus supplied by the sediment is critically important. The deepwater areas are those where rice has been grown the longest. The depth of soil over the period of 2000 years and more for which rice has been grown has usually increased by at least a metre, as sediment has accumulated. The corresponding annual nutrient additions calculated from the mean sediment concentrations given by Whitton and Rother (1988) are 25 kg ha^{-1} of N and 5.5 kg ha^{-1} of P. Unfortunately the K concentration in the sediments is not given. The quantities of N, P and K provided annually by 4000 mm of rain and floodwater of the concentrations reported by Whitton and Rother are 1.6, 3.6 and 7.6 kg ha^{-1}, sufficient to sustain a crop of about 1 t ha^{-1}. However, the figure they use for the phosphorus concentration in standing water in deepwater rice fields, 0.09 mg l^{-1}, is exceptionally high. It includes what they refer to as 'filterable phosphate,' which should be attributed to sediment rather than dissolved material.

The management of straw and stubble is obviously of critical importance to the sustainability of the system. Complete removal would soon lead to major nutrient deficiencies. Deepwater rice farmers must have learnt this lesson long ago, as the normal practice is to take only part of the straw for domestic animals, most remaining on the ground to be grazed and nutrients returned through the grazing animal (Catling, 1993).

For upland rice even a yield as low as 1 t ha^{-1} could not have been maintained without mining the soil supply of nutrients. Few soils are endowed with sufficient nutrient reserves to maintain supplies at the rate of withdrawal required. Hence the normal requirement to allow upland soils in the tropics to revert to a natural or restorative fallow for several years after a short period of cropping. The calculation of the nutrient balance for upland rice has been made assuming that there are no losses due to erosion. Some erosion losses are normally incurred in upland production systems, often comparable with or greater than the rates of addition from sediments in the lowland soils. Soil degradation then becomes serious. Ramakrishnan (1992) has reported a detailed study of an upland rice-based farming system in northeastern India, including the nutrient balances. Without erosion control measures the removal of nutrients in the eroded soil led to massively negative balances. In a terraced system where erosion losses were negligible (Table 5.13), annual losses of nitrogen and potassium were still very large, and the loss of phosphorus still significant. The phosphorus loss would have also been large were it not for the use of animal manure which was reported to return 7 kg P ha^{-1} $year^{-1}$.

5.3.2 Changes in the nutrient balance post-1960

What may appropriately be termed a watershed in the sustainability of rice farming systems dates from about 1960, when the great expansion of water storage systems for irrigation started. This was followed quickly by the release of the new stiff-strawed, high-yielding semi-dwarf rice varieties and greatly

Table 5.13. Nitrogen and phosphorus balance for terraced upland-rice-based farming system, Meghalaya, northeastern India.

Inputs kg ha^{-1}	N	P	Outputs kg ha^{-1}	N	P
Rain	4.3	1.1	Fire	239.5	0.9
Residues	65.5	0.8	Runoff	2.2	0.5
Organic manure	11.7	7.0	Percolation	1.0	0.2
Weeds	8.5	0.1	Weeds	43.8	0.4
			Crops	73.9	10.2
Total	90.0	9.0		260.4	12.2

Balance, kg ha^{-1}: N, –170.4; P, –3.2 and (details not given) K, –339
Data are for the 12th year of cropping of a terraced system, with upland rice intercropped with a range of other crops. In comparable non-terraced systems, N balance given as –430.7, with 172.9 kg ha^{-1} lost in eroded soil
Source: Ramakrishnan (1992).

increased use of inorganic fertilizers. Among the important changes in relation to sustainability were the expansion of double, and in a few instances, of triple rice cropping, the trapping of silt which would otherwise have been deposited on the rice fields behind the dams built to hold back the river waters, and the reduction in the use of manures and increase in the use of fertilizers.

A hypothetical NPK nutrient balance for a rice–rice–mungbean system is presented in Table 5.14. It has been assumed that the wet season rice crop yielded 3 t ha^{-1} and the dry season crop 5 t ha^{-1}, and that the harvest index was 0.5. The mungbean crop has been assumed to produce 1 t ha^{-1} of harvested beans, and the residues returned to the field as green manure. The fertilizer rate applied has been assumed to be 100 kg N and 20 kg P for both rice crops, but none for the mungbeans. Inputs from rainfall compared with those pre-1960 have been assumed to be slightly higher because of the effects of atmospheric pollution. A loss by volatilization of 45 kg of N from each fertilizer application has been assumed. For high yielding rice soils such as these, drainage of 10 mm per day would be desirable from transplanting until the soil is allowed to dry shortly before harvest. To enable a triple crop system to operate short duration rice varieties must be grown. Thus the irrigation would only be needed for about 75 days per crop rather than the 100 days assumed for the rice mungbean system, giving total percolation of 750 mm per rice crop. The greater solubility of the nitrogen in the fertilizer compared with that in organic manures will lead to at least temporarily higher concentrations of ammonium in drainage waters. The amount of nitrogen lost through percolation has been adjusted slightly to allow for this. The nitrogen fertilizer has been assumed to depress nitrogen fixation by blue-green algae, but the phosphorus fertilizer to stimulate not only the growth of the mungbean crop but also nitrogen fixation by the rhizobia associated with it.

Table 5.14. Typical annual nutrient balance for storage irrigated rice after 1960. Cropping system: rice–rice–mungbeans. Balances calculated from probable inputs and outputs, in kg ha^{-1}.

	N	P	K	Notes
Inputs				
Rainfall	15	0.3	15	2000 mm
Irrigation water	12	0.2	35	1500 mm
Nitrogen fixation	108	-	-	BGA 30, legume 76, bacteria 2
Fertilizer	200	40	0	100 : 20 : 0 applied to each rich crop
Total	335	40.5	50	
Outputs				
Grain	96	20	24	8 t ha^{-1} harvested from 2 crops
Straw	48	6	160	8 t ha^{-1} removed from 2 crops
Bean crop	36	8	16	1.0 t ha^{-1} harvested
Percolation	18	3	45	1500 mm
Volatilization	90	0	0	Ammonia loss from fertilizer
Total	288	37	245	
Net balance	+47	+3.5	–195	Straw removed
	+113	+9.5	–35	Straw retained

A positive balance indicates addition to soil reserves, a negative balance depletion of soil reserves. BGA, blue-green algae.

With no additions from sediments the net result (Table 5.14) is an annual gain in the soil store of N and P but a decline in the store of K of 195 kg ha^{-1} year^{-1} if straw is removed, or 35 kg ha^{-1} year^{-1} if straw is retained. The net balances can of course be changed significantly by use of larger amounts of inorganic fertilizers, and the K balance particularly by retention of the rice straw. A lower application of nitrogen and an application of fertilizer potassium of 35 kg ha^{-1} plus retention of the rice straw would lead to a sustainable system in the circumstances assumed, producing 8 t ha^{-1} year^{-1} of rice and 1 t ha^{-1} year^{-1} of mungbeans.

This calculation of a 'typical' nutrient balance for a high yielding system from a rice farm growing two rice crops and a mungbean crop each year shows clearly the importance of how the rice straw is handled. Common practice is to remove part of the straw as the grain is harvested and to burn the pile of straw produced after threshing. If the ash is spread over the rice field the benefit of the potassium contained would pass to subsequent rice crops. Equally if the straw is grazed the nutrients contained will be distributed by the animal.

Without the application of the inorganic fertilizers soil reserves of nutrients would not be able to sustain yields at the levels assumed in preparing Table 5.14. Major responses to applied P and K have indeed been widely observed (De Datta, 1983; Von Uexkull, 1985; De Datta *et al.*, 1988). Even in soils with initially high levels of available phosphorus and potassium, responses have been obtained after several years of intensive cropping. One example is given by the long-term experiments in the Philippines (Fig. 5.5). In the 17th crop year at Maligaya the phosphorus deficiency was sufficiently severe that yield of the dry season crop only just exceeded 3 t ha^{-1}. When phosphorus was included the yield reached 5.8 t ha^{-1} without potassium, and rose further to 6.6 t ha^{-1} if potassium was added. At this site the phosphorus content of the soil was initially sufficient to allow two crops with a total yield of about 7 t ha^{-1} to be obtained without addition of fertilizer, but as cropping was continued yields fell as the supply of phosphorus from the soil could not be maintained. Responses to phosphorus applied as fertilizer then increased until in the dry season the responses were over 3 t ha^{-1} (Fig. 5.6).

The increasing importance of potassium to the maintenance of rice yields as intensive cropping is continued is shown by the change in responses to potassium measured in trials in China (Lin Bao, 1985). Of 62 trials conducted on wetland rice in 1958 only 18 (29%) showed positive responses, whereas in 1980 out of 260 trials 164 (62%) gave positive responses.

5.4 Secondary and Micronutrient Balances

5.4.1 Calcium, magnesium and sulphur

Of the secondary nutrients, calcium (Ca), magnesium (Mg) and sulphur (S), only S is commonly removed from rice soils in larger amounts than are supplied in irrigation water and other inputs. In the colluvial and alluvial soils in which rice is most commonly grown the amounts of Ca and Mg deposited in sediments and irrigation and floodwaters normally exceed the crop removals by a substantial amount. The amounts of Ca and Mg in solution in flooded conditions tend to be higher in anaerobic than aerobic conditions (Robinson, 1930; Islam and Islam, 1973; Yu Tian-ren, 1981). Thus calcium and magnesium deficiencies in flooded rice are uncommon.

As discussed in Chapter 4 the problem of acidity, which is common in upland soils, seldom arises in flooded soils, because of the neutralizing effects of the reduction of iron. Thus positive responses to liming are very rare in paddy soils. Some early reports of positive responses (from +7% to +12%, which were unlikely to have been statistically significant) have been discussed by Takahashi (1965). The responses were believed to be due to secondary effects, such as stimulation of the mineralization of nitrogen in soils with rather severe nitrogen deficiencies, or a response to phosphorus contained as an 'impurity' in the lime that was applied. Although 37 field experiments on the liming of paddy rice in Japan were conducted, some continuing for more than 50 years, the extent of positive responses has been too small to merit recommendation of the use of lime as a soil amendment, except in a few restricted areas.

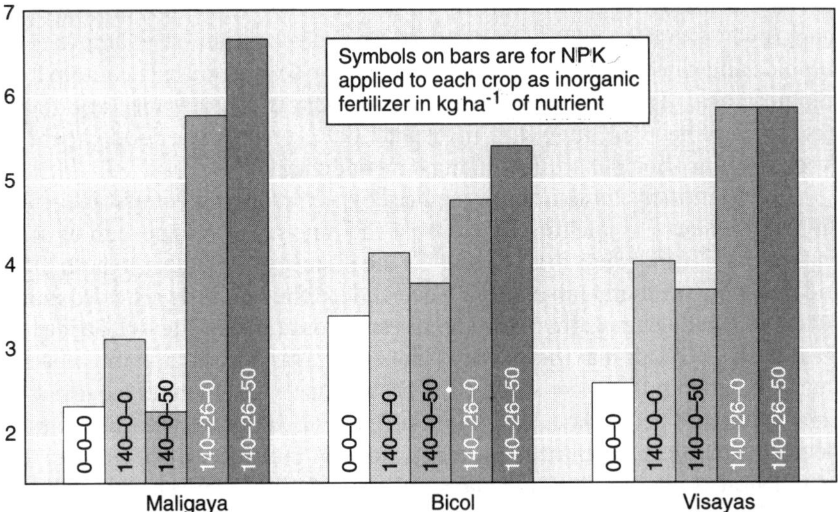

Fig. 5.5. Response of rice to NPK at three experiment stations in the Philippines; dry season crop in 17th year of a continuous 2 crops per year experiment. Results shown are for the mean of two high-yielding varieties (De Datta, 1985).

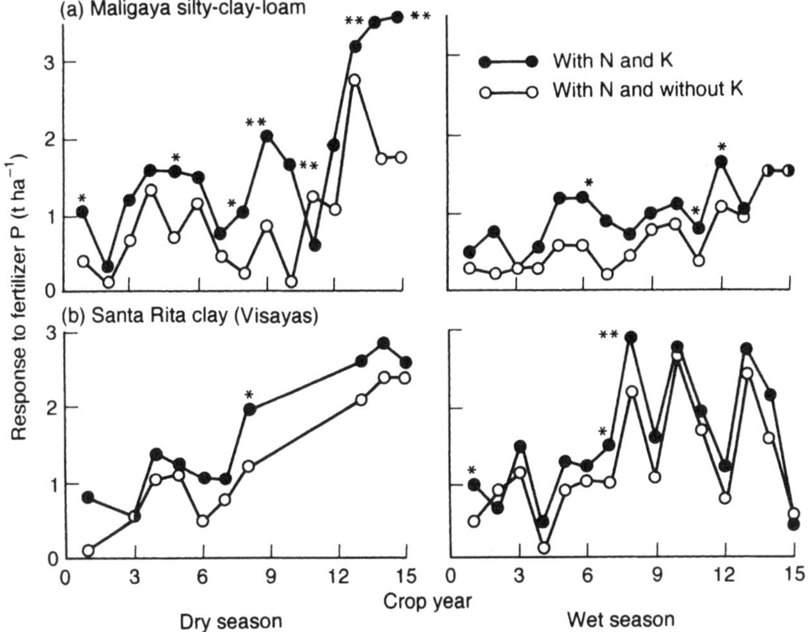

Fig. 5.6. Trends in the response of rice to phosphorus and potassium fertilizers at the (a) Maligaya and (b) Visayas Experiment Stations in the Philippines (De Datta, 1983).

An exceptional acidity problem occurs when acid sulphate soils or cat clays are drained for agricultural use. These soils occur in swampy areas where mangrove vegetation existed in the past. The soils contain large amounts of sulphide and if drained this will oxidize so that the soils essentially contain free sulphuric acid. The total area of cat clays in Asia is considerable, and there has therefore been an interest in their potential use, but their development is expensive, and their subsequent management difficult.

The quantities of magnesium required by rice crops are comparable with the requirement for calcium. By contrast the amount of magnesium in soils and waters is usually less than one half of the quantity of calcium, in both total and available (exchangeable) form. Nevertheless the amounts provided in inputs to rice paddies, together with the reserves contained in the soil, appear to have been adequate for the needs of the crop. This situation could well be changing as irrigation from storage reservoirs and replacement of magnesium from sediments decreases, and the crop removals increase with the higher yields. Responses to applied magnesium have in fact recently been reported from Tamil Nadu, India (Vijayalakshmi and Mathan, 1991) and Taiwan, China (Lin, 1992).

Problems of acidity and calcium and magnesium deficiency are of course common in upland crops of the tropics, and such problems occur with upland rice as they do with other upland crops.

Brady (1982) commented that sulphur deficiency in rice had been reported in some areas in India and Indonesia but 'as yields of tropical rices are increased, the need for sulphur fertilizers will become more common'. This prediction has been fully substantiated, widespread deficiencies having now been reported for paddy rice in Bangladesh (Bhuiyan and Islam, 1986), the Provinces of Zhejiang, Jiangxi, Fujian and Yunnan in China (Liu Zhonqun, 1986), Pakistan (Rashid *et al.*, 1992), and the Philippines (Mamaril and Gonzales, 1988). Tandon (1992) has reported that the deficiency of sulphur on paddy rice in India is now much more widespread than it was previously. It has not so far been recorded in the field in Korea, but a sulphur balance for rice production (Shin and Han, 1988) indicated a negative balance of 12 kg ha^{-1}, and a future deficiency was predicted. At that time rainfall inputs were given as 5 kg ha^{-1}, and irrigation inputs as 15 kg ha^{-1}, with net crop removals of 18 kg ha^{-1}, and percolation losses of 20 kg ha^{-1}. Adventitious additions from sulphur in N, P, K fertilizers and pesticides kept the balance to a deficit of 12 kg ha^{-1}. Increasing amounts in rainfall as the extent of fossil fuel use increases may add to the amounts in rainfall. Kawai (1988) reports a range of values for Japan in 1986 from 16 to 29 kg ha^{-1} $year^{-1}$. Lefroy *et al.* (1992) collected data from the Philippines, Sri Lanka, China, Malaysia and Korea and found a range from 5 to 29 kg ha^{-1} $year^{-1}$.

5.4.2 Micronutrients

The micronutrients required by plants are iron (Fe), manganese (Mn), zinc (Zn), copper (Cu), boron (B) and chlorine (Cl). In general terms the amounts removed by crops are almost always very much lower than the quantities present in the soil (Table 5.15). The amounts in the soil are of the order of tonnes or kilograms per hectare, and the crop requirements of the order of grams or milligrams per tonne. Often the inputs from irrigation water, rainfall and sediments are more than sufficient to offset crop removals. Problems may arise when soil conditions are such that the concentration of the element in the soil solution is too low to enable sufficient uptake by the rice plant. Problems of toxic levels in irrigation water can also arise.

Zinc is the micronutrient that has most commonly been reported to be deficient (Mikkelsen and Kuo, 1977; Randhawa *et al.*, 1978). Occurrence of deficiency is associated with soils of pH above 6.8 and organic matter contents above 3 per cent (Fig. 5.7). Some deficiencies have been observed where the absolute amount in the soil is low, and this type of occurrence may be expected to become more common as yields in excess of 10 t ha^{-1} are taken over a long period. Deficiencies of copper might also be expected to occur in this way. Zheng and Huang (1986) and Xu and Dong (1989) have reported responses of rice to copper in the southern areas of China and Sheudzhen (1991) in the southern part of the former USSR, and there have also been reports of copper ameliorating the effects of iron toxicity. A special case is copper deficiency in peat soils. Many crops grown on peats suffer from copper deficiency because the copper is strongly complexed by organic compounds and unavailable to plants; rice is no exception, as Widjaja-Adhi (1988) has reported in a study of the peat soils of Indonesia.

Table 5.15. Total micronutrient contents of soil and rice.

Micronutrient	Amount in (0–25 cm) soil (kg ha^{-1})	Amount (kg) removed per tonne		
		Grain[a]	Straw[b]	Total
Fe	54,000–210,000[c]	0.06	0.19	0.25
Mn	5,000–250,000[d]	0.05	0.46	0.51
Zn	70–300[e]	0.016	0.025	0.041
Cu	12–375[d]	0.006	0.006	0.012
B	2.4–2,500[f]	0.022	0.014	0.036
Mo	0.95–8[g]	-	0.0001[d]	
Cl	45–2,000[f]	-	0.003[d]	-

(a) From Juliano and Bechtel, 1985; (b) from Table 5.8; (c) from Kyuma, 1991; (d) from Jones *et al.*, 1982; (e) from Mikkleson and Kuo, 1977; (f) from Russell, 1988; (g) from Tang Li-hua, 1981.

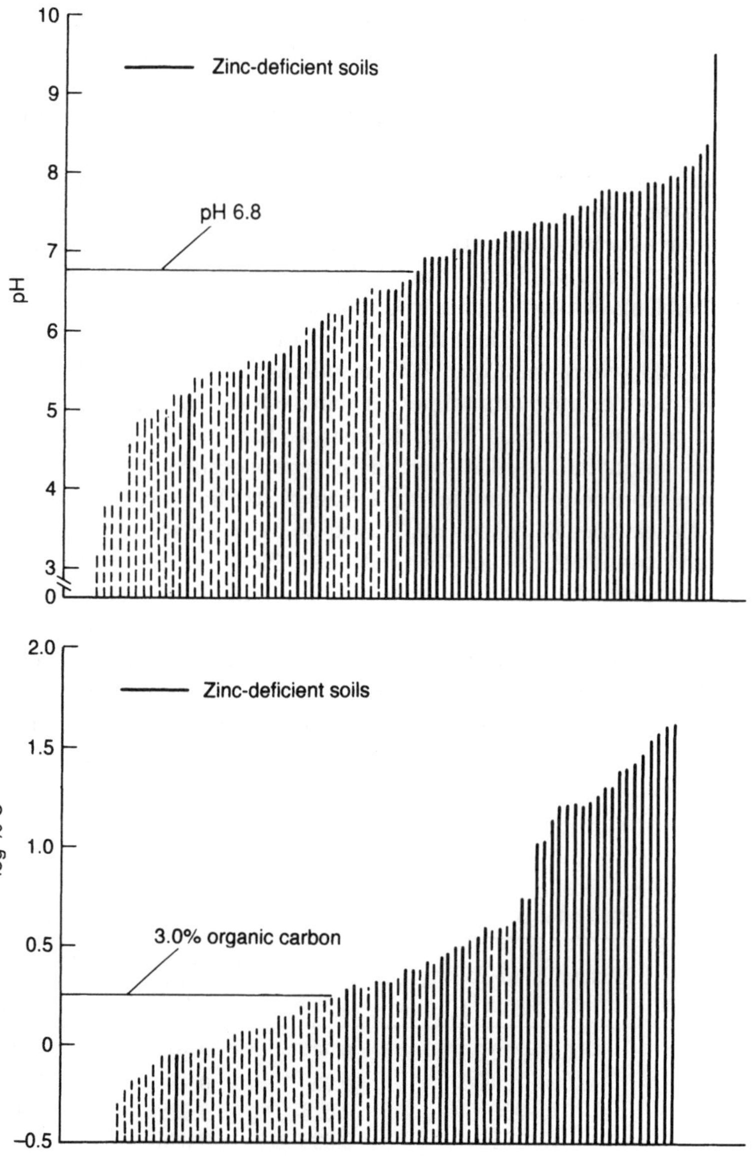

Fig. 5.7. Relation between soil properties (pH and carbon content) and response of rice to applied zinc, for 300 Philippine soils (IRRI, 1981b).

5.5 The Long-term Sustainability of Nutrient Supplies for Rice Production

Table 5.12 which summarizes the macronutrient balances before 1960 indicates that there would have been a steady drain on the soil reserves of phosphorus to maintain yields above 1 t ha^{-1} in the diversion irrigated and rainfed areas. Rice production can have persisted at this level only where the supplies of phosphorus were greater than has been assumed in the calculations from which Table 5.12 has been developed. Almost certainly the source has been the sediments; yields have been maintained where the amount of sediment deposited was greater than 1 mm $year^{-1}$, or the sediment was richer than the level assumed in making the calculations used to prepare Table 5.11. The limited data on actual nutrient contents of sediments (Table 5.3) suggest that this is not unlikely. In some areas there would also have been higher rates of addition of organic manures than those assumed, and a more efficient recycling of nutrients. But there would also have been areas where yields fell below the assumed 1 t ha^{-1} because of phosphorus deficiencies.

In the post-1960 period, in storage irrigated systems, the major source of nutrient additions from sediments has been lost. This loss has had to be made good from fertilizer additions. As shown by Table 5.14 an annual addition of 20 kg of phosphorus per crop, rather than per year, will often be necessary, as well as considerable additions of potassium and nitrogen. The potassium reserves may have been increased substantially in the past by the sediments deposited, sufficient to sustain a high yield for decades, but seldom will they sustain high yields for longer than decades. Single rice crop yields of 15 t ha^{-1} have recently been reported from Yunnan Province in China (Cassman and Pingali, 1995b). They have been obtained with very high inputs to both the rice crop and the preceding watermelon crop (Table 5.16). The annual additions of inorganic fertilizers are 235 kg N ha^{-1}, 80 kg P ha^{-1}, and 215 kg K ha^{-1}, plus the nutrients contained in 74 t ha^{-1} of (presumably wet) manure. As the source of the manure is undefined, as well as the yield of water melons, it is not possible to estimate a nutrient balance, but it would seem probable that using inorganic and organic manures at these levels is sufficient to sustain the very high rice yields. Whether such intensive farming systems are sustainable in terms of their relation to other internal and external factors will be discussed in Chapter 8.

In Spain, Italy and Australia, where little if any organic manure is used, national average yields already exceeded 6 t ha^{-1} by 1960, and have continued at this level or higher until 1995. They have been sustained by using very high levels of inorganic fertilizers. In the European countries agricultural subsidies have been a necessary factor in making the high inputs profitable. In Australia there have been no direct subsidies, but the production system has been extremely labour-efficient, with very large and highly mechanized farms.

Table 5.16. Inputs to a rice–watermelon cropping system giving rice yields of 15 t ha^{-1} and a watermelon crop at Tauyuan Township, Lijiang Prefecture, Yunnan Province, China.

Crop	Management practice	Inputs and operations
Watermelon	Seedbed preparation	Ploughing 3 times
Dec–April		Harrowing 3 times
	Manuring	37 t ha^{-1}
	Seed treatment	Fungicide
	Plastic mulching	To cover ridges
	Fertilization:	Total inputs:
	3 applications N	100 kg N ha^{-1}
	2 foliar applications	55 kg P ha^{-1}
		130 kg K ha^{-1}
	Irrigation	5 times
	Pest management	
	insecticides	3 to 4 applications
	fungicides	2 to 3 applications
Rice	Land preparation	Ploughed twice
May–Oct	Soil drying	20 days
	Manuring	37 t ha^{-1}
	Harrowing	Once
	Wet cultivation	Puddling
		Harrowing
	Irrigation	Continuous flooding
	Fertilization:	Total inputs:
	3 applications N	135 kg N ha^{-1}
	3 foliar applications	25 kg P ha^{-1}
	(included growth regulator)	85 kg K ha^{-1}
	Pest management	
	herbicide	1 application
	insecticides	1 or 2 applications
	fungicides	2 applications

(Data provided by Mr Gaoqun Yang, Agronomist, Lijiang Agricultural Research Institute, and Mr Qingrui He, Rice breeder, Yunnan Academy of Agricultural Science, and published in Cassman and Pingali, 1995b.)

An indirect subsidy has been provided through the provision of water at little or no cost to the farmer. The high yields have been attained because of high radiation intensity, and low night temperatures, which reduce respiration at night and help to avoid losses of photosynthate. While the average yields have continued to increase, the rice production has not been sustained on all the rice farms. In some areas significant problems have arisen because of salinization. As will be discussed in the following chapter, inadequate drainage is a factor threatening the sustainability of rice production in many arid and semi-arid areas. Egypt is another country where rice is produced under rather similar conditions to Australia, with high daytime and low night temperatures. Although lower levels of fertilizer are used, yields have exceeded 7 t ha^{-1} since 1990. A legume, berseem, grown in the ricefields in the winter months, will have contributed some nitrogen to the nutrient balance of the system. Probably more important is the good quality of the irrigation water from the Nile, and the excellent irrigation and drainage system installed almost a century ago.

6 Maintaining Water Supplies for Rice

Water is essential for all plant growth, but the quantity of water needed to produce rice is greater than that required for any other major crop (Bhuiyan, 1992). Revelle (1963) noted that 4000 t of water were used to grow 1 t of rice, whereas wheat used only 1000 t. Although rice will grow as an upland crop, yields are almost always considerably less than when it is grown under flooded conditions. Most experimental studies have shown that it makes little difference to rice yields whether the soil is flooded to 1 cm or 10 cm, but even a few days when the soil is not covered by water can cause a significant loss of yield (Wickham and Sen, 1978). Excess water can also be a serious problem. Floods have destroyed homes as well as rice crops throughout the great centres of rice cultivation for as long as rice has been grown. Thus control of water deficits and excess is a major biophysical factor in sustaining rice yields.

6.1 The Water Requirement of Rice

The amount of water consumed in producing a rice crop depends upon weather, landscape position, crop duration, soil drainage characteristics, and the management of the water supply. Transpiration by the crop is usually of the order of 5–8 mm day^{-1}, and percolation in the range 1–10 mm day^{-1}, depending on landscape position, soil characteristics, and how effectively the soil has been puddled. Thus a transplanted rice crop which is in the field from transplanting to harvest for 100 days will require 540–1620 mm of water, assuming that terminal drainage is initiated 10 days before harvest. Many older varieties are slower maturing, taking perhaps 130 days, and require correspondingly more water, 720 mm at the lower limits of evaporation and percolation, and 2160 mm at the upper limits. These figures may be compared with the water use for evaporation, transpiration and percolation by an upland crop in the humid tropics of 300–500 mm.

In addition to the water requirement during the time that rice is in the field, where irrigation water is available a considerable quantity of water is usually used in land preparation. This normally involves about 2–3 weeks of 'land soaking' before cultivation, a further week or more for ploughing, harrowing, puddling and land levelling, and up to a week for transplanting. The last three operations have all to be conducted in the saturated and softened soil. Thus it is unusual for these operations and the initial land cultivation to be concluded in less than a month. The water requirement for land preparation is commonly of the order of 300–700 mm, part being lost by evaporation and percolation and part used to saturate the soil, which will usually be in a dry and cracked state at the start of land preparation.

In addition to water applied to the field all irrigation systems are subject to water losses in the distribution process. These losses may exceed the water use in crop production, and are seldom much less than half of the crop water requirement. They involve evaporation and seepage from the storage and distribution canals, and problems in the timing of distribution. If farm requirements and deliveries are not synchronized there is a considerable loss as runoff. Much water might be saved if the management of irrigation systems could be improved (IRRI, 1980).

Similarly much water might be saved if the time required for land soaking and preparation could be reduced. Mid-season drainage may also reduce water use slightly, and as discussed earlier is much used in China and Japan to enhance yield. The yield and any water-saving advantages of mid-season drainage are however lost if the soil is allowed to dry to the stage when cracks appear in the soil. The crop suffers from inadequate water supply and much water can be lost by deep percolation when water is reapplied. The usual system of mid-season drainage is to allow water to drain laterally to lower outlets, and then to withhold water from the field for no more than 5 or 6 days.

In terms of yield per unit area, flooding to 5–10 cm has mostly been found to be optimal, but in terms of yield per unit of water used, relatively dry conditions are best, provided that the stress does not affect yield too severely. Data are given in Table 6.1. for two sites and longer (IR8) and shorter (IR72) maturing rice varieties. Direct comparison between the two sets of data is not appropriate, as percolation at the CLSU site would have been significantly greater than at IRRI. Maintaining flowing water in the paddy field is often regarded as the best practice, but can be very demanding in water requirement. It can also lead to serious losses of top-dressed fertilizer.

The change in water management practice in Japan from traditional continuous flooding to intermittent irrigation in which only sufficient water is applied to keep the soil just covered is illustrated in Fig. 6.1. Some form of intermittent irrigation similar to the Japanese system is now practised in many

Table 6.1. Effect of water management practices on water use and yield of (a) IR8 at IRRI (De Datta, 1981) (b) IR72 at the Central Luzon State University experiment station (Bhuiyan *et al.*, 1995). DS, dry season.

Water management	Yield (kg ha^{-1})	Water use (mm ha^{-1})	Yield (kg ha^{-1} mm^{-1})
(a) IRRI, DS, 1966			
Continuously flooded, 10 cm	8400	910	9.23
Continuously flooded, 2.5 cm	8200	804	10.20
Flooded, 10 cm, with mid-season drainage	8500	867	9.80
Continuous flowing irrigation	8300	4581	1.81
Irrigated at signs of stress (simulated rainfed)	3200	197	16.24
(b) CLSU, DS, 1989			
Continuously flooded, 5–7 cm	6600	1255	5.26
Saturated soil –			
2 cm applied on alternate days	6400	1040	6.15
Partial drying allowed –			
2 cm applied every 4th day	4200	725	5.79
5 cm applied every 7th day	3400	1015	3.35

Asian rice production schemes in which water supply is limited. These systems can maintain yields close to the maximum if weeds can be satisfactorily controlled. As much as 30–50% of the water used when standing water is maintained can be saved. When weed pressure is high, a shallow-flooded water regime maintained from transplanting to full canopy cover can achieve significant water saving, and provide some measure of weed control (Tabbal *et al.*, 1993; Bhuiyan *et al.*, 1995). Intermittent irrigation and shallow flooding require accurate land levelling, and more skilful and intensive handling of water than is possible from the current public irrigation supply systems in most of tropical Asia.

The water that needs to be provided from irrigation will be partly offset by that supplied from rainfall, and replaced by rainfall in entirely rainfed systems. For the wet season crop in monsoon Asia the rainfall during the season may be as high as 2000 mm, but somewhat less during the cropping period. Not all of the rainfall will be effective as it is normal for much to arrive in heavy sustained downpours, particularly in the areas subject to typhoons. Much is then lost in runoff. Rainfall in excess of 50 mm or 75 mm in a day is often excluded from effective rainfall when a water balance is calculated for upland crops, but because the bunds around rice paddies are able to retain much of the water that falls it has sometimes been assumed that rainfall for flooded rice is 100% effective (e.g. Oldeman and Suardi, 1977). In fact the bunds around paddy fields are usually built to retain no more than about 10 cm of water, and as typhoons often deliver more than that in a day, and the fields will normally

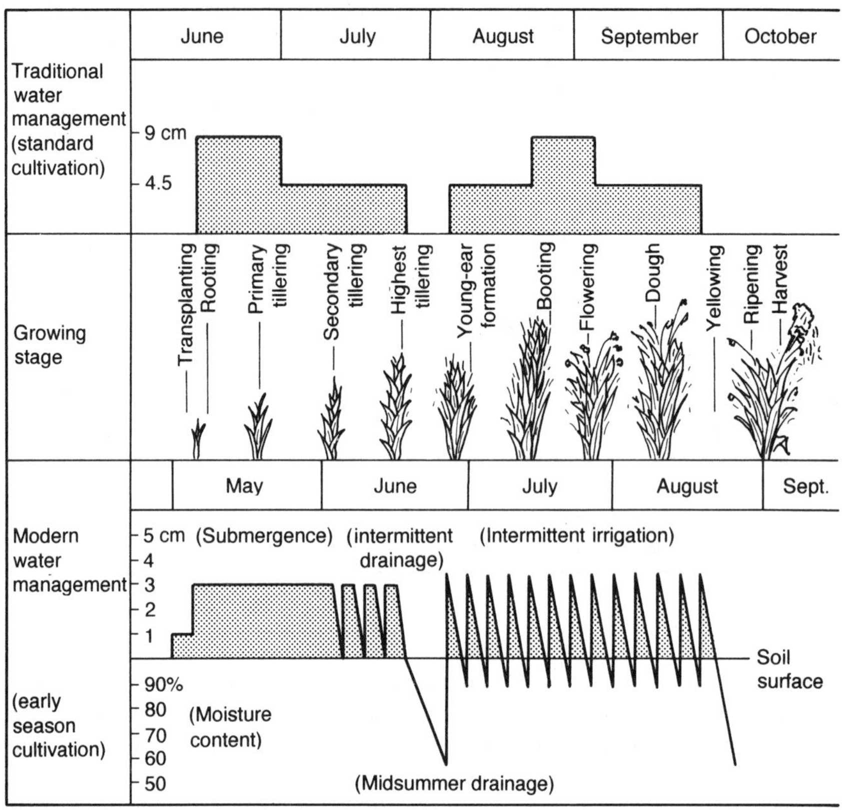

Fig. 6.1. Traditional and modern water management for each growth stage of rice in Japan (Yukawa, 1989).

contain some water at the onset of the typhoon, effectiveness may be well below 100% (Bhuiyan *et al.*, 1979).

The total water used to produce a rice crop can vary widely, both between wet and dry seasons and between years, as well as depending heavily on crop and system management. This is illustrated in Fig. 6.2. For the dry season crop the amount of water used can be as high as 5000 mm or as low as 1000 mm. In a wet season with an effective rainfall of 2000 mm the irrigation requirement can still be as much as 3000 mm, or if the rainfall is well distributed it may be nil. Revelle's figure of 4000 t of water for 1 t of rice quoted at the beginning of this chapter corresponds to 2000 mm to produce 5 t of grain. Yields are frequently less than 5 t ha^{-1}, for the same delivery of water, and are then more expensive in terms of water required per tonne of grain produced.

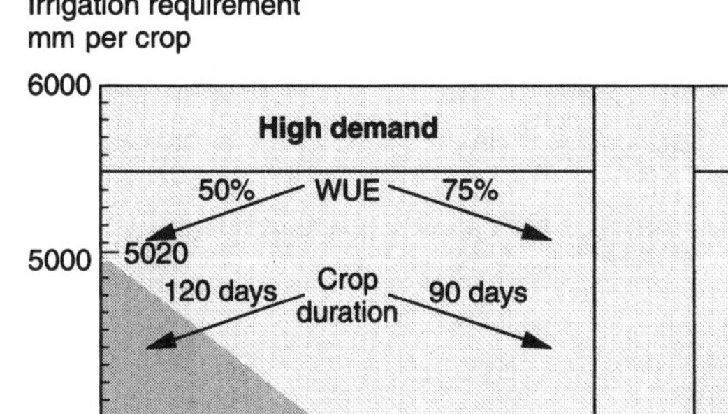

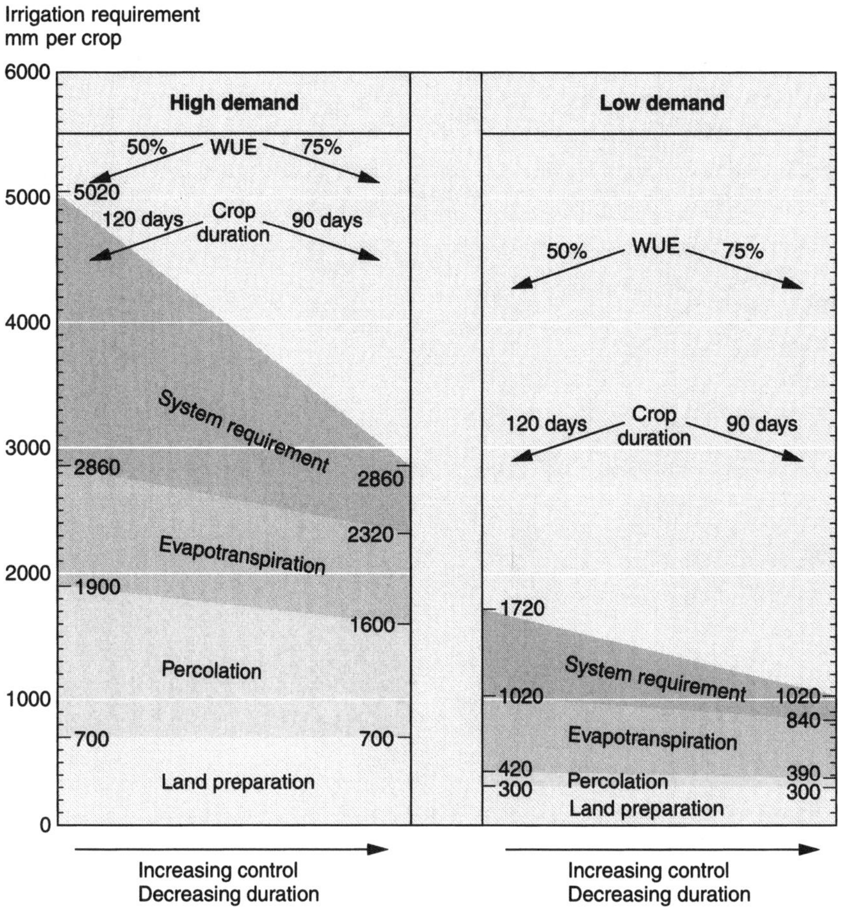

Fig. 6.2. Water used for irrigated rice production.

6.2 Water Supplies for Rice in Rainfed Systems

Yields of rice in rainfed systems are always subject to the vagaries of the monsoon. The yields are lower than for irrigated rice, and often no dry season crop is possible. Differences between average regional yields of rainfed lowland rice and irrigated rice are between 1 and 3 t ha^{-1} (Table 3.3). Differences between yields of rainfed rice obtained at the same site in different seasons can be as large. The magnitude of the yield loss due to water shortage depends on the degree of stress, and on the time at which the stress occurs. Drought at flowering time causes much larger losses of yield than similar stress at the vegetative or ripening stage (Fig. 6.3).

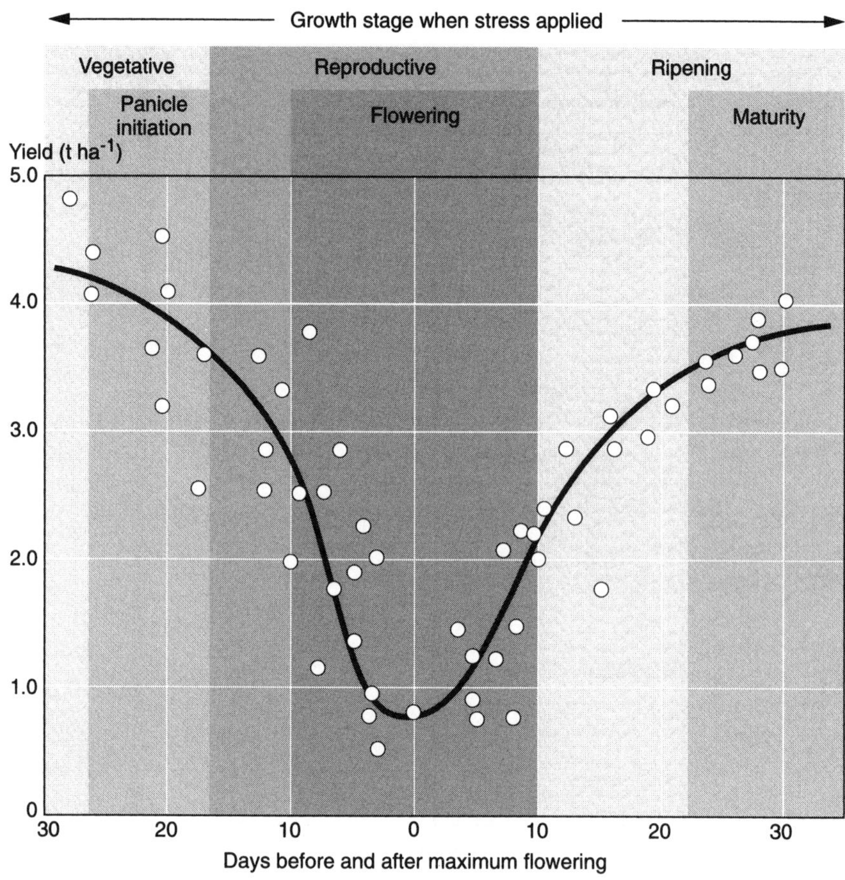

Fig. 6.3. Effect of water stress on the yield of rice (O'Toole, 1982).

Initially rice cultivation was practised in flood-prone areas which were naturally flooded for the growing period of the crop. Yields would have been very dependent on the timing of the initial flooding of the land, the rate at which water depth increased, the depth the water attained, and how long the flood persisted. As sea level rose during the period following the earliest cultivation of rice, the cultivators would have been pushed further into the inland valleys, cultivating rice in the backswamps of rivers rather than behind the levees near the junctions of river and sea. The first step towards water control was probably determined by the need to move to slightly higher areas, as sea level continued to rise, and to retain water where the rice was sown by the building of bunds around the rice paddy. This move from the stagnogleys to the pseudogleys would have been accompanied by higher yields.

In the backswamp areas, rainfall and floods and the sediments carried by them have sustained rice production, albeit erratically and with the occurrence of occasional disasters, for several thousand years. The system continues to this day, with the sustainability improved by some degree of flood control introduced by water diversion systems, and the strengthening of levees to contain rivers in excessive flood, and, in coastal areas, the sea at times of exceptionally high tides. Continued selection of flood tolerant rice varieties and breeding programmes to increase their yield have gradually improved the productivity of these areas.

The sustainability of production from the rainfed, drought-prone, alluvial terraces is particularly erratic in areas with under 1500 mm of annual rainfall, although areas such as southern Bangladesh with a high risk of major flooding also show a greater than average variability (Fukui, 1982). The variability in the drought-prone lowlands is not confined to the rainfed areas. Relatively few streams in these areas are perennial, and flow is often erratic, so that diversion irrigation systems often fail. In these areas the distinction between what land is irrigated and what is not is often a slender one. Most of the rice land may be irrigated in a wet year, but in a dry year none or at best only part of the area benefits from irrigation. The need to optimize water use is then particularly important.

By changing from transplanting systems to dry-seeding, rice can be established during the light, pre-monsoon rains. This ensures that the monsoon rains are used efficiently from their commencement, and in favourable areas may allow a second crop to be established after rice. This second crop is usually an upland crop, but in some areas dry seeding has enabled two rice crops to be taken in rainfed areas (Morris *et al.*, 1986; Shepard *et al.*, 1988; Saleh and Bhuiyan, 1995). The second crop in many of these areas has made a major difference to the local economy, notably at Iloilo in the Philippines (Greenland, 1990). The most difficult and as yet imperfectly solved problem associated with dry-seeding is weed control.

6.3 The Development of Diversion Irrigation

The need to provide water to supplement rainfall became essential for the sustainability of rice farming as soon as there was a major move from the flood-prone wetlands to the higher river terraces. Irrigation by stream diversion was probably initiated in Asia about 4000 years ago. Initially this would have involved little more than the building of simple earthworks to divert the flow of small streams, and perhaps the building of simple brush dams across the stream to bank up the water and ensure flow under gravity. Brush and basketwork dams have been used since at least the 13th century. They have the advantage that they are washed away when the river is in flood and so do not create the hazard that a larger and stronger dam may do (Ishii, 1978).

Construction of larger diversion systems had to wait for the development of better implements than the si-spades first used for cultivation. Spades made of wood or bone were probably sufficient to enable ditches to be dug in wet soil, so that water could be diverted to the rice fields from nearby streams, rivers, natural ponds and lakes. The extensive canals and water storage systems which were built in China and in the Mekong delta, in areas covering parts of what are now Vietnam, Thailand and Cambodia, would have required metal implements. Their construction dates from about 2500 years ago. The importance attached to the development of irrigation systems at this time is illustrated by a quotation from Thai history reported by Ishii (1978):

> Non-participation in the cooperative work of maintenance of the irrigation system while secretly taking water for private use was considered a grave crime against society . . . punishment for a first offender was clubbing about the head . . . for a second offence it was death.

Many of the larger diversion systems were served by major rivers, but the digging of small reservoirs was also initiated about 2500 years ago. Tanks, barays and ponds are characteristic of the regions of less intense monsoonal rains, such as northeast Thailand, the Dry Zone of Sri Lanka and the drier areas of southeast and south India of similar rainfall (Brohier, 1975; Sarma, 1986). These storages offered some escape from the vagaries in water supply caused by dry spells, and allowed rice cultivation to develop on the higher river terraces, as well as the floodplains and lower terraces which were regularly flooded, and which received sufficient water naturally. They also gave much greater stability to the existing production systems in times of low rainfall and weak floods.

The construction of larger ponds encouraged the development of settled communities, and it became a mark of power and prestige to have built a major tank or pond to store water supplies. The tanks or barays at Angkor in Cambodia are perhaps the best known examples. The western baray at Angkor built around AD 1040 is 8 km long and 2 km wide and could hold about 70 million m^3 of water (Bronson, 1978; Chandler, 1983; Higham, 1989).

In Japan water diversion systems with wood-lined ditches were operating from about AD 200 and the system of building small dams and creating ponds for water storage was developed from AD 300 (Yukawa, 1989). The practice of constructing ponds to serve a small community of farmers grew rapidly and there are now more than 280,000 ponds in Japan providing irrigation water for the rice crop in the traditional way. The great majority of these are much smaller than the barays at Angkor, and supplement rather than replace river and canal diversion systems. Similarly, tank systems in India are usually small, irrigating at most about 100 ha. The number of these is considerable. Alagh (1990) reported over 1 million in the Gangetic plain of West Bengal. It is probable that many of these were in fact silted and had fallen into disuse, as Pal *et al.* (1994) give a total of 224,000 for eastern and southern India. Whichever figure is nearer to the true number it is clear that the use of tanks was and is a significant source of water for rice production in southern and eastern India. Alagh in fact notes the problem of tank siltation and its effects on the continued utility of the tanks. He recommends the removal of the silt and its use as a soil amendment. A major expenditure of energy would be required but the work involved could be undertaken as a communal effort.

China is well supplied with streams and rivers, and historically almost all of the water used to irrigate the rice crop was drawn from these (Buck, 1937). Simple water wheels and other lifting devices have certainly been in use for several millennia. Excess water rather than too little has been the greatest threat to sustainability. The hugely damaging floods of the Yangzi and Huanghe Rivers and many others have continued to recur until the present day.

6.4 Managing Floodwaters

The huge efforts made in China over the last 2500 years to control the waters of the Huanghe and the Yangzi (Greer, 1979) reflect the importance of flood control. Hillel (1991) notes that the Huanghe is also called the 'Sorrow of China', a name associated with the damage done when it is in flood and when it changes course. The huge load of silt which it carries is derived from the great plateau of loess through which it passes in northern China (eastern Gansu, Shaanxi and Shanxi Provinces). The loess soils are easily eroded, and the plateau is characterized by massive erosion scars and the many silt laden streams and rivers which feed the Huanghe. Greer quotes Kuo Ching-hui (1956) who records silt contents of the Huanghe and its tributaries ranging annually between 2 and 160 kg m^{-3} of water (0.2% to 16%). Immediately before the floodplain and above the delta the annual range of flow rates and silt contents range from 300 to 3500 m^3 s^{-1} and from 0.4% to 5.0% respectively. As the river slows and the silt is deposited the river creates a barrier to its own path. This has caused the frequent and highly damaging changes in its course.

Maps of the channels of the Huanghe in the final 500 km of its length before reaching the sea are available from the Chou Dynasty (600 BC) through the Tang Dynasty (AD 600–900) and the Sung Dynasty (AD 960–1280) to the present. From the time of the Chou Dynasty canals were constructed to improve the flow of the river and make it less damaging in time of flood. Under the Tang Dynasty the design of the 'Grand Canal' to link the Huanghe and the Yangzi was conceived and about 200 km were built. Its length is now almost 1800 km and it links Beijing in the north with Hangzhou in the south (Yao and Chen, 1983).

There are five other rivers in Asia besides the Yangzi and Huanghe which are over 2000 km long, and with drainage basins of over 400,000 km^2 (Table 6.2). All of them rise in some part of the Himalayan massif, and carry

Table 6.2. The major rivers of Asia.

River	Length (km)	Drainage area ($10^3 \times km^2$)	Discharge (10^3 m^3 s^{-1})
Yangzi	4990	1940	53.9
Huanghe	4340	673	8.1
Mekong	4020	802	27.3
Indus	2735	926	13.7
Brahmaputra	2700	934	49.0
Ganges	2480	1058	46.2
Irrawaddy	2011	430	33.5

Source: Leeden *et al.* (1990).

substantial silt loads, although none so great as the Huanghe. Their floods and those of other major rivers have presented the greatest threat to the sustainability of rice production not only in China but in the rest of Asia.

Bangladesh lies at the junction of the Ganges and Brahmaputra and before they reach the sea they are joined by another large river, the Meghna (Fig. 6.4). The average rate of discharge of the three rivers is more than twice as great as the sum of the discharge rates of the Yangzi and the Huanghe. It is hardly to be wondered at that Bangladesh is regularly subject to floods, and devastated by severe floods at irregular intervals. In spite of the floods, and partly because of them, Bangladesh has the highest concentration of population of any country in the world with a primarily rural population. The floods which afflict the country are not only a danger to sustainability; they are the source of the water and nutrients on which rice production depends, as well as replenishing the lakes and ponds as homes for the fish to supplement the rice diet. In India the Ganges is known not as the Sorrow but as the Mother of India.

Nevertheless the scale of the flooding and the losses of life and livelihood associated with the major floods are a cause for continuing concern. In an average year, when about one-fifth of Bangladesh is flooded, the floods are regarded as entirely beneficial. The serious damage comes from the major floods which have tended to occur about one year in ten (Dempster and Brammer, 1992). The more severe floods sweep away homes as well as rice crops. Around 1900 the Brahmaputra shifted its course dramatically to a new more westerly path. The new channel, known as the Jamuna, is in places more than 10 km wide. In 1988 near the junction of the Ganges and Brahmaputra, the combined river, the Padma, shifted its course by 500 m and cut a new channel 45 m deep (Pearce, 1991). The devastating floods of 1987 and 1988 covered more than half of the country. Two thousand died and hundreds of thousands were made homeless. Much the worst disaster occurred in 1970 when a cyclone at the coast met a major river flood, and the death toll was estimated at around half a million (Pearce, 1991).

Naturally flooded river backswamps were the initial sites of rice cultivation, and over many years the people of the flood-prone areas have learnt to adapt to their environment. Production in these areas would always have been

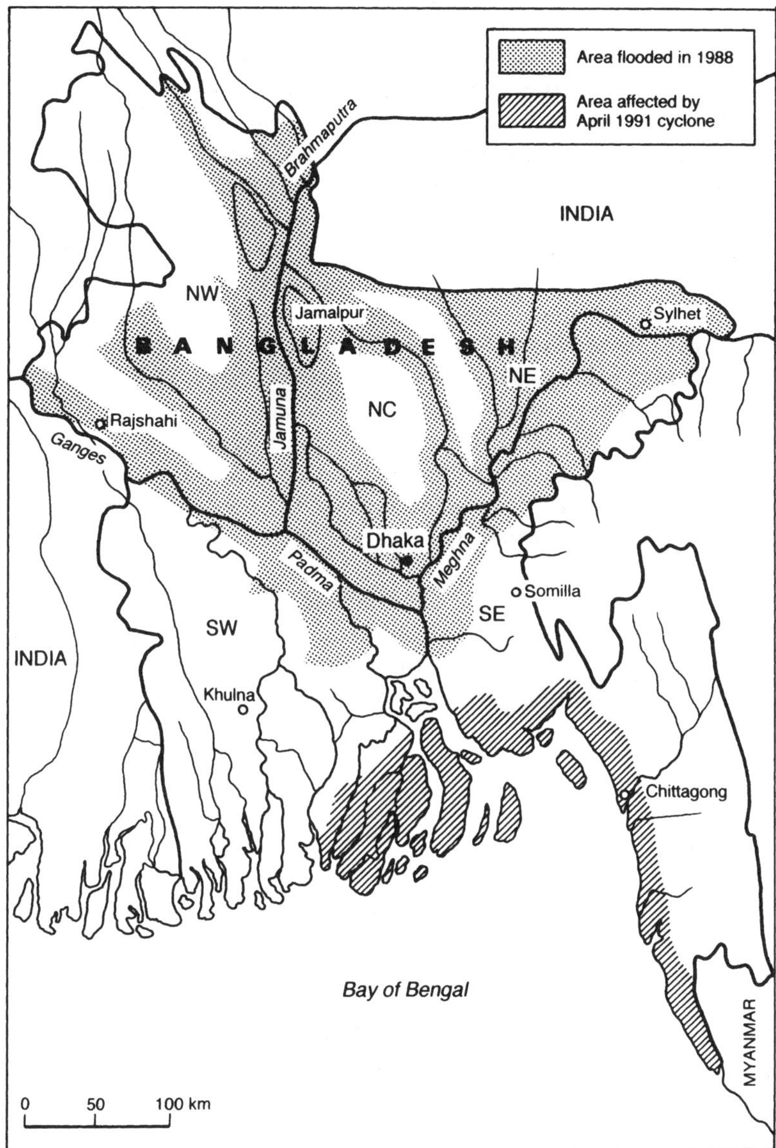

Fig. 6.4. Bangladesh – areas affected by the 1988 floods and the 1991 cyclone (Dempster and Brammer, 1992).

precarious, threatened by floods which were uncontrolled and might wash away the crop, or drown it, or bury it under a load of silt. Some rice varieties have evolved which are tolerant to as long as 10 days of submergence, and some are able to elongate by as much as 10 cm day^{-1}. These constitute the so-called floating rice varieties. These varieties are adapted to survive in areas where flooding to a metre or more is the norm. The great majority of rice varieties suffer yield losses if they are even partially submerged for more than 3 or 4 days, and may be killed by complete submergence (Table 6.3).

Strengthening of the river banks and levees and the construction of embankments to reduce the impact of the more severe floods are practices that have been in use for at least 5000 years (Pereira, 1989). The embankment of rivers was extensively developed by the rulers of India in the Mogul period of the 16th century. They built more than 1000 km of embankments in India and what is now Bangladesh. By 1947 there were more than 3000 km, and by 1978 more than 10,780 km of new embankments had been built (Pereira, 1989).

Construction of canals to divert floodwaters when the monsoon was particularly heavy dates back at least 2500 years (Parry, 1992) and possibly further towards the beginning of the bronze age (Chaudhuri, 1991). The Grand Canal in China, construction of which started over 1000 years ago, is the prime example, but many others built earlier and on a smaller scale were also important, not least for leading some of the floodwaters into channels where they could flow harmlessly to the sea, or to areas where they would be less damaging.

A serious problem associated with the building of embankments is that the natural deposition of silt in the floodplain of the river is largely prevented. Much of the sediment load is deposited in the river bed and so raises the level of the river. Eventually, as with the Bahmaputra, the river is forced to divert to a new

Table 6.3. Reduction (%) in yield of rice (IR30) as affected by inundation treatments.

Crop growth stage	Average crop height (cm)	Type of submergence	Period of submergence (days)			
			1	3	4	5
2 weeks after transplanting (early tillering)	30	Fully submerged	24.6	61.2	63.7	84.2
		Partially submerged	6.9	–3.9	10.3	12.2
4 weeks after transplanting (max. tillering)	48	Fully submerged	24.5	37.6	81.5	94.5
		Partially submerged	0.1	7.6	8.7	4.8
6 weeks after transplanting (panicle initiation)	68	Fully submerged	74.2	94.0	96.4	100
		Partially submerged	10.2	8.0	9.5	11.3

Partially submerged means crop was submerged at half its height.
Source: Bhuiyan and Undan (1986).

course where it develops a new and deeper channel. Many rivers in Asia are now contained by high embankments and flow above the level of the surrounding paddy fields. Unless elaborate drainage systems are built and pumps installed to return the water to the river and prevent it stagnating, soils which had been productive pseudogleys become less productive stagnogleys. In China in the lower reaches of the Huanghe vast settling basins have been dug to collect the silt. These are 3–5 m deep. Once filled they form fertile rice fields. There are now more than 500,000 ha of such resilted land in the lower floodplain of the Huanghe (Greer, 1979).

In spite of the control measures that have been installed, floods in India and Bangladesh have been more frequent and more damaging in the past 30 years than previously. The extent of the annual floods in India in the 1950s affected on average 6.86 Mha, and this increased to 16.57 Mha in the 1980s (Centre for Science and the Environment, 1991). The cause is the increasing rate of erosion in the headwaters of the rivers.

6.5 The Development of Storage Irrigation

In contrast to the antiquity of the diversion and tank storage systems, storage of water behind high dams is a very recent development. A famine in Java in 1848/49 is believed to have caused the death of 200,000 people. It provoked the Government to build a dam on the Tuntong River, and distribution canals to deliver water to 12,000 ha. A further famine in 1872 led to an expansion of the storage systems until by 1940 the total area in Java served by a reliable irrigation system was 1.5 Mha (Grist, 1975). The question of the reliability of the supply of irrigation water is all-important. Thus Buck (1937) reported that all rice in China was irrigated. It was, but a great deal of the irrigation water was supplied from streams and rivers which contained little or no water in times of low rainfall. Of the 16,786 farms in 22 provinces which were studied by Buck and his associates, only 20% reported no failure in the water supply. The commonest frequency of failure was one year in three. As the population increased the ability of the water systems to meet the growing demand fell. In 1929–1933 diversion of water from ditches, streams and rivers still accounted for 82% of the water used for irrigation on the farms in Buck's survey. The water required to produce the rice needed to feed the population of China by 1900 could not be obtained from such systems. They fail in times of drought and provide little control of floodwaters. The need to ensure the reliability of measures to control water supplies can be gauged from the historical records for China reported by Yu and Buckwell (1991). The number of major floods between 206 BC and 1949 is given as 1092, and the number of major droughts as 1056. This means that on average some part of China was affected by a major flood or drought each year. The record of the series of natural calamities which have affected the Chinese Province of Hupei between 1644 and 1911

(Table 1.2) illustrates the scale of the problem. 'Flood, excessive rain, and typhoon' were by far the most common cause of disaster.

China went through a long period of severe hardship before a massive programme of dam building was started in 1950. The number of dams in China able to hold more than 1 million m^3 of water was eight in 1950, and 18,595 in 1982 (Leeden *et al.*, 1990). The increase in total reservoir capacity (Fig. 6.5) was correspondingly spectacular, allowing not only an increase in the total irrigable area but much greater reliability of supplies, and the capacity to deliver water for two and sometimes three crops per year. Of the 48 Mha irrigated in China in 1978 about 17 Mha were in reservoir irrigation districts, 11 Mha in pump (groundwater) irrigated districts, and 20 Mha were irrigated by other means, principally diversion systems.

In Japan the expansion of reliable irrigation started in the 19th century (Hayami, 1975) and in Korea and Taiwan the period of most rapid expansion was from 1940 to 1960, influenced strongly by Japan (Kikuchi, 1975). Elsewhere in Asia the years after 1960 also saw a great surge in dam building and expansion of the irrigated area. Most but not all was due to the increase in storage irrigation. The expansion in South Asia was almost as rapid as in China. In India 600 major storage dams and many smaller ones were

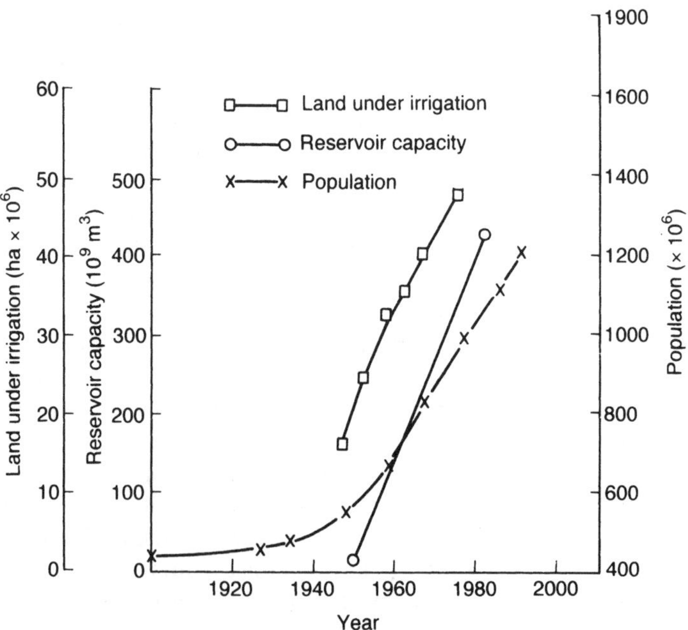

Fig. 6.5. Growth of land under irrigation, reservoir capacity and population, China, 1900–1990 (data from IRRI, 1993a, 1995c; Barker *et al.*, 1985; Tang and Stone, 1980; and Leedon *et al.*, 1990).

Table 6.4. Average annual growth rates (%) of irrigated agricultural area in Asia, 1960–1988.

Years	South Asia	Southeast Asia	East Asia
1960–65	1.8	1.6	2.2
1965–70	2.9	1.2	2.4
1970–75	1.8	2.8	2.1
1975–80	2.8	3.6	1.2
1980–85	1.8	4.1	–0.3
1985–88	0.1	1.5	0.3
1960–88	1.9	2.7	1.1

Growth rate 1940–1960 in Korea, 5.0%, Taiwan, 2.8% (Kikuchi, 1975).
Source: Rosegrant and Svendsen (1993).

built between 1950 and 1990. The 600 major reservoirs had a total storage capacity of 160 billion m^3 (Ghassemi *et al.*, 1995). In Southeast Asia the rapid increase started later but has continued longer, although at a lower rate (Table 6.4). An essential element in promoting the surge in dam building and extension of the irrigation areas was not only the availability of the high-yielding semi-dwarf varieties of rice, but also of wheat, which increased the advantage to be obtained from the water provided.

Relatively few countries provide data for rice areas separated from the total irrigated area. The average annual growth rates from 1960 to 1988 for South, Southeast and East Asia given in Table 6.4 are for all crops. The actual areas of irrigated rice in five major rice producing countries are given in Table 6.5. The data quoted are for the crop area. Thus where two rice crops are grown in a year the area is counted twice. The irrigated dry season crop by the mid-1990s had covered 28 Mha, or about one-third of the wet season irrigated area.

6.6 Other Water Supply Systems

Dams and reservoirs do not have to be large. Tank and pond systems as used in southern and eastern India, Sri Lanka and Japan are often constructed to serve areas of less than 100 ha. Recently small farm reservoirs in which the area of the reservoir is of the order of 1500 m^2 and the embankment no more than 4 m high have become increasingly common (Bhuiyan, 1994; Maglinao *et al.*, 1994).

These on-farm reservoirs have mostly been constructed to provide water to small farms in areas lacking access to major irrigation systems, and where there is a pronounced dry season and dry periods during the monsoon are not uncommon (Ghalang *et al.*, 1994). The construction and use of small on-farm reservoirs, often built at a cost below US$200, is spreading rapidly in rainfed

Table 6.5. Irrigated rice areas and growth rates, 1950 to 1990, selected countries.

	Bangladesh		India		Indonesia		Philippines		Thailand	
Year	Area	% $year^{-1}$	Area	% $year^{-1}$	Area	% $year^{-1}$	Area	% $year^{-1}$	Area	% $year^{-1}$
1950	120		9,844		–		–		–	
		9.8		2.4						
1955	179		11,035		–		596		–	
		10.7		2.6				12.2		
1960	285		12,461		4100		960		1404	
		14.2		0.7				0.2		4.0
1965	488		12,909		–		971		1744	
		21.4		2.2		2.8		9.5		0.2
1970	1011		14,339		5236		1432		1758	
		4.1		1.2		1.4		0.8		1.0
1975	1218		15,220		5608		1492		1845	
		0.7		1.5		1.1		1.5		1.7
1980	1259		16,340		5928		1609		2006	
		5.6		1.6		3.2		2.8		3.7
1985	1613		17,677		6489		1838		2376	
		11.3		1.3		2.8		2.0		–1.0
1991	2706		19,039		7403		2060		2206	
1992	2710		–		–		1980		–	

Areas × 1000 ha.
$\%^{-1}$year is the average yearly increase in area as a percentage of the area in the initial year.
The Thailand data in the 1960 line are for 1961; the Indonesian data in the 1985 line are for 1983.
Where more than one irrigated rice crop is grown in a year the area is counted twice if two rice crops are grown, and three times if three rice crops are grown.
Sources: IRRI (1987d, 1993a, 1995c).

areas of the Philippines, eastern India, Bangladesh and Indonesia (Bhuiyan, 1994).

In addition to the increase in the irrigated area due to the building of major and other dams, pump irrigation systems have also been extended. Manual, animal and water and wind driven pumps all have a long history in Asia (Elvin, 1982). Their significance was initially small compared with diversion systems in which water was led under gravity from rice terrace to rice terrace. Manual and foot operated pumps of various types have been in use in Asia for many years. However with the arrival of petrol, diesel and electric pumps in the 20th century the use of pumps has expanded considerably.

Soon after such pumps became available 'dragon boats' could be hired in China. These plied the rivers and canals and carried a small pump. They could be hired to pump water into or out of a paddy. More commonly the pumps have been installed to serve a small community of farmers. The greatest use of pump systems has been in South Asia, where the fall of the great rivers once they emerge from the foothills of the Himalayas is very limited, so that gravity fed systems cannot serve as large an area as is desired.

By using small pumps to lift surface water in the dry (boro) season 600,000–700,000 ha can now be irrigated each year. Extensive use of deep tubewells (DTWs) able to lift water from about 40 m started in Bangladesh about 1965 and of shallow tubewells (STWs) a few years later. This encouraged an increase in the use of hand-operated tubewells (HTWs). During the wet season when much of the lower land is flooded, and at the start of the dry season when the water table is no more than a metre or two deep, hand-operated low lift methods can be used to bring water to the surface. By 1982 there were 12,000 DTWs and 78,000 STWs in operation, and in addition some 150,000 HTWs (Bhuiyan, 1984).

The area that can be served by the manually operated systems is around one-quarter of a hectare, compared with 5 and 25 ha for the STWs and DTWs respectively. The area which it is possible to irrigate from pumps was over 80% of the total irrigated area of Bangladesh in 1992/93 (Table 6.6). The rapid increase in irrigated area in Bangladesh in recent years has been almost entirely due to increased use of shallow tubewells, using small pumps to lift water a few metres, or hand operated wells. The costs in energy, human or purchased, are significant, and not all pumps are operated to capacity.

Elsewhere in Asia there have been substantial developments of tubewell systems in Pakistan and India. By 1976 it was estimated that in India, of a total irrigated area of 34 Mha, 13 Mha were irrigated from groundwater, and in Pakistan of 13.6 Mha, 1.2 Mha were supplied from groundwater. The number of irrigation wells in India in 1950 was 90,000. In 1990 it was 12 million

Table 6.6. Areas (1000 ha) of tubewell, canal and hand powered irrigation, Bangladesh, 1970 to 1993.

Year	Power pumps	Tubewells	Traditional methods	Canals	Others	Total
1970/71	418	48	417	104	181	1168
1975/76	552	95	450	91	219	1400
1980/81	666	222	452	150	149	1640
1985/86	609	964	254	163	109	2099
1990/91	675	1784	397	173		3029
1992/93	687	2014	396	159		3255

Power pumps are small pumps used to lift surface water into rice paddies.
Tubewells include deep, shallow and hand operated pumps, used to pump water from the underlying water table.
Canals lead water from reservoirs and rivers and other water channels to paddy fields by gravity flow.
Source: Bangladesh Statistical Yearbooks.

(Alexandratos, 1995) of which 3.36 million were private tubewells, 46,000 state operated tubewells, and the remainder private hand operated wells (Reddy, 1989). Elsewhere there have been increases in pump irrigation wherever there are high-yielding aquifers, but, except in China, their use has been on a more limited scale.

Pump irrigation has the advantage that at least the immediate costs of the water are readily apparent to the user. No loss of land and displacement of people is involved as it is in the construction of dams and entrapment of water in reservoirs. The long-term costs in depletion of aquifers by tubewells where the aquifers are not regularly and fully recharged are less apparent. The long-term sustainability of systems dependent on the use of groundwater is obviously directly related to the extent and speed with which groundwater supplies are recharged.

The coastal wetlands offer both challenges and opportunities for rice production. The opportunities arise because there are considerable areas which might be used for rice which are not yet developed (IRRI, 1984b). The challenges come from the problems of intrusion of saline water, and the occurrence of particularly difficult soils such as peats and acid sulphate soils (Dost and van Breemen, 1982) as well as the flood problem. Near the coast three types of flooding are superimposed, diurnal flooding due to tidal movement, fortnightly 'moon' floods at the time of maximum tidal depth controlled by the phases of the moon, and seasonal flooding when the rivers reach their peak flows. Diurnal effects are only important in areas immediately adjacent to the sea. Moon floods are progressively dampened with increasing distance from the coast, and further still from the coast the tidal swamp rice areas merge with the flood-prone floating rice and deepwater areas. Diurnal floods and moon floods are normally predictable so that agronomic operations can be timed to take account of them, and rice varieties used that are tolerant to the submergence problems. Exceptional tides caused by typhoons and other weather factors create additional problems, as when the typhoon that struck Bangladesh in 1970 coincided with high flood levels and half a million people died. Sea water salinity also influences rice in the tidal swamps, and varieties which have some degree of tolerance to salinity must be grown.

6.7 Waterlogging and Salinity

The recent rapid expansion in the number of small and large reservoirs, rehabilitation of older and degraded systems, and utilization of groundwaters by pump systems has made an enormous difference in the sustainability of water supplies for rice production. No longer is the availability of water universally threatened by poor rainfall. The measures taken have however also introduced new problems of sustainability, as have some of the recent flood control measures. Most of the problems are associated with the effect of the measures on the groundwater table. Storage of water on the land establishes a higher level of the water table, and unless drainage measures are introduced waterlogged areas are formed, and salts in subsoils and groundwaters may be brought to the root zone. Access to the fields may also be impeded.

Waterlogging for rice production is not the disaster that it is when other crops are being produced. Nevertheless as discussed earlier poor drainage reduces the productivity of ricelands. Lack of control of the depth of flooding prevents the use of other than the lower yielding flood-tolerant varieties, and the yields of these varieties will be reduced by the accumulation of toxic concentrations of iron, manganese and other compounds. Most importantly waterlogging means that whatever is brought into the rice paddies in the floodwater accumulates there. As there will always be some salt dissolved in the water, gradual salinization of the land is inevitable. Even with good quality water containing only 200 mg l^{-1} of soluble salts, the amount added per year, if 1000 mm of irrigation water are used, is 2 t ha^{-1}.

Some areas may be flooded for part of the year, but drain naturally after the floods subside. Percolation and lateral drainage before the floods of the following year enable any salt which accumulates during the dry season to be washed away as the rains commence. This natural recovery of the land may be prevented if the water table remains above or close to the soil surface throughout the year. This situation may arise when a large reservoir is established at a level higher than the flood-prone area, or when an unlined earth canal that transports large volumes of water is built on permeable soils. No seasonal renewal of the land occurs as the soils remain waterlogged. The situation may be alleviated by lining canals with impervious materials, or by ensuring that adequate drainage facilities are provided. This has not always happened as the cost of drainage facilities and lining canals is often high.

Salinity problems are found more frequently where irrigation is practised in arid and semi-arid areas. Not only is there inadequate rainfall to remove the salts but it is more common to find groundwaters which are saline in such areas. With continuing irrigation the rise in the level of the groundwater will bring this salt into the root zone, so that the problem of salt accumulation is accentuated. There is no doubt that the extent of the soil salinity problem is increasing. In part this increase is due to the further expansion of rice production into coastal areas where groundwaters are affected by sea water intrusion. However secondary salinity associated with expansion of major irrigation networks is the major cause of recent increases in the extent of areas affected by salinity. The significance of the problem in Australia has led to the initiation of on-farm restrictions to try to alleviate the recharge of groundwaters from rice fields (Humphreys *et al.*, 1994).

Salinity is not a new problem. Between 3000 and 4000 years ago accumulation of salt in the irrigated areas of Mesopotamia and other parts of the Middle East was causing the decline and disappearance of kingdoms. Some 4000 years ago Akkad and Sumer, where much of the earliest development of irrigation had occurred, declined in importance and power moved to areas less affected by salt in higher reaches of the Tigris and Euphrates (Boyden, 1987; Hillel, 1991). Salinity has also been a problem in China for many years. Hseung and Liu (1990) quote a book written in China about 2500 years ago which states 'The strongly saline and alkaline soils taste very salty and bitter. They are low-yielding land but suitable for cropping rice'. This quotation is important not only because it indicates that salinity is a long-standing problem in China, but because it points to a special relation between rice and salinity. Although rice is often perceived to be a plant which has at least moderate tolerance to salinity, it is in fact more sensitive as a plant to salt than are some other cereals (Flowers and Yeo, 1981; Yoshida *et al.*, 1983). Rice grows in soils which are not able to support dryland crops because under flooded conditions salts in the soil are diluted by the floodwater. Provided the floodwater is of good quality the salt concentration to which the rice roots are exposed will be significantly lower than that in the soil solution when an upland crop is grown. This has led to the common perception that rice is tolerant to salt. There are a few tolerant varieties, but most rice is sensitive to salinity, and yields are reduced by quite moderate concentrations of salt (Fig. 6.6). Bhatti and Kijne (1992) quote data of the Drainage and Reclamation Institute of Pakistan which show that rice yields in slightly, moderately and severely saline soils are reduced by 32%, 63% and 79%, respectively, compared with non-saline soils.

The extent of lands adversely affected by salt is at least 45 Mha, or 20% of the irrigated land area of the world (Ghassemi *et al.*, 1995). Several higher figures have been given. The difference in area quoted depends on the criteria used to determine what is and is not saline. The salt content of the soil varies considerably in time and space, depending on the position of the water table at the time measurements are made, the antecedent weather conditions, and whether the sampling procedures used to measure the extent of the problem are sufficiently intensive. Oldeman (1994) gives the global extent of soil degradation due to salinization as 21 Mha where it is 'strong to extreme', 20 Mha where it is 'moderate' and 35 Mha where it is 'light'. Ghassemi *et al.* (1995) give a detailed analysis of salinity problems in several countries. They give the salt-affected area in China as 20 Mha, in India as 7–20 Mha, and Pakistan as 4 Mha. In the United States of a total of 17 Mha of saline land, 5.6 Mha are in irrigated areas, representing a quarter of the total irrigated land. Whatever the exact figures are for the total global extent of saline land, there is no doubt that salinity poses a growing threat to the sustainability of crop production, including rice production.

Studies in India (Joshi and Jha, 1991) and Australia (Beecher, 1991) have shown that in an irrigation area many farmers will have land adversely affected by a water table higher than it was before the establishment of the irrigation system. Even if the area is not regarded as suffering from salinization the yields these farmers obtain are in some years much reduced by salinity, even though the levels are generally low.

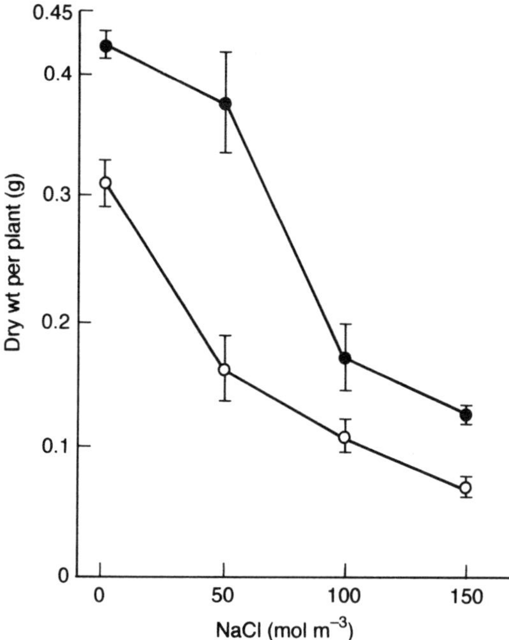

Fig. 6.6. Salinity resistance of rice varieties (Flowers and Yeo, 1981). The effects of sodium chloride on dry matter production by two cultivars. Dry weight per plant (mean ± S.E.) for cv. IR28 (○) and IR2153-26-3-5-2 (●) at a range of additions of sodium chloride to the culture solution. Seedlings were salinized commencing at age 14 days and harvested age 24 days.

Another problem associated with waterlogging and salinization is that of sodicity. When the proportion of sodium ions associated with the soil is higher than about 15% a soil may be dispersed, i.e. its aggregation is lost, and it dries to large, tough clods. Salts that accumulate in groundwaters often contain a high proportion of sodium, so that even when the salt content is not damaging, the soil may become less suitable for crops. This is generally not important for rice, but may be very important when an upland crop is grown after rice. With a sodic soil, dispersion of the soil by puddling is very easy, and such soils are generally impermeable. Thus they readily retain water for the rice crop, although the permeability may be less than optimal even for rice. Preparing a seedbed for an upland crop to be grown after rice may be quite impossible, and if it does rain the waterlogged soil will inhibit growth of the upland crop.

In areas where pump irrigation is practised other problems arise. These are due to lowering of the water table. If the aquifer is not recharged sufficiently to replace water removed by pumping, the water table inevitably falls, so that the costs of pumping increase. In the drier parts of India irrigated by pump systems groundwater levels have generally fallen by 5–10 m (Ghassemi *et al.*, 1995). In the lower parts of the landscape which were naturally flooded this may mean that rice production is seriously affected by drought. It can also mean that water from a deeper aquifer which is of lower quality has to be pumped to maintain the water supply. A further problem arises in coastal areas, where lowering of the water table can allow sea water intrusion into an aquifer, causing salinization of land and water supplies. Reddy (1989) reports that in Gujarat, India, salinity has penetrated to 7 km from the coast as the water table has been lowered by pumping.

As well as the general problem of salinity associated with the high osmotic potentials created, and those of sodicity associated with excess sodium ions, other problems can arise owing to the presence of specific toxic ions in the irrigation water. This occurs mostly when groundwater is pumped from considerable depth. Boron has been the ion most commonly found to cause toxicity for rice plants (Ayers and Westcot, 1989). In the United States selenium has been found to be introduced to the food chain from saline groundwater and cause mammalian toxicity (Tanji *et al.*, 1986).

Lowering the water table can sometimes be a help where saline groundwater is close to the surface. Pakistan's SCARP Project, designed to pump saline groundwater from a near-surface water table and mix it with better quality surface water, is perhaps the largest of such reclamation schemes (Awan and Latif, 1982). Pump schemes can also be introduced to remove water directly from the plant root zone, as is done in parts of India (Patnaik and Sahoo, 1971). Provided that there is an opportunity to dispose of the pumped water safely, degraded areas or naturally saline areas can be brought into production. Evaporation ponds may be used to dispose of the saline water, or it may be mixed with high quality surface water as in Pakistan, but more commonly an outlet to the sea is sought, or the saline water is allowed to run into a deep sump so that the water enters a deeper and already saline aquifer, as is done in Australia (Rhoades, 1974). Many of these schemes at best provide a partial solution; they often do no more than transfer the problem to a new area.

To achieve sustainability it is necessary to be able to manage the depth of the water table within acceptable limits. Chinese engineers and agronomists have introduced the concept of the 'four waters'. This concept is of management of surface, ground, soil and rainfall water in harmony, to achieve a controlled water table and avoid waterlogging and salinization but ensure maximum crop water availability (Shen and Wolter, 1992).

6.8 Erosion and Sedimentation

The sustainability of water supplies depends not only on the management of water resources but also on the management of the uplands from which the water is obtained. Soil erosion in the uplands affects water supplies in two ways: (i) loss of topsoil exposes subsoils with poor infiltration capacity, which reduces water entry into the soil, and increases runoff, so enhancing the likelihood of floods and (ii) eroded soil forms the sediments which cause siltation of reservoirs and canal distribution systems. When storage irrigation systems are designed it is normal for the life of the reservoir to be determined by the length of time it will take for the reservoir to become so full of silt that it is no longer useable. It is a sad fact that over the last 50 years while the great expansion of reservoir systems has occurred, the speed at which the reservoirs will be silted has almost always been seriously underestimated (Pereira, 1989; White, 1990). One of the largest reservoirs, the Nizamsagar in India, has had its life expectancy reduced to 6% of the designed life, and another large reservoir, the Tungabhadra, to 24% of the designed life (Alexandratos, 1995). These figures are by no means exceptional.

One reason for the excessive siltation rates is that estimates of silt load to be expected after construction of the dam have had to be made without adequate data. The estimates have sometimes been based on studies using data from temperate areas to predict erosion in tropical areas (White, 1990). There is now more than adequate data to show that erosion rates in the tropics are almost always considerably higher than in temperate areas (Greenland and Lal, 1979). An important cause is that rainfall in the lowland tropics is more intensive and erosive than rainfall of the temperate zone. Increases in the number of people living in reservoir catchment areas, and logging and agricultural activities in the catchments, cause deforestation and as a result soils are exposed to the more erosive rains. Erosion and runoff increase, and lead to much higher deposits of sediments in the reservoirs than were anticipated. The rivers rising in the Himalayas certainly carry heavy loads of silt into the reservoirs built to store irrigation water for the rice producing areas of the Indo-Gangetic Plain (Pereira, 1989). The rates at which sediments have been deposited in some of these reservoirs exceed the predicted rates by up to 22 times (Table 6.7). In the Philippines the loss of storage capacity due to siltation of the reservoirs from which the principal rice growing areas are irrigated has meant that the service areas have had to be considerably reduced (Masicat *et al.*, 1990). The major problems of siltation of the reservoirs on the Huanghe in China are even more severe. They were discussed earlier.

In India and China and elsewhere the importance of erosion control measures in the catchment areas of reservoirs is now well recognized. The relative importance of the erosion which occurs naturally and the accelerated erosion which occurs when the vegetative cover is removed has sometimes been disputed (e.g. Ives, 1991). However there is strong evidence to show that appropriate measures for erosion control in the catchment areas of major reservoirs can reduce the sedimentation rates, for instance that provided by Gupta (1980) from a series of studies in India (Table 6.8) in which the effects of soil conservation measures in the catchment areas of the reservoirs on the rates of siltation were determined. At the time the studies were reported erosion control measures had been established on less than half of the critically eroded areas in the catchments, but the siltation rates had nevertheless been reduced by a quarter. The costs of erosion control in catchments, added to those of construction of the dam, introduction of proper drainage systems in the service area, plus the costs in loss of useful land and displacement of people, often make investment in new storage irrigation systems uneconomic. The result has been the sudden decline in such investments in recent years (Table 6.9). The decline in the rate of construction of new major storages has been partly compensated by the recent rapid development of groundwater irrigation systems.

6.9 The Long-term Sustainability of Water Supplies for Rice

Although rice is very demanding of water, there are at least in theory sufficient renewable water resources to meet the demand (Table 6.10). On a global basis

Table 6.7. Predicted and actual sedimentation rates in reservoirs in India.

Reservoir	Catchment area (km^2)	Year of impounding	Year of observation	Annual average silt load, (t × 10^6)	Sedimentation rate m × ha 100 km^{-2} Assumed	Observed
Hirakud	82,650	–	–	46	2.54	3.60
Tungabhadra	28,179	1953	1972	24	4.32	6.57
Mahi	25,330	–	–	32	1.29	8.99
Nizamsagar	21,694	1931	1973	15	0.29	6.34
Panchet	9,920	1956	1974	12	2.47	9.92
Tawa	5,983	–	–	6.8	3.61	8.10
Maithan	5,206	1956	1979	7.6	1.62	12.15
Ramganga	3,134	1974	1974	7.3	4.29	17.3
Mayurakshi	1,792	–	–	4.1	3.6	20.9

Sources: Narayana and Ram Babu, 1983; Randhawa and Abrol, 1990; Abrol and Sehgal, 1994.

Table 6.8. Effects of soil conservation treatments on sedimentation in the Bhakra (Himachal Pradesh) and Ramganga (Uttar Pradesh) reservoirs, India.

Reservoir	Year of impounding	Catchment area (km^2)	Area of critical erosion (km^2)	Critically eroded area treated, as % of catchment area	Annual sedimentation rate m × ha 100 km^{-2} Before	After	% Reduction
Bhakra	1959	18,200	5,370	23	6.91	5.19	25
Ramganga	1974	3,134	1,040	40	22.0	16.8	24

Source: Gupta (1980).

Table 6.9. Index of average annual public expenditures for irrigation development (1976–80 = 100).

	Bangladesh	China[a]	India	Indonesia[a]	Philippines[a]	Sri Lanka	Thailand
1971–75	97[b]	70	60	20	25	37	88
1976–80	100	100	100	100	100	100	100
1981–85	143	74	94	192	125	92	151
1986–90	103	54	80	170	45	55	109

[a]For China, Indonesia and the Philippines the successive time periods are 1969–73, 1974–78, 1979–83, 1984–88 (1974–78 = 100).
[b]1973–75.
Source: Rosegrant and Svendsen (1993).

Table 6.10. Proportion of world water resources used for agricultural irrigation.

	Renewable water resources (m^3 person^{-1} year^{-1} 1990)	Estimated utilization for agriculture m^3 person^{-1} year^{-1}	% utilization for agriculture
World	7,690	436	5.7
East and South Asia	3,729	448	12.0
Central and South America	3,566	428	12.0
West Asia/North Africa	1,446	659	45.6
Sub-Saharan Africa	8,010	120	1.5
Korea	1,450	196	13.5
Brazil	36,070	107	0.3
Malaysia	26,300	275	1.0
Thailand	3,210	581	18.1
Philippines	5,180	471	1.9
Indonesia	14,020	76	5.0
India	2,450	498	20.3
Nigeria	2,730	21	8.0
Bangladesh	20,391	224	1.1
Tanzania	2,780	25	0.9

Source: Abernethy (1994).

the use of renewable water resources for agriculture in 1992 was less than 6%. In Asia it was under 20%. The problem is to get the water into the right place, and without creating conditions which damage the production potential of the soil. The methods for providing the water must also be economically sustainable and environmentally benign. A further problem in relation to agricultural use is the rapidly growing competition for available supplies from urban and industrial users. One prediction of the likely increase in the number of persons living in water-stressed environments is illustrated in Fig. 6.7. Much the

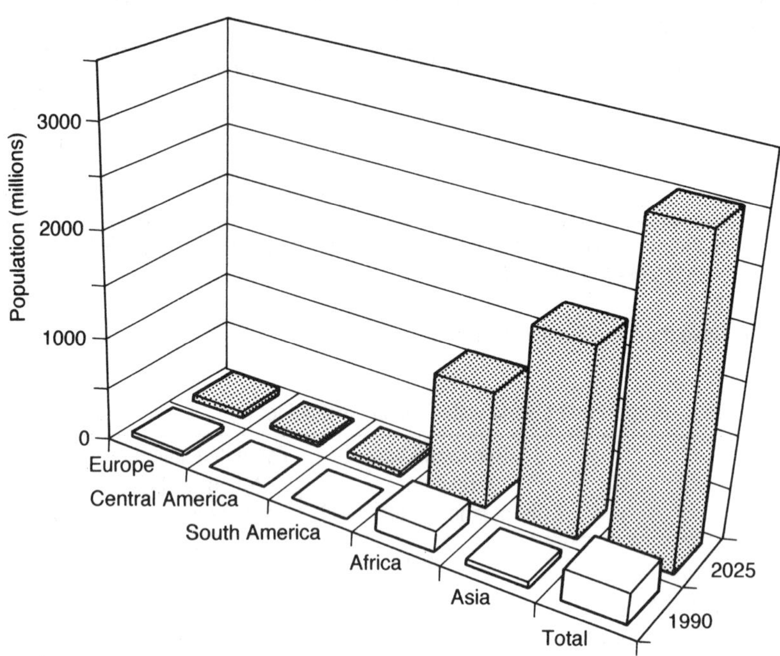

Fig. 6.7. Number of people living in water-stressed regions in 1990, and projected for 2025 (Falkenmark, 1994).

greatest increase is likely to be in Asia. Water requirements for agricultural production to provide basic food requirements are of the order of 500 to 600 m^3 per person per day (Alexandratos, 1995). In the areas where flooded rice is the primary source of staple food it is higher, at least 1000 m^3 per person per day. Urban and industrial requirements are of the order of one-third of the agricultural requirement, greater in more developed countries, and less in agriculturally dependent countries. Both Table 6.10 and Fig. 6.7 are based on inadequate data, but they illustrate the importance which must be given to the problem of water availability and use, as Bhuiyan (1993), Abernethy (1994) and others have emphasized.

The evolution of the supply system can be traced historically. Rice was initially confined to naturally flooded wetlands. The ability to divert water to paddy fields, and to hold it where the rice was grown, by means of shallow earth barriers evolved first. Later came ditches and canals to guide the water where it was wanted. Then came simple brush and basketware barriers to retain water at the start of the wet season, then simple earth dams, and much later large and permanent dams of concrete. About the same time that the brush and basketware dams evolved simple lifting devices started to appear, to move water from natural surface accumulation sites on to rice terraces, and again much later pumps of various types to replace or supplement the manual labour used to move water, and eventually to raise it from the groundwater levels.

As with rice production, where the evolution of rice production changed to a green revolution, evolution in the methods of water storage and delivery has changed to a 'blue revolution'. The quantities of water which can be stored and delivered to rice fields, or pumped from groundwater, has increased dramatically in the past 50 years (Table 6.5, Fig. 6.5). As with the green revolution, or any other revolution, there are questions about the sustainability of the blue revolution. As described in earlier sections of this chapter they include:

1. The problems of storage and delivery, where sustainability is threatened by siltation of canals and reservoirs, increasing costs of operation and maintenance of the system, and decreasing efficiency in operation of the system.

2. In drier areas, the problems of rising and falling water tables, rising levels often bringing salts to the root zone and creating waterlogged conditions, and falling levels increasing the costs and energy requirement to raise water to the surface, and in coastal areas allowing sea water penetration and salinization of groundwater.

3. In wetter areas, increasing flood hazards, due to erosion in the areas from which the flood waters are derived. Less of the rainfall enters the eroded soil, but is immediately added to runoff, and so creates a sudden surge in water flow in the rivers which flood the downstream areas. These problems may be further aggravated by rising sea level associated with global warming.

4. Paradoxically, the reverse of the above, when dams and weirs prevent flooding of downstream areas, so that rice production is restricted by droughts.

5. Environmental problems, most commonly associated with changes in hydrology caused by water storage and extraction. This can lead to deterioration of coastal lagoon environments, and drying of inland waters, on scales from the Aral Sea and Lake Chad to small farm ponds.

Social and Economic Factors and the Sustainability of Rice Farming 7

7.1 Rice in Asian Cultural Systems

Some of the earliest written records about rice concern its place in the cultural and religious systems of China and India. The development of strong and organized societies depended on the production of rice in excess of the requirements of the family, so that there was a surplus to support not only the rulers and administrators but also the armies needed to protect the community and maintain the authority of the rulers. Once the ability to produce a surplus above subsistence needs had been established the potential was created to develop complex social structures, and well organized states. The states in turn spread their influence through trade and conquest.

As the religious, political and economic significance of rice grew over the course of centuries, many rituals and festivals became associated with its cultivation. These extended from the family to the village and higher levels. The Chinese ceremony in which the Emperor plants rice was observed through many dynasties. The Hindu scriptures require that rice be offered to the gods as an expression of gratitude and happiness. In Hindu and many other marriage ceremonies the throwing of rice at the bride and groom is a significant part of the proceedings. The religious use of rice is important not only in India but also in China, Thailand, Indonesia, Sri Lanka, Malaysia and elsewhere. In Sri Lanka the words for food and for rice have been synonymous for many centuries (Brohier, 1975).

The importance of rice to trade and financial dealings in India is reflected in the fact that the Hindi word for unhusked rice, dhan, is the same as the word for money, and related to that for wealth, dhani (Sharma, 1987). Until a monetary system evolved, the social importance of rice and continuing family needs for food were all that was needed to ensure that every effort was made to sustain production. As community organizations developed rice gradually took on a wider economic importance. Family producers were required to provide a portion of their produce to support those in the community assigned special roles. In return they received various services, and protection from others. The services often included access to water supplies, for which they were required to provide the necessary labour. Once water supplies were assured the provision of other inputs to boost the rice harvest became important – the use of manures of many kinds, and simple tools for weeding and cultivation, and the use of water buffalo to enable more land to be prepared than could be cultivated without animal power. In time the use of money to pay for other foodstuffs and necessities of life evolved. A value for land and water as well as for labour emerged. Local, and then district, regional and national markets for rice developed, establishing the monetary value of rice.

For many Asian rice producers with insufficient land to produce much beyond their family needs rice farming was a way of life, and for many it remains so. But as urban communities grew and rice became an important trading commodity, the demand to produce a surplus for sale and to pay taxes to whatever authority was in power increased. Thus rice came to have a position of political and commercial as well as social importance. Rice continues to hold a revered position in life in Asia. Governments can still expect to fall if the rice harvests are bad or if the price of rice is too low to the producer or too high to the consumer.

7.2 Rice in the Asian Economy

Food crop production has always been demand driven, and on a long-term and global scale production and population have largely grown in parallel with each other (Chapter 1). Nevertheless within political boundaries there have been periods when demand has outstripped supply, and major importation of food has been necessary to meet demand. In Asia this has occurred in several countries. Within China and India different provinces or states have experienced famine when other parts of the country have had a surplus of food. The most productive areas have also become the most populous, as in the lower Yangzi region of China. Failure of the rice supply in these densely populated areas has, from time to time, led to major disasters, and a rapid decline in the population (see p.9). The need for rice imports to alleviate potential famine in parts of China and eastern India was most pronounced in the period between 1900 and 1970 when wars and civil disturbances disrupted production and technology failed to keep pace with the need for higher productivity (Hayami and Otsuka, 1994).

Those countries without serious land constraints such as Burma and Thailand extended the areas of rice production and became major exporters of rice to other areas of Asia. India, which for most of this period included what are now Bangladesh and Pakistan, was a major importer until 1977, since when it has become an exporter in most years. Other major importers between the 1930s and 1970s included Sri Lanka, Indonesia and Malaysia (Table 7.1). The island of Bali within Indonesia provides an interesting example of an area which is highly productive, but where the population grew as a result to the point where it became an area of outmigration. Fertile soils and well managed community irrigation schemes allow three rice crops, each of which was often well over 5 t ha^{-1}, to be harvested every year. Nevertheless the growth of population and the restricted availability of land has meant that many people have had to leave Bali and seek land in less productive but also less populated parts of Indonesia. The Philippines, like Indonesia, has experienced periods when it has been a major importer, but since 1970 has been able to export in years of good harvests while importing relatively small amounts in less favourable years. Vietnam, a traditional exporter of rice, needed to import between 1940 and 1990, but recently returned to being a significant exporter.

In the 1930s Japan was a major importer of rice from its then colonies of Korea and Taiwan, but a rigorous policy of farm-price support and investment in research and technology has enabled it to create the potential to produce well in excess of the needs of its population. The cost is high, the price paid to the Japanese rice farmer exceeding the world market price for rice by a factor of ten or more for much of the 1980s and into the 1990s.

The position of China is rather more complex. Between 1850 and 1950 China had to import considerable quantities of rice during periods of recurrent internal strife and the resultant famines. The principal limitation on its imports was the ability to pay for them. Since 1950 China has mainly been an exporter of rice while at the same time importing considerable quantities of wheat (Fig. 7.1). The ratio of imports of wheat to exports of rice has depended on their relative prices, that of rice often being more than twice that of wheat (Fig. 7.2). Similarly in the 1990s as the ability to produce large surpluses of rice in China has increased, so China has produced increasing quantities of the more valuable *japonica* varieties for export, at the expense of reducing the amount of hybrid rice grown. The margins between surplus and deficiency in cereal grains is small, and in the face of increasing demand not easy to manage for optimum advantage (Yu and Buckwell, 1991).

Rice farmers in Asia produce rice to meet two needs, for their own and their families' food security, and to realize a profit. They will not invest in the inputs necessary to increase yields unless they are reasonably assured that there will be a market for their produce, and that the price they receive will show a profit above what they have invested. Most Asian governments seek to be self-sufficient in rice. Hence they have generally adopted policies to encourage farmers to produce more, and managed markets and prices to give the farmer a profit. These policies allied to green revolution technology have enabled many Asian countries to become rice exporters rather than importers, so that in both 1991 and 1992 the exports of rice from Asia exceeded imports by more than 4 Mt.

Table 7.1. Annual exports/imports milled rice, '000 t, 1909–1992.

Region/country	Years	1909/13	1925/26	1939/40	1961/62	1971/72	1981/82	1991/92
World	Exports	5,849	7,218	8,749	6,739	8,621	12,567	14,434
Asia	Exports				4,798	5,278	7,652	9,581
	Imports				4,745	5,762	5,807	4,918
China	Exports	0	6	6	648	1,517	731	926
	Imports				68	0	255	127
India	Exports	248	0	261	75	16	751	619
	Imports				767	418	37	27
Pakistan	Exports				150	190	1,098	1,359
Bangladesh	Imports				361	515	172	28
Burma/Myanmar	Exports	2,169	2,439	2,991	1,655	668	688	190
Indochina	Exports	885	1,187	1,497				
Vietnam	Exports				135	3	4	1,475
	Imports			12	116	785	105	4
Cambodia	Exports				0	20	0	0
	Imports				0	37	139	51
Indonesia	Exports	59	31	43	19	0	0	22
	Imports				1,080	620	424	391
Japan	Exports	28	128	80	0	548	587	0
	Imports	573†	1,253†	1,736‡	169	27	71	18
Korea	Exports	48	717	?				
S. Korea	Imports				0	879	1,448	0
N. Korea	Exports				0	101	237	8
Malaysia	Exports	552	265	153	22	0	0	0
	Imports	409	770	833	410	225	360	422
Singapore	Imports	469			343	342	188	223
Philippines	Exports				0*	3	41	23
	Imports	187	86	61	94	413	0	1
Sri Lanka	Imports	373	468	576	440	303	134	185
Thailand	Exports	792	1,100	1,425	1,423	1,852	3,405	4,742
Africa	Imports				561	935	2,936	3,610
Egypt	Exports	25	?	138	198	486	57	170
Madagascar	Imports	0	0	0	0	48	274	42
Nigeria	Imports	11	10	0	1	1	598	240
South America	Exports				266	295	543	662
	Imports				49	28	250	1,153

Table 7.1. *Continued*

Region/country	Years	1909/13	1925/26	1939/40	1961/62	1971/72	1981/82	1991/92
Brazil	Exports	0	52	53	98	76	31	2
	Imports	13	24	0	0	5	141	720
Argentina	Exports	0	0	0	25	58	75	129
Uruguay	Exports	0	0	1	22	57	235	299
USA	Exports	7	27	144	943	1,758	2,837	2,204
Australia	Exports	0	0	0	57	143	439	472

*Refers to 1911/20; †to 1921/30; ‡to 1931/38. From Huke and Huke, 1990b, and IRRI, 1995c

The advantage to the individual rice farmer has often been that he is able to afford to employ more labour. He can then give his wife the option of devoting more time to the family and household and less to the drudgery in the mud of the paddy field, and his children can attend school and later seek better paid employment outside the rural sector (Smith and Gascon, 1981). The wider effect has been to keep rice prices low for the urban populations, and to decrease the extent of rural poverty among the landless labourers (Hayami and Kikuchi, 1981; Herdt, 1987).

From a position of threatening widespread famine in the 1960s Asia has moved to general rice self-sufficiency, and some measure of greater prosperity. Herdt (1987) has concluded that greatest advantage was to the urban consumer of rice, although all involved in its production have derived some benefits from the changes which have occurred over the past 30 years. Regional problems due to floods and typhoons have been largely mitigated by national relief programmes, without need for external support except in the case of the most serious disasters. Rice food aid fell from 0.69% of production in 1971 to 0.25% of production in 1991.

The problem now is whether the momentum can be sustained, and the benefits extended to less advantaged areas and communities. Many critics of green revolution technology have emphasized the negative effects of falling rice prices on the less advantaged producer (e.g. Griffin, 1975; Lipton, 1979; Lipton and Longhurst, 1985; Norland *et al.*, 1986; Hazell and Ramasamy, 1991). A falling price for his produce undoubtedly hurts the producer who is unable to increase his yields. More attention needs to be given to the social and economic problems of the disadvantaged areas. But much the greater problem for governments of rice dependent countries is to ensure that production keeps pace with demand. The rate of increase in production has fallen from 3.37% per year in the period between 1965 and 1975 to 1.7% per year in the period

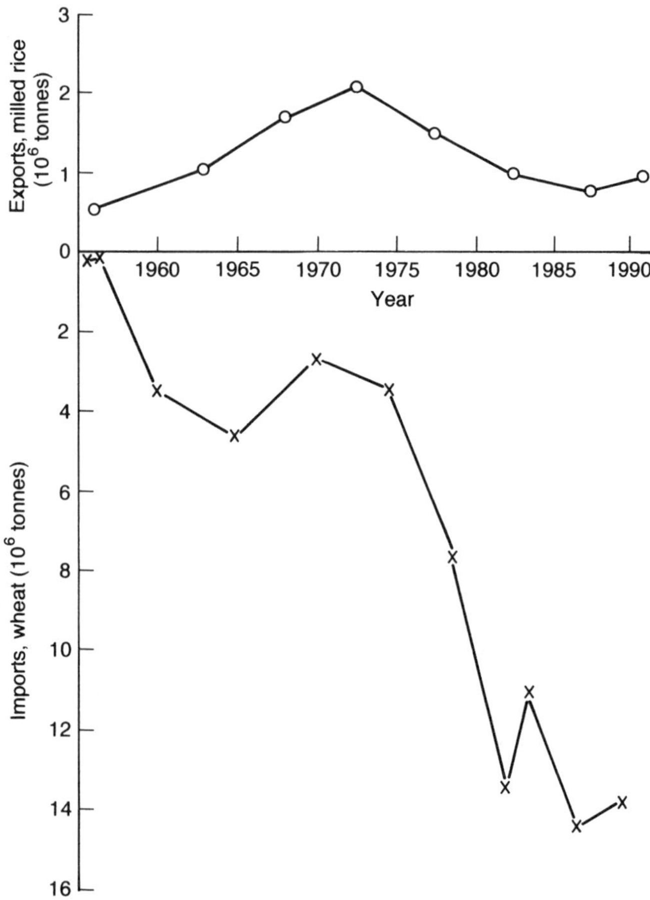

Fig. 7.1. China. Exports of milled rice (5-year means), (from IRRI, 1995c), and imports of wheat (selected years) (from Yu and Buckwell, 1991).

between 1985 and 1993 (Table 1.3). In several countries the production growth rates fell below 1% per year in the period between 1988/90 and 1993/95, in spite of population growth rates well over 1% per year (Table 7.2).

In addition to the production and equity problems the further spectre has been raised of degradation of the resource base, particularly by more intensive production practices. Some of these problems are very real, such as the problem of declining water quality discussed earlier, and health risks associated with pesticide use (Rola and Pingali, 1993). The significance of others, such as land contamination by pesticides and intensive use of inorganic fertilizers, has still to be properly evaluated. There is always a cost to the protection of natural resources, and it is important that this cost is not forgotten (Spendjian, 1991).

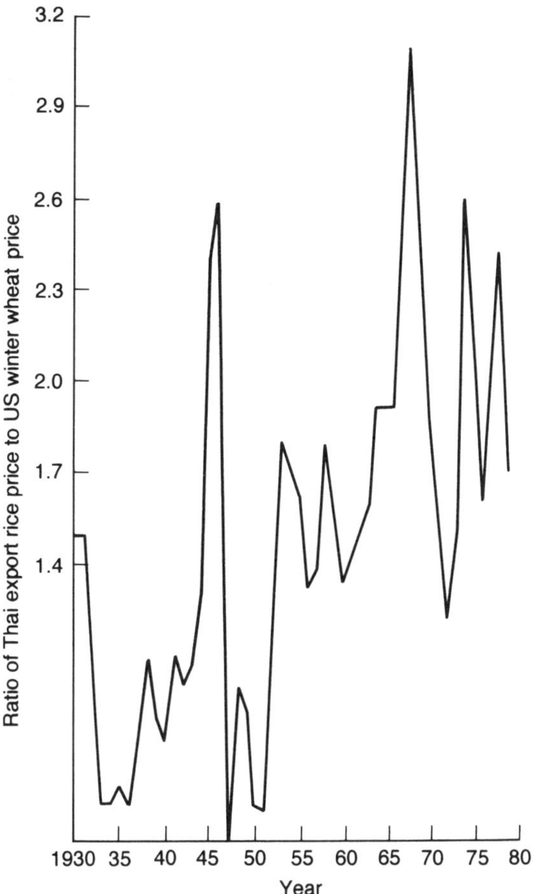

Fig. 7.2. Ratio of average Thai export rice price to the US No. 2 winter wheat price, FOB Kansas City, 1930–1979 (Barker *et al.*, 1985).

Economic sustainability requires that the costs of inputs are less than the value of outputs. Several books on the economics of rice production deal with these aspects in considerable depth (e.g. Barker *et al.*, 1985; Bray, 1986; David and Otsuka, 1994; Pingali and Hossain, 1997). In the next two sections an attempt is made to summarize briefly the more important aspects of the costs of inputs and the value of products.

7.3 The Costs of Inputs

The inputs include labour, land, water and seed as primary essentials, and manures and fertilizers, implements and pest control measures as secondary

Table 7.2. Rice production and population growth rates, 1988/90 to 1991/93, selected Asian countries.

	Rice production (Mt year^{-1})		Growth rates (% year^{-1})	
Country	1988/90	1991/93	Production	Population
Bangladesh	25.6	27.6	2.50	2.57
China	181	187	1.08	1.43
India	110	110	0	2.05
Indonesia	43.9	46.9	2.25	1.96
Malaysia	1.86	2.10	4.04	2.51
Myanmar	13.7	15.1	3.24	2.23
Nepal	3.39	2.97	–4.4	2.25
Philippines	9.25	9.44	0.68	2.15
Sri Lanka	2.36	2.39	0.40	1.42
Thailand	19.7	19.7	0	1.70
Vietnam	18.4	21.2	4.7	2.59

Source: IRRI (1993a, 1995c).

requirements. Initially labour was the economic determinant of rice production. It then became land, and the costs of water and other inputs have become increasingly important as the need to obtain greater yields per unit of land and labour has grown.

7.3.1 Labour

Rice production has always been labour intensive. The number of man-hours needed to produce a tonne of rice were well over 100 before 1900, but with modern production methods and high yields may now be as low as 16 (Table 7.3). As labour has become more productive through the use of improved implements and machinery and better management of other inputs, the advantage to be derived from greater employment of labour on small Asian rice farms has grown – even on farms as small as 1.6 ha up to 76% of the labour used has been hired for the purpose of producing rice (Barker and Cordova, 1978; Barker *et al.*, 1985).

The cost of labour depends not only on the number of man-days required to produce a tonne of rice but also on the alternative work available. Industrialization in Asia has tended to raise demand and wage levels for labour. The proportion of the labour force engaged in rural work has fallen sharply in each of the past three decades (Table 7.4) although less so in South Asia than elsewhere. Nevertheless a large pool of landless labourers available for seasonal employment still exists in much of tropical Asia. Thus there is often a conflict between the use of labour-saving techniques and the need to provide work

Table 7.3. Labour productivity in man-days per tonne of rough rice produced (day t^{-1}) and labour used in man-days per hectare (day ha^{-1})

Country	Productivity (day t^{-1})	Labour (day ha^{-1})
Japan		
1888–91	104	271
1956	55	229
1971	28	141
Central Taiwan		
1926–27	39	96
1936–37	41	126
1961	34	138
1972	16	84
Java, Indonesia		
1875–78	228	225
1920–30	111	204
1968–71	59	158
1977–80	39	144
Philippines		
1966		
Central Luzon		
Local variety	37	60
Modern variety	28	82
Laguna		
Local variety	35	88
Modern variety	30	106

Data recalculated from Barker *et al.* (1985) who give full details of sources of original data.

opportunities for the landless (Lipton and Longhurst, 1985). There has been a great deal of debate about the effects of green revolution technology on equity issues, from which it has gradually become clear that there are two opposing factors involved: (i) an increased demand for labour, as irrigation has enabled more rice to be grown, and multiple cropping has become much more common (multiple cropping provides work in the dry season when previously there was none, as well as greater employment opportunities in the management of each crop); and (ii) a decreased demand as wage rates have risen and farmers have employed machinery to displace labour, or used less labour intensive production methods, such as direct seeding rather than transplanting of rice. Where innovations have increased production, there has often been greater income from rice production to enable more labour to be employed, but where the yield advantages have been small, then labour has tended to be displaced by new technology with little advantage in extra employment to compensate (Herdt, 1987; David and Otsuka, 1990, 1994).

Table 7.4. Proportion of Asian labour force in agriculture, 1960–1993 as % of total labour force.

Region/country	1960	1970	1980	1990	1993
Asia	74.8	70.2	65.7	59.8	58.0
East Asia					
China	82	78.3	74	67.5	65.2
Japan	33	19.7	12	6.4	5.5
North Korea	62	53.4	49	33.6	31.1
South Korea	66	49.1	34	24.6	21.8
Southeast Asia					
Burma/Myanmar	67	59.5	53	46.7	45.3
Cambodia	82	77.4	75	70.6	67.6
Indonesia	75	66.4	58	48.5	45.9
Laos	83	78.6	75	70	71.4
Malaysia	63	54.1	50	32.8	28.9
Philippines	61	54.7	52	46.9	45.5
Thailand	84	83.2	76	64.4	62.4
Vietnam	80	76.4	71	60.6	58.5
South Asia					
Bangladesh	87	81.5	74	68.6	66.5
India	74	71.7	69	66.5	65.5
Nepal	95	94.1	93	92.4	90.6
Pakistan	61	59.1	57	49.6	48.2
Sri Lanka	56	55.8	54	51.6	51.5

Adapted from: Barker *et al.* (1985) and IRRI (1995c).

Many local factors affect the costs of labour, and determine the extent to which family and other labour is employed. These include such factors as tenure arrangements, farm size, water availability, crop duration and timing in relation to other labour demands, and the economic advantage of employment on the family farm as opposed to other opportunities. The net effect of changes in productivity of rice differs from location to location. In general it has increased labour use per hectare harvested, but reduced labour input per tonne of rice produced.

The increasing effectiveness of labour in rice production is an important contribution to the sustainability of production, particularly in the face of competing demands for labour. Had a labour requirement of over 100 days to produce a tonne of rice persisted labour costs could have more than doubled per tonne of rice produced. Purchase of other inputs would have been inhibited and other income producing activities restricted. Rice yields and rice production would not have reached a level where the price of rice has fallen sufficiently to enable yields to be sustained at the levels necessary to meet current demand.

7.3.2 Land

When land was abundant in relation to the number of people that had to be supported its ownership was not an issue. It was deemed no more possible than ownership of the sea or the sky. With the advent of agriculture and the development of settled communities the right to the use of land steadily assumed greater and greater importance. Traditional rights, initially of kinship communities and later of wider communities, became the norm. Later still legal land tenure, with rights to individual ownership, became the economic basis for trade in land, but individual rights are not yet universally established. In the past rights to the use of land have often been determined by force of arms, so that warlords became overlords and landlords for many small farmers. Colonial rule in several countries in Asia served to consolidate land holding in few hands. Only recently through revolutions, the granting of independence, and land reform, has individual title of small rice farmers to their land become the norm. In many areas the process of reform continues (FAO, 1991).

Individual or community ownership and confirmed right to the land are essential both to increase production from the land, and to ensure that the increases are sustainable. Firm title to the land places responsibility for its future productivity in the hands of the farmer, who has the primary interest in ensuring that the land will continue to support him and his family. The farmer with secure title knows that any surplus he produces will be primarily to his own benefit, and not to the benefit of a landlord. Without firm title and a continuing relation to the productivity of the land, measures to enhance productivity through investment in irrigation systems and the fertility of the soil are less likely to be given the support of the farming community.

The absence of a right to continuing tenure often arises when population pressure forces the displacement of people from their traditional homes. They may settle on land where they have no more than 'squatters rights'. The land may then be used as a temporary source of food, and little concern given to its long-term productivity. Often the damage is caused unwittingly, because the land is new to the settlers. It is more common for damage to occur in upland than in wetland areas, because the former are more likely to be subject to irreversible soil damage by erosion.

The continuing pressure on land resources throughout Asia has meant that farm sizes have tended to decrease, with many countries showing a large increase in the number of farms of less than one hectare (Table 7.5). As a consequence land values have risen. The actual value and costs of land are affected by many factors, so that generalizations can be misleading. Nevertheless studies of a single village over time give an impression of the changing

Table 7.5. Total number of farm holdings, number less than 1 ha, and total which are rice farms, selected countries and years.

Country	Year	Total no. holdings	Holdings < 1 ha	Total no. rice holdings	Total area rice holdings
Bangladesh	1960	6,139	3,170	5,654	8,548
	1983	10,045	7,066		
India	1960	48,882	19,897		33,820
	1970	71,011	36,200		
	1980	88,883	50,122		
	1985	97,731	56,748		
Indonesia	1963	12,237	8,577	9,202	5,766
	1973	14,374	10,115		
	1983	15,927	10,482		
Japan	1960	6,056	3,911		
	1970	5,354	3,639	4,654	3,274
	1985	4,376	3,046		
Korea (South)	1961	2,332	1,655	2,066	1,054
	1969	2,421	1,600	2,099	1,185
	1979	2,162	1,407		
Nepal	1961	1,540	852		
	1971	1,735	1,335	1.203	995
	1981	2,194	1,455	1,035	1,394
	1991	2,736	1,878	2,281	1,481
Pakistan	1960	5,691	1,598		
	1972	3,762	521	1,014	1,789
	1980	4,400	878		
Philippines	1960	2,166	250	1,042	3,112
	1971	2,354	319	982	2,661
	1980	3,420	776	1,867	3,650
	1991	4,610	1,685		
Sri Lanka	1963	1,170	764	568	459
	1971	1,645	1,162		
	1982	1,791	1,390	781	
Thailand	1963	3,214	595	2,651	5,957
	1983	4,471	662	3,623	11,824
	1988	9,532	867	3,797	10,067

Numbers of farms in thousands, areas in thousands of hectares.
Source: IRRI (1995c).

values. San Bartolome village in Tarlac Province, Luzon, the Philippines, is one village that has been intensively studied. Land values have tended to be related to the expected yield of rice. They have therefore increased as the productivity of the land has increased, with large increases when irrigation became available, and high-yielding rice varieties were introduced. Prices paid for the freehold right to land in San Bartolome were 3200 pesos per hectare in 1957/59, and 22,700 in 1979/80 (Huke *et al.*, 1982). Expressed in terms of rice yield, these values represented about 2 pesos per kilo and 8 pesos per kilo respectively, corresponding reasonably well to changes in the cost of rice in the local market. The land reform act in the Philippines stated that rice land owned by an individual had to be sold if his holdings exceeded 10 ha. The price the large landholder received in the 1970s was 2.5 times the value of the expected annual rice production, or something of the order of 15,000 pesos per hectare.

7.3.3 Water

Like land, the value of water is dependent on its availability. In the past when most irrigation works were developed by communal labour, the cost of the water was the labour contributed to ensure its availability. In many places in Asia the communal development and management of water resources has continued. Where government schemes to provide water from reservoirs held behind large dams, and distributed by canals managed by an irrigation authority, have been introduced, a water fee may or may not be charged. Where the supply of irrigation water is regarded as a social good, as has been the supply of drinking water in many countries, it is expected that no charge will be made. As experience in the economics of water management and distribution systems has increased, it has become an accepted economic fact that water fees should be levied. Where such fees have been introduced the collection has often been resisted by farmers, particularly in areas where charges had not previously been made. But inefficient use of water has been widely recognized as a major factor in limiting the economic viability of irrigation schemes. Hence a clearer recognition of the real costs of water collection and distribution, commonly involving farmer participation in the management of the fee collection as well as of the system, would undoubtedly be advantageous in ensuring sustainable use of water supplies (Rosegrant and Binswanger, 1994; Svendsen and Rosegrant, 1994).

The appropriate pricing of irrigation water has to be done in the light of the wider problem of resource use efficiency. Financial autonomy of the irrigation management authority is one way in which the linkage between fees and provision of irrigation water can be properly made (Small and Carruthers, 1991). To some extent this would be comparable with the charges levied by the operators of the 'dragon boats' in China, and has a close parallel in the ways in which charges for the use of privately operated pump irrigation systems have developed in Bangladesh and elsewhere.

The rapid decline in irrigation investments was discussed in the previous chapter. Part of the reason has been the rapidly escalating capital costs for the construction of new irrigation systems (Table 7.6) Even these costs may be too low, as adequate account may not have been taken of the costs of drainage, avoiding too rapid siltation of reservoir and distribution canals, and of avoiding

Table 7.6. Real capital costs for construction of new irrigation systems, 1966–88 (US$ ha^{-1}).

	India (1988 prices)	Indonesia (1985 prices)	Philippines (1985 prices)	Sri Lanka (1986 prices)	Thailand (1985 prices)	Unweighted average
1966–69	2698	1521	1613	1470	1419	1744
1970–74	2368	1681	1882	2056	2584	2114
1975–80	1656	3187	2263	2909	2366	2476
1981–85	4033	3283	2688	5288	2276	3514
1986–88	4856	4096	na	5776	2812	4385

Source: Rosegrant and Svendsen (1993).

environmental damage and providing compensation for social disruption. There are of course several mitigating factors which have also to be recognized, such as the value of secondary effects of irrigation schemes in stimulating economic development (Rosegrant and Svendsen, 1993).

7.3.4 Seed

Rice seed has traditionally not been a cost to the producer. Most farmers saved their own seed, and many continue to do so. Only as new varieties evolved, and were selected for multiplication by farmers did a value become attached to seed, as it did to *Azolla* after its beneficial effects were discovered (p.89). With the increasing availability of new and further improved rice varieties in recent years, and the introduction of hybrid rice in China, a market for improved seed has become a small but regular component of the costs of rice production. For about 3500 years before 1960 rice yields evolved slowly in the subtropics, and even more slowly in tropical regions. Then, when the introduction of the semi-dwarfs established a much higher yield potential for tropical rice the availability of high quality seed of the new varieties assumed major importance. Farmers quickly recognized the value of seed of the new varieties; one, in Tamil Nadu, India, named his son 'IR8'. Production and distribution of high quality seed was often government sponsored. Demand from farmers helped to establish a system of certified seed producers. Their contributions have helped to ensure the continued availability and rapid dissemination of new varieties, particularly pest resistant varieties, the availability of which was important when new problems arose. In economic terms there has also been considerable importance in the provision of varieties with a shorter time to maturity, which enables double cropping to be developed and less water used per crop.

Essential support to the seed producers has come from the national rice improvement programmes. Their efforts have been allied with the efforts of IRRI in varietal improvement through the International Rice Testing

Programme (IRTP), which was supported by the United Nations Development Programme (UNDP) to ensure that a partnership was created between all those involved in rice varietal improvement. The IRTP has been succeeded by the International Network for Genetic Evaluation of Rice (INGER), in which greater emphasis is given to sharing material between plant breeders for rice improvement. The Network ensures that all new material can be evaluated by national programmes, and distributed freely and rapidly to the licensed seed producers wherever it offers an advantage to local farmers.

Rice is self-pollinating, so that once a variety has been grown it will normally produce seed suitable for replanting if it is decided to continue using the same variety. Hybrid rice seed, produced by outcrossing a male sterile rice variety with a suitable high-yielding second parent, is a special case. Male sterile lines were first identified in China in 1970 (Lin Shih-Cheng and Yuan Loung-Ping, 1980). The hybrid vigour gave a yield advantage of more than 20%. The hybrids produced were planted on about a quarter of the rice area in China in 1985 (Yuan Loung-Ping and Virmani, 1988) and almost half of the rice area (16 Mha) by 1991 (IRRI, 1993a). The extra yield from hybrid rice is obtained at the cost of labour and land used in the production of the hybrid seed. Nevertheless it has been calculated (He Guiting *et al.*, 1988) that the rate of net return to total cost in 1985 was 30%.

7.3.5 Manures and inorganic fertilizers

Use of manures and other means of restoring soil fertility has long been an essential part of obtaining high yields from the wetlands. It has always had a significant cost. This was initially in labour required to collect and transport manure or more fertile silt to the rice fields. As discussed in Chapter 5 many forms of organic manures have been used (Table 5.6). In China human and animal faeces were long used, usually mixed with straw or other waste material. There are early records of the sale of briquettes made from night soil (Bray, 1986) and of a trade in river mud, silkworm waste and animal manures. As noted in Chapter 4 (p.89) live *Azolla* plants were sold as early as AD 540.

Since the 1950s however the use of inorganic fertilizers in rice production in Asia, and particularly tropical Asia has increased dramatically (Fig. 5.3). Use of inorganic fertilizers (N, P, K) on all crops in China increased from 728,000 t in 1961 to over 29 Mt in 1991. At the same time use of organic manures has decreased, although not to the same extent. Fertilizer prices differ considerably between countries, depending on whether they have the resources necessary to manufacture fertilizers themselves. Indonesia and Malaysia have from time to time provided farmers with inorganic nitrogen fertilizers without charge, or at highly subsidized prices. The fertilizers were produced as an offshoot from their oil production facilities. Countries unable to manufacture fertilizers and where foreign exchange was not readily available tended to charge higher prices, so that fertilizer use has been less, as for example in the Philippines compared with Indonesia (Fig. 5.3). The Philippino farmer actually spends more on fertilizer than the Indonesian farmer (Table 7.7) although he applies considerably less to his crop.

7.3.6 Machinery, pesticides etc.

The amount of money invested in machinery and farm implements and used on pesticides and sprayers has become a significant cost for some farmers. Rice production in China and tropical Asia has been largely characterized by the fact that it has been labour intensive. Only as increasing prosperity and alternative avenues of employment have become available has the use of machinery become a significant factor in the costs of rice production. This is most apparent in Japan where even on farms of 1 or 2 ha a heavy capital investment in sophisticated mechanical transplanters and combine harvesters is common. In tropical Asia the low cost of labour and value of the rice harvest compared with its highly subsidized price in Japan has meant that only relatively simple machines have been used. Simple power tillers, axial flow threshers, and water pumps manufactured in village workshops are now an increasingly common feature of small irrigated rice farms (Fig. 7.3).

Pesticides and sprayers have also become an increasingly common feature. Their use started to spread rapidly when the early high yielding varieties with limited pest resistance were first introduced, but the sales of insecticides have tended to diminish in recent years as a wide spectrum of pest resistance has been bred into the rice varieties released (Table 4.4). Consequently the costs of pesticides have been only a small component of expenditure for most small rice farmers in tropical Asia. The importance and potential

Fig. 7.3. Small farm machinery for rice production. (a) Power tiller, (b) transplanter, (c) reaper, (d) axial-flow thresher, (e) seeder

Table 7.7. Some costs of inputs to rice farming (US$ ha^{-1}), and rice (paddy) and urea fertilizer N prices (US$ kg^{-1}).

Country	Year	System or region	Hired labour	Fertilizers	Chemicals (pesticides)	Others	Rice (paddy) price	Urea price
Bangladesh	1993	Irrigated	236	76	17	19	0.14	0.27
		Rainfed, MV	90	68	9	16		
		Rainfed, TV	84	19	3	23		
China	1991	Irrigated	—	96–124	16–26	28–62	0.09–0.13	
India	1987/88	Punjab	80	78	13	11	0.14	0.65
	1984/85	West Bengal	73	26	2	13	0.13	
	1986/87	Assam	26	2	0	14	0.11	
Indonesia	1987/88	Irrigated, W. Java	150	34	20	5	0.13	0.18
		Upland, Lampung	73	26	10	12		
Japan	1993	Irrigated	63	799	710	836	2.49	1.2
Madagascar	1993	Irrigated, MV	112	23	0	6	0.22	
		Irrigated, TV	87	0	0	10	0.22	0.98
		Rainfed, TV	34	0	0	8	0.24	
Philippines	1991	Irrigated	122	45	21	39	0.15	0.41
		Rainfed	87	25	13	29		
		Upland	41	10	5	13		
Thailand	1992/93	Irrigated	57–66	47–58	19–24	19–23	0.15	0.39
		Rainfed	6–16	14	1	6	0.15	0.40
Vietnam	1990	Irrigated	32	84	10	35	0.09	0.38

'Others' includes costs of seeds, manures, and water fees, but not of land, capital, or family labour. MV, TV = systems using modern or traditional varieties.
Source: IRRI (1995c).

of biological control has also been more widely recognized. As a result pesticide sales, and in particular insecticide sales in several tropical Asian countries have fallen by more than 50% between 1990 and 1993 (IRRI, 1995c).

The case for reduced use of insecticides is strong. The fall in sales of fungicides and herbicides has a less satisfactory scientific basis and is largely a response to a general negative reaction to the use of chemical pest control. The need to educate farmers in the rather complex arts of integrated pest management (IPM), in which chemicals are appropriately used in combination with varietal resistance and biological control techniques, is important (Kenmore *et al.*, 1987; Escalada and Heong, 1993; Heong *et al.*, 1995).

7.3.7 Total costs

Table 7.7 lists the production costs which require an immediate outlay from the farmer. Most of the outlay will be in the form of cash, although many small rice farms in Asia continue the practice of part payment to hired labour in the form of a share of the crop. In addition to these costs the concealed or opportunity costs of family labour, of credit, and of capital investments have to be added. Capital investments include the return on capital expected from land values, and where land is not held on a freehold basis it will usually include a rental cost. Costs of the hire of animals or machinery or of capital invested in machinery may also be significant. Thus total costs to the rice farmer (Table 7.8) will usually be considerably greater than those listed in Table 7.7. Total costs of production for several countries shortly before 1990 have been discussed by Yap (1992) and Hossain and Fischer (1995).

7.4 The Value of Outputs

The principal output of rice farming is of course the rice grain, but the straw also has value. In addition the use of the land when rice is not occupying it can produce substantial extra income. So can family labour when it is not engaged in work on the rice crop. In irrigated areas the water in the ricefields can be a major source of income if it can be managed to produce fish.

7.4.1 The rice grain

The governments of the rice dependent countries have a fine balancing act to perform. They have to balance the need for rice farmers to receive a sufficient return for their crop to ensure that rice farming remains economically sustainable, against the demands of consumers to have sufficient rice at prices they can afford. In most instances this has meant balancing the needs of the city

Table 7.8. A comparison of the cost of producing rice in selected countries, 1987–89.

Country	Paddy yield (t ha^{-1})	Cost of production (US\$ ha^{-1})	Cost of production (US\$ t^{-1})	Share of labour in total cost (%)	Price of rice unmilled (US\$ t^{-1})
Bangladesh					
Irrigated	4.60	633	138	30	180
Rainfed	2.63	419	160	32	190
Indonesia					
Irrigated	5.76	677	118	25	120
Rainfed	3.49	414	119	44	110
Thailand					
Irrigated	4.00	580	145	24	150
Rainfed	1.84	221	120	35	141
Vietnam					
Irrigated	4.60	461	100	17	100
Rainfed	2.60	304	117	22	130
Philippines	4.18	679	162	15	170
Korea	6.61	4,347	658	17	957
Japan	6.51	12,943	1,986	28	1,730
USA	6.27	1,223	195	5	157

Source: Hossain and Fischer (1995).

against those of the rural community. As shown in Table 7.4 there has been a continuing movement of people from country to town in almost all of Asia since 1960. Correspondingly the political importance of the urban communities has grown and there has been a tendency to manage the price of rice so that the consumer has been the principal beneficiary of increased rice productivity (Herdt, 1987). Possibly as greater economic prosperity is achieved there will be a move to make the returns to the farmer higher, as happened in Japan. Certainly such a move favours the biophysical sustainability of rice farming. The size of the subsidy to rice farming in Japan is more than most governments would wish, or be able to, pay. Nevertheless it is important that the price of rice is managed so as to ensure a fair return to the farmer. He is then able to manage his land and water supply in such a way that rice production is biophysically sustainable. If the biophysical sustainability is lost the option of favouring rural or urban population is lost, together with a major natural resource of the country.

7.4.2 Straw

Rice straw has limited value, even by comparison with other cereal straws. It has high silica and lignin contents, and is not readily digestible by animals.

Considerable work has been done on measures to improve its digestibility, mostly involving treatment with ammonia or caustic soda (Doyle *et al.*, 1986; Schiere and Ibrahim, 1989). Although the digestibility has been improved the measures have not yet been shown to be economic. Paper making and other possible uses have also been explored (Munck and Rexen, 1985) but for these purposes the high silica content, which commonly is more than 10% by weight, is a major drawback. Returning the straw to the land is frequently advocated as a measure to improve or help maintain soil fertility (e.g. Ponnamperuma, 1984b). It may contribute to the soil nutrient content, but this will be at the expense of adding to the carbon released as methane, which is a much more powerful greenhouse gas than carbon dioxide. Phosphorus and potassium and other non-volatile nutrients are also added to the soil with the straw, but are equally obtained from the ash if the straw is burnt and the ash spread. As discussed in Chapter 4, the physical effects of organic matter which are so important in upland soils can be disadvantageous in wetland soils.

Incorporation of straw involves a significantly greater energy input during soil cultivation than if the straw is burnt and only the ash incorporated. Hence most rice farmers in common with other cereal producers prefer to burn the straw. Some straw will usually be retained as bedding for animals and, together with other straw consumed, eventually end up in the soil as part of the farmyard manure. This may well be its most valuable function. In terms of sustainable land management recycling of the nutrients in straw is important. For most rice farmers with very low incomes the decision on whether to follow the recycling route or use fertilizers will be made on the basis of the relative costs of fertilizers and the labour required for recycling. The more perspicacious farmers will follow the route of integrated nutrient management, using some organic residues together with inorganic fertilizers, to keep the advantage of a balanced nutrient supply and slow release of the nutrients from the manure.

7.4.3 Other crops grown in rice-based farming systems

Most rice farmers endeavour to maximize their income by using their land throughout the year. If adequate water is available then two or three rice crops may be grown. Two rice crops are grown on about 28 Mha of rice land at present; the triple cropped area is probably less than 1 Mha (from IRRI Social Sciences Division Database). The area on which an upland crop is grown after rice is probably greater than the area where a second rice crop is taken, and an upland crop following two rice crops is more common than growing three rice crops each year. In South and Southeast Asia the most common practice is to plant a grain legume after the rice crop, usually mung bean, cowpea or pigeon pea (IRRI, 1977, 1982b). The dryland crop receives little attention after the seed has been sown, and yields seldom exceed 1 t ha^{-1} and are often much less.

Nevertheless a return is given for very little input, and the provision of extra protein for consumption by the farm family may be important.

Wheat and vegetable crops grown after rice are usually more profitable than legumes, but require more care and attention, and most importantly usually some supplementary irrigation. Thus their production tends to be confined to areas where there is a possibility of some water being available. The use of small farm ponds to provide irrigation for vegetable crops in the dry season has been strongly advocated as a means to provide extra income for rainfed or partially irrigated farms (Bhuiyan, 1994). Where a market exists in reasonable proximity vegetable growing is often very profitable.

Rice–wheat systems have been practised for many years in northern India, including parts of what are now Bangladesh and Pakistan, in the Terai and Hill regions of Nepal, and in north-central China. In many of the rice–wheat areas there is a problem in establishing a satisfactory seed bed after the soil has been puddled to produce the rice crop, and yields are not always satisfactory. Nevertheless the value of the wheat crop is such that the system can be expected to expand wherever conditions are suitable (Harrington *et al.*, 1992, 1993).

7.4.4 Animals in rice-based farming systems

The majority of small rice farmers in tropical Asia own at least one water buffalo, and in China the majority of rice farmers own at least a pig, but more commonly several animals, kept for food, manure and profit. The water buffaloes are kept primarily as working animals, but also yield milk and meat, and some are used as breeding stock. While they need considerable care – for instance a water wallow once every 2 hours – and consume considerable amounts of feed, they return the care with excellent service. Consequently provision of the necessary grazing and feed is often a significant part of the operation of a rice farm. Some feed is usually obtained from the surroundings of the rice farm. Many farmers will have some non-irrigated land where the animals can be allowed to browse, and a common use of ricelands in the off-season is to provide grazing for the farm animals. The ability of water buffalo to move through the mud of a flooded rice field is an important attribute, not only to make cultivation of the wetlands possible, but to contribute to the puddling process, and to deepen the depth of the puddled soil layer and so promote better root growth of the rice crop. While power-tillers have replaced the water buffalo in many areas for the initial land cultivation, because they can complete the operation more rapidly, many farmers continue to use a water buffalo for the subsequent harrowing of the soil, and some farmers continue to use a water buffalo from time to time for the initial cultivation, to maintain the depth of soil above the cultivation pan.

Animal manure is also an important component of the rice farming system; chicken manure has a high reputation as a manure for rice fields and can still be bought in many country markets. Use of pig manure in Chinese rice farms is a very long-standing practice, and the practice of growing vetch or other legumes after the rice crop as an animal feed as well as acting as a green manure is still common.

7.4.5 Rice–fish systems

Aquaculture has been more successfully developed in Asia than in other parts of the world. For much of Asia fish are a prime source of protein in the diet. Traditionally fish have been caught in the streams and canals which provide the irrigation water. As the population has increased in the rice areas, the intensity with which such sources have been exploited has meant that the natural fish population has declined sharply. Although there has been a tendency to blame the use of chemicals, and specifically pesticides, for the decline in the natural fish population, the increased exploitation of the limited natural resources is much the most common reason. In aquaculture systems the fish may be raised in special ponds, but they can also be raised in the rice fields, if precautions are taken to ensure that if the water level in the field is temporarily low, there are areas where sufficient depth of water persists to preserve the fish population. This is usually achieved by digging a small trough within the rice field which does not drain with the rest of the field.

The potential of rice–fish culture to add to the food produced from rice fields, and to the income of the rice farmer, is considerable. Additional labour is required, and initially a modest capital outlay, but the economic returns can be substantial (Table 7.9). Many claims have also been made that some fish species will contribute to the control of weeds and insect pests in rice fields, so that rice production is enhanced by the fish. Not surprisingly there has been a considerable expansion in rice–fish cultivation; in China alone the area increased from 120,000 ha in 1981 to 987,000 ha in 1986 (Cai Renkui *et al.*, 1995). The same authors believe that about one-tenth of all rice land in China could be used for rice–fish cultivation, which would produce about 350,000 t of fish annually. The potential in other parts of Asia will be less because the availability of water supply to the rice fields is less, but it is certainly considerably greater than at present.

7.4.6 Alternative employment opportunities

The survey of San Bartolome village in the Philippines (Huke *et al.*, 1982) revealed that the apparent prosperity of the village, with an average farm size of 1.7 ha, was largely dependent on income returned to the village from

Table 7.9. Economic benefits from different forms of fish culture in Hunan and Sichuan in 1984.

	Area (× 1000 ha)	Raising days	Avg. fish harvest (kg ha^{-1})	Avg. rice harvest (kg ha^{-1})	Average inputs per ha		Avg. income (CNY ha^{-1})	Income ratio (%)*
					Labour (Days)	Net invest. (CNY)		
Hunan								
Rice only	46.7	—	—	4883	240	450	639	—
Rice–fish	4.8	65	113	5250	285	480	720	113
Rotation fish and rice	0.5	200	600	6750	311	750	2001	313
Intercrop, rice and fish	0.7	35	Raising fry	5250	285	858	998	156
Sichuan								
Rice only	8.2	—	—	5273	240	675	1434	—
Rice–fish	0.3	70–90	83	5438	300	837	1635	114
Rotation fish and rice	0.2	200–240	128	5513	315	930	1658	116
Fish culture in half-dry fields	0.08	90–100	879	8513	375	1320	4706	328
Fish culture (ponds + canals)	0.08	90–100	885	7575	27	1335	4881	340

*Net income from fish culture in ricefields as a percentage of income from planting only rice. Labour in Hunan is CNY2.5 day^{-1}. CNY = China new yen. Source: Jiang Ci Mao and Dai Ge (1995).

external sources. These included salaries from employment in part-time occupations outside the village, from larger farms which members of the family had acquired by migrating to less populous parts of the Philippines, pensions from the Philippine constabulary and United States Navy, and remittances from children who had acquired professional qualifications and were now employed elsewhere in the Philippines or overseas.

The importance of remittance of money from non-agricultural sources to rural families is almost certainly greater in the Philippines than elsewhere in Asia. Nevertheless as the process of change in the rural areas continues, and the proportion of the population finding employment other than on the family farm increases, so the income derived externally tends to increase. While the actual amounts remitted will differ considerably between countries, and between different areas within a country, there is no doubt that it is an important contribution to rural family incomes throughout Asia.

7.4.7 Total farm income

In 1985 as part of the activities to mark the 25th anniversary of the establishment of the Institute, IRRI invited 14 'Outstanding Rice Farmers', nominated by their governments or FAO, to IRRI. They were asked to record their views on rice farming and the sources of their success. The results of the interviews with the farmers were reported in an IRRI publication 'Insights of Outstanding Farmers' (IRRI, 1985f). Some were able to provide quantitative data on the sources of their income (Table 7.10). The information offered showed that the income from rice farming ranged from less than 10% of the total to slightly more than half. This was typical of the group, although probably not typical of rice farmers throughout Asia. The successful rice farmers are generally those with the initiative to develop interests wider than production of the rice crop. But even if rather atypical the importance of income from activities other than rice farming is clear. Except for two Japanese farmers included in the group, the annual income of these rice farmers was less than US$5000 per year. This was in spite of the fact that they owned farms rather greater than the average size in much of Asia, and were considered to be among the best farmers in Asia. An annual income of less than $5000 per year would be considered by many in developed countries as little more than the poverty level. Their farms also provided income for employed labour. Some labourers were employed on an annual basis, primarily to look after animals; most of the others were engaged for seasonal work and often paid in kind, particularly those engaged for work with the rice crop.

It has been true for several millennia that farm activities contribute to the wider social and economic well-being of the rural community, or at least its survival. The sustainability of the rice farming system has depended as much on its contribution to the life of the community as a whole, as it has on the

Table 7.10. Sources of income of some 'outstanding rice farmers'.

Farmer	Country	Source of income	Farm size (ha)	Annual net income (US$)
Abul Kalam Azad	Bangladesh	Rice	7.4	1300
		Sugarcane		1300
		Turmeric		926
		Potato, jute, wheat		456
		Jack fruit, mango, coconut, bamboo		370
		Chicken and eggs		222
		Net annual income		4574
Qu Yong Shou	China	Rice	0.3 plus	400
		Oranges	?	2600
		Pigs		600
		Annual salary		16
		Received from relatives		1440
		Net annual income		5056
Serapio San Felipe	Philippines	Rice	3.8	1216
		Fish		278
		Fruit and vegetables		500
		Animal products		361
		Net annual income		2355
		plus remittances from 3 sons and 1 daughter		

Source: IRRI (1995f).

biophysical components of sustainability. The new factor which has appeared in the course of the past century is the development of alternative employment opportunities, affecting both the social and economic structure of the rural communities. This has been part of the worldwide trend to move from subsistence farming to semi-commercial systems to commercial systems (Pingali and Rosegrant, 1995). While rice farming in Japan has generally reached the fully commercial stage, in most of the rest of Asia it is still at the semi-commercial stage, with isolated communities living at subsistence level. The semi-commercial stage is characterized by a mix of production for family consumption and for sale, a mix of traded and non-traded inputs, a mix of family and hired labour, and a mix of farm activities. This is well illustrated by the activities of the 'outstanding farmers' discussed above. Increasing commercialization at the farm level leads to increasing specialization as the demands on technical and managerial skills increase, as well as the increased capitalization of the production process. The two 'outstanding rice farmers' from Japan, who were among those invited to IRRI in 1985, both retained a part of their farms for non-rice production. One produced half of his annual income of US$57,200 from mushroom growing in the forested part of his land.

His farm worked as an extended family unit covered 8.3 ha of ricefields. In addition he owned 20 ha of forest, which is unusual in Japan.

7.5 Policies and Profitability

The sustainability of rice production in economic terms requires that the value of outputs exceeds the cost of inputs. Through the long period of the green evolution in Asia, and while land was readily available, this was not a significant problem. The social and religious importance of rice tended to safeguard the position of the rice farmers, who were in any case the great majority of the population. As they evolved from the subsistence stage to the semi-commercial stage government interference in taxing the rice farmer steadily increased. Initially taxes were levied by requiring that some of the crop be handed over. Later the farmer was required to sell the crop, beyond that needed for family use, to the government at a price well below the market value. Alternatively or in addition a monetary tax was levied.

Technology has greatly increased the productivity of rice, following the impact of the green revolution. This has meant that government policy in controlling the price of rice to farmer and consumer has become increasingly the dominant factor in determining the profitability and hence the economic sustainability of rice farming. Commercialization of rice farming will undoubtedly tend to increase as the industrial revolution continues to make its impact on rural Asia, and more advantageous opportunities, or the promise of them, become available in the cities. Thus pressures to reduce labour use in rice farming will intensify, and farms will need to be increasingly specialized in rice production, with an increasing premium on efficient production (Pingali and Rosegrant, 1995). If this process goes too quickly and too far there is a real danger that many rice farmers will struggle to survive. To ensure the economic sustainability of rice production there must be an adequate return to the farmer. As the costs of labour rise it will be increasingly important to assist farmers in the adoption of more efficient rice production practices. Governments may need to find ways to subsidize rice production, by controlling the price to the consumer while paying the farmer an adequate price for his production, or subsidizing the price of inputs, or in other ways, including investment in research and extension.

Many economists have argued strongly against the need for subsidies to the rural sector. However the question of ensuring the long-term productivity of the resource base is a prime government responsibiliy. All developed countries have built their success on a firm foundation of sustainability of their agricultural production by subsidizing farming, whether directly as in Europe and Japan, or indirectly through support services and other ways as in the USA. The profitability and sustainability of rice farming should not be dependent, as it is at present in many tropical Asian countries, on subsidies from other members of the family employed outside the rural sector; nor should its economic viability depend on production of non-rice crops from the family farm.

The effect of Japanese policies has been to increase greatly the capitalization of individual farms, so that labour requirements are reduced, while yields and productivity per unit of labour, land and water have increased. The value of land for rice production has been inflated, and not only is all land suited to irrigated rice production cultivated, but some unsuitable land is used, for instance peat land in Hokkaido, and many urban gardens and plots. The production of rice is sustained to meet the social demand for national self-sufficiency in rice.

In stark contrast to the situation in Japan, Thailand has exported rice in increasing amounts since 1900 (Table 7.1). The price has been kept low by not subsidizing farmers for their rice production, so that they have had to sell to the government for export at a controlled low price, or into an over-supplied local market. The rice farmers who have been able to find more land with natural flooding have been able to increase their production and maintain their economic position. To maintain their income this has often meant that that those with poorer land have had to seek more land to farm. This has usually meant that less suitable land has had to be used. From about 1900 to 1950, while production and exports increased, yields fell as the less suitable land was brought into production (Fig. 7.4). The less favourable areas used to produce rice in Thailand are subject not only to unpredictable floods but also droughts. Thus yields vary considerably from year to year. To give greater stability to production the Thai government therefore encouraged irrigation development

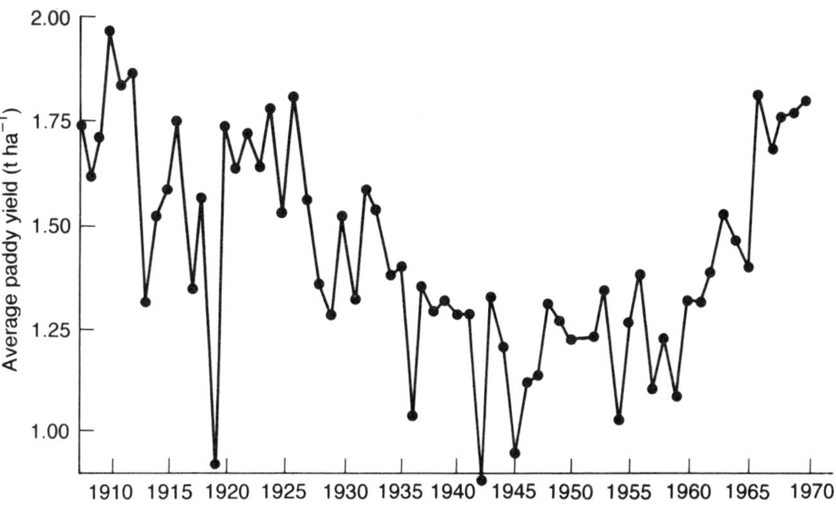

Fig. 7.4. Average paddy yield for the whole of Thailand, 1905–1971 (Ishii, 1978).

to supplement the wet season floods when necessary, and to enable dry season crops to be produced. Flood control measures have also been undertaken. These measures have reversed the downward yield trend, and have helped to boost and stabilize production, so that exports of rice have continued to grow. Modern high yielding varieties have become available for the irrigated areas with the grain quality required for Thai markets. The quality of Thai rice has been an important factor in ensuring that demand for the rice produced remained strong.

For both Japan and Thailand government policies have supported rice production. In Japan strong economic incentives have been provided by the Government in the form of subsidies to the price paid to the rice farmer, and rigid policies to prevent import of rice produced more cheaply elsewhere. In Thailand Government policy has been to keep the price low to enable Thai rice to be strongly competitive on the world market. In Thailand, although farm size has been declining, most farms are still over 2 ha, significantly larger than the average size in Asia. Sufficient rice can be produced to support a family at a little above subsistence level. Provided that the rice is produced with minimal inputs the farmer makes a profit from his rice sales. Government policy, while not always appreciated by the farmer, has in fact helped to sustain production by ensuring that there was a stable and expanding export market for the rice produced.

The Japanese and Thai examples represent extremes – the Japanese government procurement price for rice in 1993 was 230,000 yen t^{-1}, or a little over US$2000, whereas the Thai government price was 3900 baht t^{-1}, or US$156. Rice production in both countries has been sustained, but different environmental and economic conditions in the two countries have led to very different policies.

8 Concerns about the Sustainability of Rice Farming

Many concerns have been expressed about the sustainability of rice farming. Several of these have been mentioned in earlier chapters. Although the long history of rice farming indicates that the system is basically sustainable, what is less certain is whether the dramatic increases of rice production which followed the green revolution are sustainable. Agricultural scientists are concerned about the biophysical problems of maintaining the high yield levels which have been reached in the past two decades; engineers are concerned about the problems of maintaining existing water supplies as well as the increasing costs of developing new supply systems; economists are concerned not only about the rising costs of water, but also about the overall costs of inputs and whether the right policies are being followed to ensure the long-term sustainability of rice production; environmentalists are concerned about the effects of increasing use of agricultural chemicals on biodiversity and human health, and the effects of global warming; and sociologists about the problems of growing inequity and changes in the way of life associated with traditional farming systems.

The level at which sustainability must be achieved has also to be addressed. Sustainability at traditional levels of production, or even at today's levels, will not be adequate even for the present population, let alone the extra population which will be added to present numbers before the world population stabilizes at some time in the second half of the 21st century. Conservative estimates of the population increase are that by 2030 there will be more than 8 billion people to be fed. At the present inadequate nutritional levels this requires at least a 50% increase in rice production, as has been discussed in Chapter 1. If a start is to be made towards the desirable improvement in nutritional standards in areas such as South Asia, where they are at present well below optimal, and likely dietary shifts in Africa and elsewhere are taken into account, the increase in production will need to be closer to 70%, as IRRI has estimated, than the 50% based solely on a conservative estimate of the increase in the population. Thus there are serious concerns to be faced not only about maintaining present production levels but also about whether they can be greatly increased and sustained at the higher level required. In the present chapter the bases of these concerns will be reviewed in the light of current knowledge of rice farming and its traditional stability as discussed in previous chapters.

8.1 Concerns Arising from Current Production Trends

Rice production is dependent on the area of land used for rice each year, the number of rice crops produced each year, and the yields obtained from that land. The current yield trends are those which give rise to most concern and will be discussed before the contributions that may be made by increasing the land base and cropping intensity.

8.1.1 Yield trends

The evidence from yield trends has to be considered from a national and regional viewpoint, from trends in farmers' fields within a given local environment, and from trends in carefully controlled long-term experiments (Cassman and Pingali, 1995a). For Asia as a whole the rate of increase of rice yields fell from 2.86% year^{-1} between 1973/75 and 1981/83 to 1.32% year^{-1} between 1985/87 and 1991/93 (Table 1.3). The decline has been more dramatic in China, where the national average yield reached 5.96 t ha^{-1} in 1993, than elsewhere. If China is excluded from the statistics, and all developing countries considered, the decline is less dramatic (Fig. 8.1). Southeast Asia follows the general trend, but South Asia, where national average yields have not so far exceeded 3.5 t ha^{-1}, continues to record yield increases above 2% year^{-1}.

There are various factors influencing the yield changes, of which economic factors are dominant. High levels of yield depend on high levels of inputs. If adequate water is available the response to inputs is reasonably assured, and the high yields obtained in East Asia (Tables 2.1 and 3.3) reflect the good irrigation services provided. But the law of diminishing returns also operates, so that the cost of each further increment in yield tends to cost more. This can be seen from the decreasing response to fertilizer nitrogen in Indonesia (Fig. 8.2), where yields increased by 70% between 1976 and 1986, while nitrogen fertilizer consumption increased by 440% (Cassman and Pingali, 1995a). The response to nitrogen fell from 75 to 28 kg grain kg^{-1} of inorganic nitrogen fertilizer applied. This is not really a fair comparison because at the start of the period considerable use was being made of organic manures, and

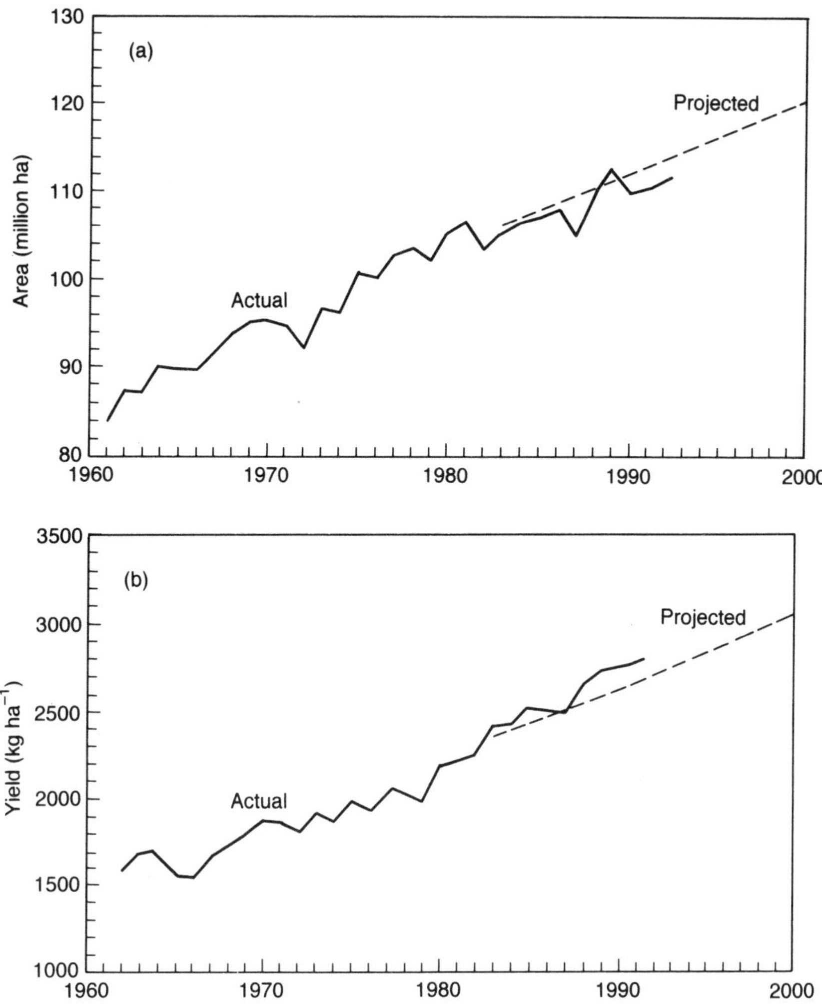

Fig. 8.1. Comparison of actual trends and FAO projections of (a) rice area (b) rice yield, for all developing countries excluding China. Projections 1982–2000 from Alexandratos (1988, Table 4.1). Comparison from Alexandratos (1995).

the application of organic nitrogen was almost certainly declining while that of inorganic nitrogen was increasing. Other factors were also changing. However Cassman and Pingali (1995a) support this evidence with calculations of nitrogen factor productivity from experiments at research stations and from studies of farmer groups. These also show a declining rate of return to nitrogen inputs.

Another important economic factor in the decline in the rate of yield increase has been the achievement by several Asian countries, such as Indonesia

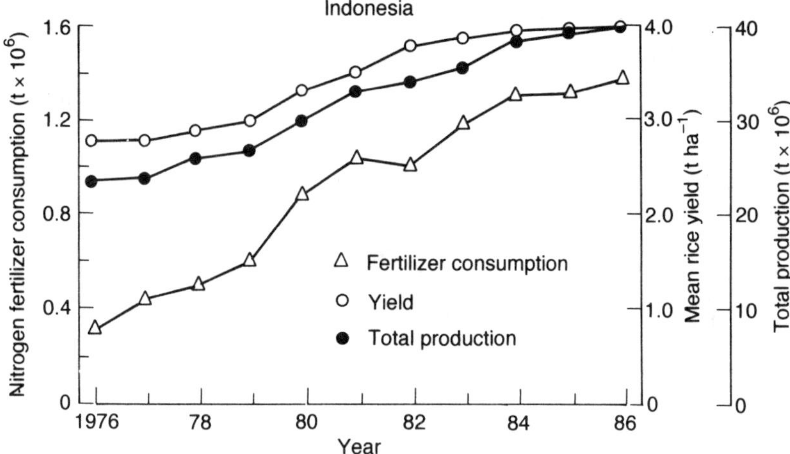

Fig. 8.2. Trends in nitrogen fertilizer use, average rice yield, and total rice production in Indonesia, 1976 to 1986.

and the Philippines, of their goal of self-sufficiency in rice production. Others such as Vietnam, Myanmar and Thailand are exporting rice at the low world price and are reluctant to increase levels of inputs until the price of rice on the world market increases. The same problem of balancing costs of inputs against value of the rice produced is also central to the dilemma of how to maintain sufficient supplies to meet expanding domestic demand at a price the poor can afford. All Asian governments wrestle with the problem of maintaining a plentiful supply of rice at low prices while ensuring that the farming community is sufficiently rewarded to want to produce more, as discussed in the previous chapter.

To maintain supplies to the domestic market, and to maintain production of a surplus for export, farmers must be given adequate incentives to seek higher yields. National policies have to be adjusted to take account of this, or not only will the rate of increase in yields fall, but production will start to decline. The escape from this dilemma has in the past been the development of new technology which has enabled yields to increase at a greater rate than the cost of inputs. In areas where yields are already relatively high, yield increases have only been obtained as a result of the application of greater inputs. Plant breeders have tended to concentrate on stabilizing the yield potential of the semi-dwarfs by improving varietal resistance to pests, grain quality, and decreasing the time required for the grain to mature. If the yield potential is to be raised, and the efficiency of higher levels of inputs improved, plant breeders have to develop a new plant type with greater yield potential than the stiff-strawed semi-dwarfs.

The failure to increase yield potential is clearly shown by the studies on farm level trends in yields conducted by IRRI since 1966 in two provinces of the Philippines. Inputs other than labour have tended to increase throughout the period (Table 8.1) so that the gap between the research station yields at IRRI and those obtained by the top one-third of farmers in the survey has disappeared (Fig. 1.9). Present yield levels obtained by the farmers in the survey, of around 6 t ha^{-1}, have been made sustainable by continuing efforts to produce pest resistant varieties that mature more quickly. A slight reduction in pesticide use has been possible since 1982, and in some areas greater cropping intensity has been achieved, but no new technology has emerged to raise the yield ceiling. As Khush (1995) comments, 'During the 30 years since the development of this (IR8) plant type, however, only marginal improvements in the yield potential of rice have occurred'. To achieve higher yields other than by increasing inputs requires a new yield potential to be realized (Cassman, 1994). Hybrid rice represents a step in this direction, but the need for specialized facilities to support hybrid seed production introduces an additional cost.

The difficulties that are likely to be encountered in sustaining rice yields at levels that are needed are well exemplified in the results of long-term experiment station trials conducted in the Philippines and India. The Philippine experiments were started at four sites between 1964 and 1968. Without exception they show either a non-significant or downward trend (Flinn *et al.*, 1982; Flinn and De Datta, 1984; Cassman and Pingali, 1995a). The downward trend was observed in spite of the addition of what was believed to be more than sufficient fertilizer, carefully managed and sufficient irrigation, and full crop protection measures. The yield decline is shown by the highest yielding variety as well as others included in the trial. Yields obtained at the Maligaya Research Station of the Philippine Rice Research Institute are shown in Figure 8.3. Two crops were grown under fully irrigated conditions each year. The yield decline has been from 14 t ha^{-1} $year^{-1}$ to 11 t ha^{-1} $year^{-1}$ over

Table 8.1. Mean input levels used by farmers on wet season rice in Central Luzon, 1966–1990.

Year	No. of farmers	N rate (kg ha^{-1})	Pesticide use* (kg ha^{-1})	Seed rate (kg ha^{-1})	Total labour (days ha^{-1})	Tractor use (days ha^{-1})
1966	91	9	0.1	45	44	0.1
1970	62	29	0.2	52	47	0.2
1974	58	39	0.5	49	62	0.3
1979	146	63	0.8	115	50	0.7
1982	136	63	1.0	129	48	0.8
1986	120	67	0.7	184	42	0.8
1990	107	70	0.9	169	42	1.3

*Based on kg active ingredient ha^{-1}.
Source: Cassman and Pingali (1995b).

a period of 23 years. At IRRI in a continuous cropping experiment in which three crops were grown each year, again receiving all perceived necessary inputs, yields of the dry season crop have fallen from 8 t ha^{-1} in 1968 to less than 6 t ha^{-1} in 1990 (Fig. 8.4). IR8 was included as a variety in this experi-

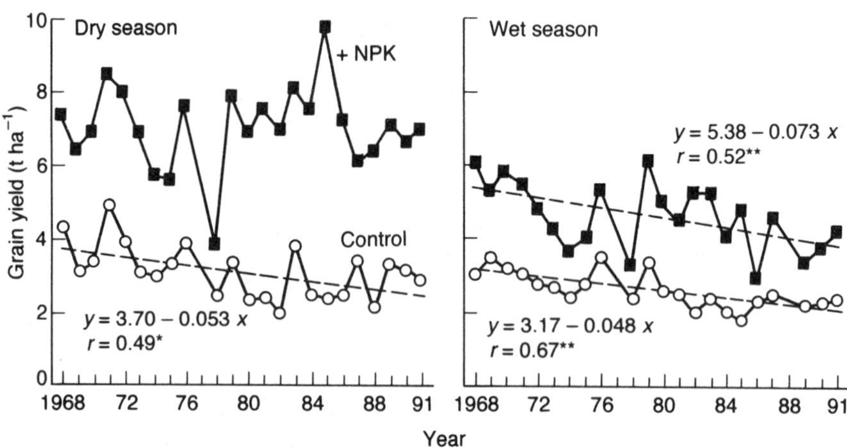

Fig. 8.3. Yield trends of the highest yielding variety in the long-term fertility experiment conducted at the Phil Rice Research Station, Maligaya, the Philippines, for treatments that receive complete nitrogen (N), phosphorus (P) and potassium (K) inputs in each crop cycle (+NPK) and in the control treatment without fertilizer inputs (Cassman and Pingali, 1995a).

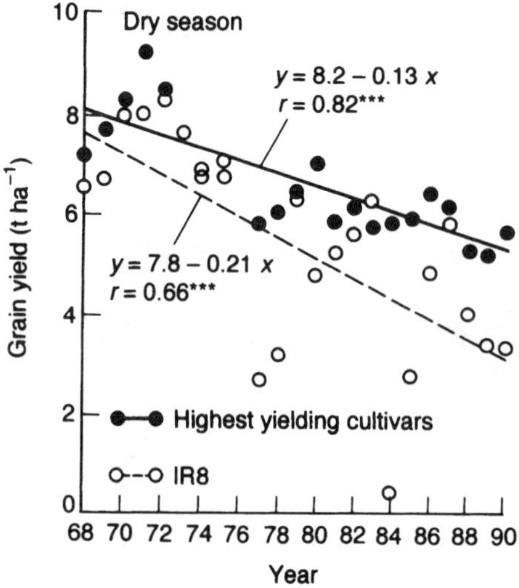

Fig. 8.4. Dry season yield trends for the highest yielding variety and IR8 for plots receiving 150 kg of fertilizer N per crop, in the 3 crops per year long-term continuous cropping experiment at IRRI, Laguna Province, the Philippines (Cassman and Pingali, 1995a).

ment as a comparison. Its yields fell from 8 t ha^{-1} to less than 4 t ha^{-1}. Total yields from the three crops grown each year fell from around 20 t ha^{-1} to 14 t ha^{-1}.

The Indian studies (Nambiar and Ghosh, 1984; Nambiar, 1989) were initiated in 1972 as part of the All-India Co-ordinated Research Project on Long-term Fertilizer Experiments. Two experiments, at Bhubaneswar and Hyderabad, studied a rice–rice cropping system, and two, at Barrackpore and Pantnagar, a rice–wheat cropping system. A wide range of fertilizer treatments were examined. Results shown in Fig. 8.5 are for the treatments using rates of nitrogen, phosphorus and potassium (NPK) fertilizers shown to be adequate from soil tests, and the same inorganic NPK treatment supplemented by an average dressing of farmyard manure. Varieties grown were those that were giving the highest yields in the region.

For the rabi (summer, dry season) crop the experiment at Hyderabad shows a consistent and high rate of decline over the 10 years that it was conducted. For the kharif (winter, wet season) crop there is also a downward trend after the first 3 years, yields falling to 2–3 t ha^{-1} by 1980, and showing only a slight increase by 1985. Part of the problem with the dry season crop was the difficulty experienced in maintaining adequate irrigation. If this was the sole cause it would have been expected to lead to an erratic sequence of yields depending on seasonal conditions, rather than the consistent downward trend. At Bhubaneswar in eastern India it has been possible to sustain yields for 14 years in the wet season, but at a low level between 2 and 4 t ha^{-1}. The initial dry season yields were close to 5 t ha^{-1} but fell in the succeeding 2 years to 2–3 t ha^{-1}, and over the following 12 years have remained between 2 and 4 t ha^{-1}. At Barrackpore and Pantnagar there have been no significant changes in the yields of either rice or wheat, rice yields being maintained at about 5 t ha^{-1}, and wheat yields at 2.5 (Barrackpore) to 4.5 (Pantnagar) t ha^{-1}. In a rice–wheat system in Nepal, a slight and statistically non-significant downward trend in wheat yields has been recorded over 11 years (Harrington *et al.*, 1993).

The experiment station results from the Philippines have raised doubts about whether, with present technology, there is a real prospect of increasing annual yields beyond about 6 t ha^{-1} on a sustainable basis. Some of the Indian experiments imply that there are still problems to be solved in sustaining yields at lower levels, at least in some tropical conditions. Although the experiments in the Philippines show a consistently declining yield trend, this is only apparent where annual production from the trial plots exceeds 12 t ha^{-1}. It is not yet clear whether yields will stabilize at or about this level. Kropff *et al.* (1994) report several studies at IRRI in which individual crop yields have been raised above 10 t ha^{-1} by using additional late additions of nitrogen fertilizer, and suggest that the yield decline is associated with decreasing nitrogen supplies from soil nitrogen, and a limited capacity of the semi-dwarf plant type to use extra nitrogen, and to resist soil borne fungal diseases.

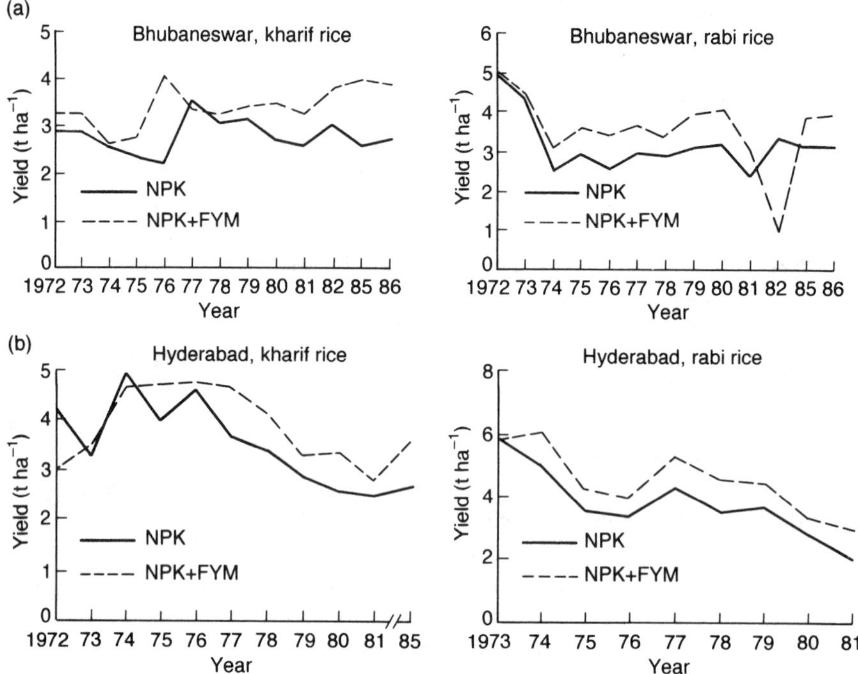

Fig. 8.5. (a)–(b) Yield trends for rice–rice (Bhubaneswar and Hyderabad) and rice–wheat (Barrackpore and Pantnagar) cropping systems in the All-India Co-ordinated Research Project on Long-term Fertilizer Experiments for treatments receiving NPK fertilizer and NPK plus farmyard manure. (Replotted from data from Nambiar and Ghosh, 1984, and Nambiar, 1989.)

In sub-tropical conditions in several Provinces in China, average yields have exceeded 5 t ha^{-1} per crop for more than 20 years. In Japan the national average yield has exceeded 5 t ha^{-1} from from 1962 to 1992. In Australia and California average yields have grown from 6 t ha^{-1} in 1962 to over 8 t ha^{-1} in 1992; only one crop per year is grown. In China several of the Provinces where high yields have been maintained are those where multiple rice cropping is practised. Thus there are grounds for believing that yields of well over 6 t ha^{-1} for a single crop, and over 10 t ha^{-1} for a multiple crop, are sustainable in sub-tropical conditions. It has yet to be established if this is so for humid tropical conditions.

8.1.2 The land base

FAO staff have attempted to estimate the probable land available for further agricultural use in the year 2010, and its production potential (Alexandratos,

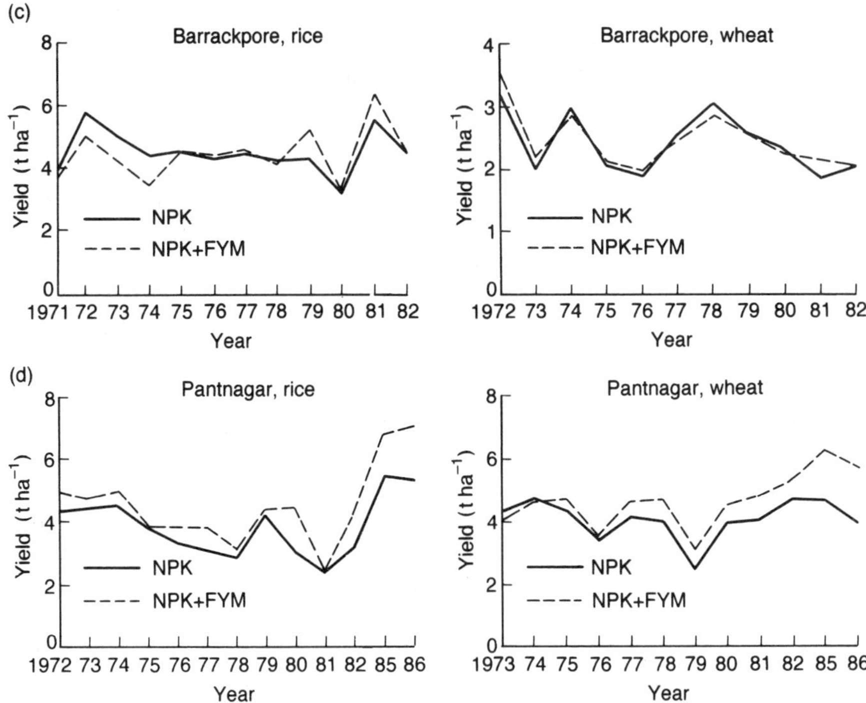

Fig. 8.5. (c)–(d) *Continued*

1995). The uncertainties about the present use of agricultural land in China caused FAO to omit China from their assessments. This of course leaves a vast gap in so far as rice production is concerned. A further major gap arises from the problems associated with estimating the areas available for irrigation development. For this purpose the land and climate characteristics have to be interfaced with those on water resources, and political, social and economic factors which can also have a strong influence on what is practicable in terms of irrigation developments. For South and Southeast Asia, including all Korea, FAO estimated that there are 134 Mha of land with agricultural potential not currently in use for agriculture. Some of this land is already occupied by roads and buildings, and some is reserved for forests or conserved or protected for other reasons. The possible addition to the irrigated rice area in Asia by 2010 appears from the FAO figures to be projected as 10 Mha, representing a growth rate of 0.8% year^{-1}, a much slower rate of growth than in the period between 1965 and 1985 (Table 6.4), but still considerably greater than would be projected from recent rates of decline.

Among the uncertainties is the problem of assessing the area of irrigated land lost to salinization and waterlogging. FAO believes this may be of the order of 2.5 Mha per year. An almost comparable uncertainty relates to the area of land which is brought back into use by rehabilitation of irrigation systems. FAO suggests it may be sufficient to compensate for the losses.

The available information is clearly inadequate for firm conclusions to be reached about the additional land that will be available for either irrigated or rainfed use. The FAO projections for the decade 1983 to 1992 in fact overestimated the actual expansion in the area cultivated to rice, and underestimated the yield increases (Fig. 8.1). IRRI (1993b) discounts the likelihood that extra land development will make a significant contribution to rice production in Asia, and Barker *et al.* (1985), Herdt (1987) and David (1991) all accept that the additional rice production that is needed must come from yield increases, with some contribution from increases in cropping intensity.

Loss of land to urban and other non-agricultural uses is an increasing source of concern. Land alienation has already contributed to the loss of rice land in East Asia, where a decrease in irrigated area of 0.3% $year^{-1}$ was recorded between 1980 and 1985.

In Africa there is scope for cultivation of new land, but much of the development will be for upland rice cultivation. Much of the land will be in areas not at present considered suitable for cultivation, and so raise concerns about soil erosion and other forms of degradation. There are also considerable areas of wetlands suitable for development. Together these have accounted for the rapid increases in land area cultivated in Africa since 1960 (Table 8.2). In Latin America there has been considerable development of upland rice in the Amazon areas of Brazil, where rice was used immediately after felling the forest as a crop that would help to control regrowth and clear the land of weeds before its use as pasture. The widespread expressions of concern about deforestation have meant that the production of upland rice in these areas has decreased considerably. The loss of production from upland rice has been more than compensated by the development of irrigated rice in southern Brazil (Table 8.2).

8.1.3 Cropping intensity

Increases in the land area cultivated to rice depend not only on the availability of irrigation water but also on the length of time for which the rice crop must occupy the ground to reach maturity. From the time of the spread of the short-duration 'champa' varieties from Vietnam about a millennium ago the advantages to be obtained from multiple cropping have been recognized. The suggestion (p.39) that famine in the Yangzi Provinces around 1850 might be relieved by growing 'the 30 to 40 day miracle rices' defied credibility. If such rices existed their yields would have been negligible. The problem in the breeding of short duration varieties is to maintain yield as the time from seed-to-seed is reduced. Thus the advantage to be obtained from short duration

Table 8.2. Rate of growth, production, area and yield of rice, Africa and South America, 1961/63 to 1991/93 and 1985/87 to 1993.

	Production (Mt)			Area (Mha)			Yield (t ha^{-1})		
Africa									
Years	61/63		91/93	61/63		91/93	61/63		91/93
	5.28		14.82	310		713	1.69		2.00
Rate (% $year^{-1}$)		3.16			2.63			0.53	
Years	85/87		91/93	85/87		91/93	85/87		91/93
	10.0		14.82	533		713	1.87		2.00
Rate (% $year^{-1}$)		6.47			4.82			1.65	
Africa excluding Egypt									
Years	85/87		91/93	85/87		91/93	85/87		91/93
	7.68		10.5	4.92		6.5	1.56		1.59
Rate (% $year^{-1}$)		3.65			3.63			0.02	
South America									
Years	61/63		91/93	61/63		91/93	61/63		91/93
	7.40		16.03	4.19		5.95	1.77		2.70
Rate (% $year^{-1}$)		2.46			1.16			1.39	
Years	85/87		91/93	85/87		91/93	85/87		91/93
	15.1		16.03	6.86		5.95	2.21		2.70
Rate (% $year^{-1}$)		0.99			–2.4			3.33	

Source: IRRI (1995c)

varieties is best measured in terms of yield per day. A further complication arises from varieties that are photoperiod sensitive, i.e. that will only flower when the daylength has a certain duration. An important aspect of the breeding of IR8 and the other high-yielding semi-dwarfs was to select for photoperiod insensitivity. The varieties produced would then be stable in their growth duration. IR8 showed advantages over many traditional varieties which required 160–170 days to mature; it was photoperiod insensitive, matured in 130 days, and produced about 70 kg ha^{-1} day^{-1}. Later IR varieties such as IR36, IR64 and IR72, which are now extremely widely grown, are also photoperiod insensitive, mature in 110 days, and produce about 90 kg ha^{-1} day^{-1} (Khush, 1987). Their advantage is that they not only allow a second or third crop to be grown where it was not previously possible, but they also use less water than slower maturing varieties. The great popularity of IR36, 64 and 72, which are also based on the semi-dwarf plant type, has stemmed not only from their shorter growth duration and high yield potential, but also from their widely acceptable grain quality and their broad-spectrum pest resistance.

It is difficult to quantify the contribution that the shorter duration varieties have made to rice production, and to savings in water use. Varietal effects are confounded with changes in management practices such as the change from transplanting to direct seeding, which also reduces the time required for the crop to mature. The advantage is not only to increase the intensity of rice cropping but also to increase the frequency of rice–non-rice crop multiple cropping. The shorter duration of the rice crop often means that the upland crop can take advantage of greater residual water left in the ground after the rice crop has been harvested.

The advantage is not only in the irrigated areas. Zandstra and his colleagues were able to show that with a short duration rice variety, and dry seeding the crop to ensure early establishment, double rice cropping became possible in several rainfed production systems (Zandstra *et al.*, 1981).

FAO (Alexandratos, 1995) estimate that the cropping intensity in East Asia (excluding China) will increase from 120% to 126% between 1988/89 and 2010, and in South Asia from 118% to 136%. FAO divides what has been referred to in the present text as Southeast Asia between South and East Asia. Their data refer to all crops and not only rice-based systems. Concerns about increased cropping intensity relate to the effects on pest problems, and the environmental effects associated with the higher levels of inputs required. These concerns are discussed further below.

8.2 Concerns about Water for Rice Production

There are several concerns about maintaining the existing water requirements for rice (Chapter 6, section 6.9). These include competing urban and industrial demand, the escalating costs, economic and social, of increasing water storages and distribution systems, the deterioration of existing systems due to siltation of reservoirs and canals, falling water tables in areas where pump irrigation systems are used, and rising water tables causing water logging and salinization where storage systems have been built.

Recent success in the increase of rice production in Bangladesh has stemmed from successful exploitation of groundwater by pump systems. Where much of the land is flooded each year so that groundwater supplies are replenished, as it is in Bangladesh, the system may be sustainable. The greater threat to sustainability is from flooding. In drier areas where exploitation of groundwaters has been a component of the green revolution it is unlikely that the technology is sustainable. Falling water tables and intrusion of saline water are widely observed. As water tables fall the costs of pumping increase.

Whatever the source of water for arid areas the problems of salinization pose a continuing threat. Seepage of water from reservoirs and canals raises water tables, often bringing saline water into the root zone of the rice crop. In arid regions major drainage works have always to be installed if the agricultural system dependent on water from the dam is to be sustainable. The seepage of water from reservoirs and canals can also cause waterlogging in the service area, with its attendant problems of difficult access to land, and development of incipient toxicities.

These concerns about the physical problems of maintaining and improving water supplies are often felt to be secondary in importance to the problems of who controls and manages existing systems. Where large water control systems have been built to manage the water from major rivers, international agreements are required. While some have been successfully negotiated, others have led to conflict. Some predict major conflicts are likely to arise in the future over control of water resources (Clarke, 1991; Carruthers, 1992). If the extent of irrigated land is to be increased, as is necessary if rice production is to grow to meet future demand, further sources of supply have to be found. As all obvious and readily developed sources in Asia have already been tapped, this requires major and expensive development of new irrigation schemes. At present rice prices the costs of such schemes (Table 7.6) are unlikely to be economic, unless the price of rice rises on the world markets (Rosegrant and Svendsen, 1993).

For many areas a problem at least as serious as a possible deficiency in water supplies is that of flood control. China and Bangladesh are particularly vulnerable, and it is probable that there will be further areas subject to serious flooding if deforestation continues in the catchments, where the rain which feeds the river systems of the lowlands falls. Land pressure and exploitation of the upper catchment areas for timber are the principle causes of deforestation. When the protective forest cover is lost the soils are exposed to erosion so that rainwater runs off into streams and rivers much more rapidly than when the forest protects the land. The accumulation of water becomes more than can be accommodated within the existing channels. The larger the river system the more devastating is the damage caused by the floods. Probably the worst problems have been in China, in the lower reaches of the Huanghe and Yangzi, but many other areas have also suffered. The beneficial effects of the floods have also to be recognized, as discussed in Chapter 6, section 6.4. Nevertheless the management of the major rivers, both to create better storages and to control flood damage, is a huge problem which will be an increasing threat to the sustainability of rice production. The costs, social and economic, as demonstrated by the intense debates surrounding the construction of dams in the Yangzi Gorges, will be considerable.

8.3 Soil Fertility

The most important beneficial effects of the floods have historically been the maintenance of soil nutrient supplies. The nutrients in the sediments deposited by the floods have been sufficient to support rice yields of 1–2 t ha^{-1} for very many years. Supplementing them with organic manures enabled yields of 2–3 t ha^{-1} to be maintained, and further supplementing them with inorganic fertilizers has enabled yields of 5–6 t ha^{-1} to be obtained. Under favourable conditions very much higher yields have been produced (Table 5.16). Sediments and manures normally contain all essential nutrients in amounts that are appropriately balanced in terms of plant requirements. Inorganic fertilizers are usually constituted to supply the elements most needed by the crop, and in most soil conditions this means nitrogen, phosphorus and potassium, but as yields are raised so the demand for secondary and micro-nutrients increases and deficiencies of these elements may then limit yields. With appropriate scientific support the additional nutrients required can be identified and fertilizers tailored to meet the needs, although the costs of the extra nutrients will increase the price of the fertilizers. Apart from a need for zinc in alkaline soils and some soils with higher than normal organic matter contents, there have been few deficiencies of elements other than N, P and K demonstrated to be necessary for wetland rice, although the incidence of such deficiencies is increasing (Chapter 5, section 5.4).

The price of fertilizers relative to the price of rice has been such that since about 1930 there has been a very substantial economic advantage to use of fertilizers for rice production, even when there have been multiple nutrient requirements to meet. Also as seen from the data collected in Chapter 5, to sustain rice yields at the levels necessary to feed present and future populations the use of increasing amounts of inorganic fertilizers is and will continue to be essential (Parish, 1993). Environmentalists and sociologists have often expressed concern about the effects of using increasing amounts of fertilizers. Environmental effects are discussed below. Regarding the social effects, as Lipton and Longhurst (1985) and Herdt (1987) have concluded, earlier fears of adverse social effects have not been realized. The evidence from many sources shows that the use of fertilizers with modern rice varieties benefits the landless and urban poor as well as the better-off (although by developed world standards still poor) rice farmers.

Concerns are often expressed about the extent of the global supplies of phosphorus and potassium, and how long these can continue to be exploited for fertilizer manufacture. Similarly concern is often expressed about the cost of energy supplies required to produce nitrogen fertilizers. Known sources of phosphorus and potassium are in fact sufficient for several hundred years at least (Stangel, 1976). At present costs of energy there is also little problem in maintaining inorganic nitrogen supplies to meet expected demand. However for nitrogen the opportunity exists to use biologically fixed nitrogen, or at least supplement applications of inorganic fertilizer with biologically fixed N.

It is of course highly desirable that maximum use should be made of biological nitrogen fixation. The case for returning organic wastes to wetland soils is less strong. All of the ‘outstanding farmers’ mentioned in Chapter 7 continued to use organic manures in substantial quantities as well as inorganic fertilizers. Nevertheless it has to be accepted that as indigenous soil nitrogen and other nutrient supplies are depleted by continued cropping, yields will only be increased to the levels needed by adding additional nutrients, rather than recycling such residues as are available. Unless adequate supplies of NPK in particular are added to the soil from external sources recycling crop residues and organic wastes generated on the farm will mean recycling poverty.

As discussed in Chapter 4 (section 4.2.3) declining soil organic matter levels are seldom a problem in wetland rice soils. There may sometimes be a problem about the quality of the organic matter, in that the nitrogen contained is less readily mineralized after several years of intensive cultivation (Cassman and Pingali, 1995b). However this observation needs much wider confirmation. Another problem arising from the use of organic manures is the enhancement of the rates at which methane is evolved. This is discussed below together with other environmental effects of rice farming.

Acidification and erosion are also seldom problems in wetland rice production, although they have their usual strongly detrimental effects when rice is grown under upland conditions. Although this is a serious concern for some of the poorest rice farmers who are dependent on upland rice production, it affects less than 1% of total rice production.

8.4 Concerns about Pests and their Control

As discussed in Chapter 4 (section 4.3) farmers have long been concerned about pest problems. The major concerns are that the widespread planting of single varieties will lead to major outbreaks of pests able to attack the variety, and that increases in cropping frequency will mean that the pest will be able to maintain itself throughout the year, and from one season to the next. The initial response to these fears was a greatly increased use of pesticides. However as varietal resistance has improved the need for pesticides has been substantially reduced. Understanding of the ecology of pests and their natural enemies has enabled insecticides to be used more judiciously and in much lower amounts without loss of yield (Ooi and Waage, 1994; Heong *et al.*, 1995).

In 1993 the total rice insecticides market was US$1114 million. However of this no less than 34% was spent in Japan, although Japan produces less than 3% of the world's rice (Woodburn, 1993). The greatly inflated value of rice to the Japanese farmer accounts for his concern to safeguard his crop against the slightest risk of loss of yield. Subsidizing of pesticide prices in Indonesia also led to unnecessary use of insecticides there, but when the resurgence problem was understood the subsidy was withdrawn, and insecticide sales fell. Yields continued to increase (Kenmore, 1991). However as the incidence of pests decreased a few farmers reverted to growing rice varieties with more desired eating qualities but susceptible to pests, notably the brown planthopper.

Consequently outbreaks have occasionally recurred in limited areas, but sufficient to renew fears of wider outbreaks.

Concerns about disease problems centre on the difficulties of managing the frequent emergence of new strains of diseases such as blast and bacterial blight, and identifying sources of resistance to viral diseases. Uniform and intensified planting provide a more suitable environment for the spread of diseases, as is the case for insect pests. Consequently disease problems have tended to become more frequent. Sales of fungicides have increased, for instance in India from US$15 million in 1981–83 to US$30 million in 1991–93. Again the market is heavily dominated by Japan, where sales of fungicides have doubled over the same period, but from US$225 million to US$450 million.

In contrast to the reduced problems with insect pests and reduced market for rice insecticides, weed problems and sales of herbicides have been increasing, in much the same way as fungicides. Expenditure on herbicides is about double that on fungicides (IRRI, 1995c) and again Japan accounts for more than half of the world market.

Thus in developing countries pest control remains largely dependent on varietal resistance, crop management, natural biological control, and for weeds and some insect pests, removal by hand. Combination of these methods with judicious use of pesticides in an integrated pest management (IPM) strategy is now the accepted way forward (Heong *et al.*, 1995). With improvements in breeding for pest resistance made possible by advances in biotechnology it is probable that much better control will be achieved in the future than in the past, in spite of the intensification of production.

Concerns about human health and environmental contamination are likely to persist as long as any pesticides are used. It is important that these concerns are recognized and that any measures taken to increase rice production do not introduce health risks or lead to serious problems arising from contamination of air, water or soils.

8.5 Global Warming and Rising Sea Level

Concern about global warming and its possible effects on sea level have not been confined to rice farming, but have been a matter of worldwide concern for the past decade (Jones *et al.*, 1988; Warrick *et al.*, 1993). In historical perspective there has been a relatively stable phase as far as temperature and sea level are concerned for the past three centuries. As shown in Table 3.1 sea level and temperature have fluctuated considerably for the past 15,000 years, but in the past 300 years temperatures have not varied by more than 0.6°C and sea level has not fluctuated by more than 0.6 m. However over the past 120 years temperature has risen rather consistently (Fig. 8.6). Several predictions of further changes in temperature and consequent sea level rise have

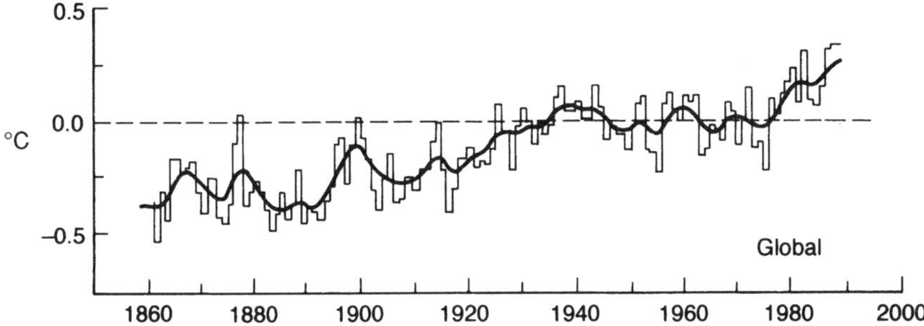

Fig. 8.6. The global temperature record 1860–1988. The values are departures (anomalies) from the 1950–1979 reference period (Warrick and Barrow, 1990).

been made. Estimates are mostly that by 2030 temperatures may rise between 1.1° and 1.9° and sea level between 0.14 m and 0.24 m (Fig. 8.7).

At least part of the cause of the upward trend in global warming is associated with the release of carbon dioxide from the burning of fossil fuels. However, given the magnitude of the fluctuations before the use of fossil fuels this remains controversial. The upward temperature change is not debatable, although some uncertainty relates to its effect on sea level. The increase in the level of carbon dioxide in the atmosphere may be expected to increase rice yields, as will the higher temperatures expected (Matthews *et al.* 1994a, b). Thus concern should be primarily related to the effects on sea level. This will further increase the depth of flooding in many of the flood-prone rainfed areas, cause increasing salt water intrusion into many coastal areas, and rising groundwater levels in many areas including some remote from the sea. Perhaps the most disastrous effects will arise from the damage caused when floodwaters and the sea meet in the way they did in Bangladesh in 1970.

There may well be other problems arising from climate change. Great uncertainty still surrounds the likely changes in rainfall patterns. Higher temperatures mean that more water will be evaporated and enter the atmosphere, and it is probable that there will be a general increase in precipitation. How much, where it will fall, and with what intensity remains uncertain, but there is significant probability that there will be increased flooding in the areas that at present receive higher rainfalls than average. In the arid areas receiving only irrigation water to grow rice the higher evaporation will also mean less water available for crops, and salinization problems will intensify.

Carbon dioxide is not the only gas causing global warming. The other significant greenhouse gases are methane and nitrous oxide. Both of these are evolved from rice fields, although only methane in amounts that have a significant impact on global warming (Neue *et al.*, 1995). Some concern must therefore be exercised about factors affecting methane emissions. These include particularly the use of organic manures, which breakdown under anoxic

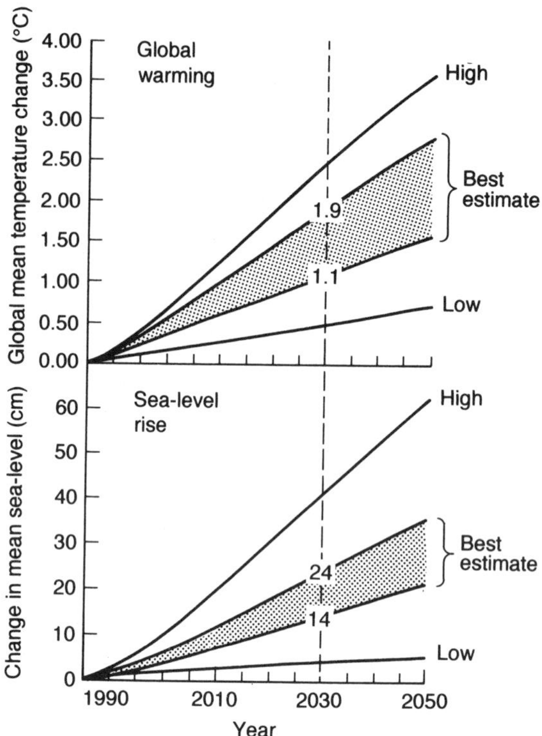

Fig. 8.7. Projected global warming and sea-level rise, 1990–2050. Based on analyses conducted by the Climatic Research Unit, University of East Anglia, UK (Warrick and Barrow, 1990).

conditions in wetland rice fields to release a large part of their carbon as methane. Present evidence suggests that about a quarter of the methane released to the atmosphere each year arises from rice fields. This problem has been given very little attention until recently, but evidence has been obtained that by agronomic practices such as mid-season drainage of rice fields, and incorporation of characteristics into rice cultivars which minimize the transport of methane to the atmosphere, the release of methane can be controlled (Sass, 1995).

8.6 Environmental Contamination and Loss of Biodiversity

Ever since the publication of *The Silent Spring* in 1962 there has been considerable concern about the effect that inorganic chemicals used in agriculture have on the environment. These fears were reinforced by the appearance of the Brundtland Report (Brundtland and Khalid, 1987). Two aspects in particular have been of continuing concern, the effects of inorganic fertilizers on the pollution of water, and the effects of pesticides on wildlife and human health.

Is there a case for serious concern about the use of inorganic chemicals on rice in Asia? Some believe that this is an unnecessary question, any use of inorganic chemicals being undesirable. However there are important differences between rice farming and farming of other crops. The background to these has been given in Chapter 4. Most important is the fact that nitrate is not normally formed under flooded conditions, and if added to flooded soils is promptly reduced to gaseous nitrogen, nitrous oxide and ammonium. Thus the problems of groundwater pollution by nitrates does not normally arise in the areas where wetland rice is grown. Most inorganic nitrogen added to rice fields is in the form of ammonium. As a positive ion it becomes attached to the exchange complex in the soil and so is relatively immobile. Some nitrous oxide may be formed during denitrification of added nitrate, or nitrate formed from soil nitrogen when the soil is allowed to dry. Much of the nitrous oxide escapes to the atmosphere, where it acts as a greenhouse gas. However its significance in relation to global warming is much less than that of carbon dioxide and methane, and it is more rapidly decomposed in the upper atmosphere than the other two greenhouse gases (Bouwman, 1990).

Phosphorus is the other element widely used in inorganic fertilizers and commonly believed to cause undesirable pollution of groundwaters. In fact phosphorus is almost always present as phosphate and strongly adsorbed in the soil. Most problems of pollution due to the use of inorganic phosphorus fertilizers have arisen when the fertilizer has been surface-applied and rainfall has caused surface runoff before the fertilizer has been in contact with the soil. Phosphate used on rice is normally incorporated in the soil before flooding and transplanting, so that little is lost in runoff. The concentration of phosphorus in flooded soils is higher than under oxidized conditions so that a potential exists for some movement to groundwaters. No significant problems directly attributable to such movement of phosphorus are known. In fact, the concentrations of phosphorus are mostly too low in paddy water to maintain an adequate population of nitrogen-fixing blue-green algae, or support growth of *Azolla* (Roger and Kurihara, 1991).

Although the amounts of fertilizers used on rice in South and Southeast Asia have been increasing, they are still much lower than the levels used in Japan and Korea (Fig. 5.3). It is generally more important to increase the use of inorganic fertilizers to raise income and hygiene standards and to improve health and environmental conditions in other ways than to seek to limit the use of inorganic fertilizer. Reducing fertilizer use is a commendable policy in most developed countries but inappropriate in most developing countries.

Pesticides have been and are a more serious source of concern. With pesticides the major pollution problem arises from their misuse and overuse. Several studies have shown that rice farmers using pesticides suffer serious health problems (Rola and Pingali, 1993; Pingali and Roger, 1994). Several factors contribute to this problem. One is the sale and use in tropical Asia of pesticides which are a known health hazard. Combined with this is the problem that farmers using the pesticides might not be able to read the instructions on the labels, or understand them, whether or not they were written in a language known to the user. Greater responsibility on the part of the pesticide companies, and greater efforts to improve knowledge and understanding of pesticide use, have alleviated these problems, at least in some regions. Nevertheless impact of pesticides on human health must remain a continuing concern. It will be many years before the level of pesticide use in tropical Asia approaches that in Japan. Hopefully before that time non-chemical control methods will have replaced use of toxic chemicals for pest control, and if not farmer education in pesticide use will be at a more advanced stage than at present.

The persistence of some pesticides in soils and waters under temperate conditions is known to pose problems in relation to environmental contamination. In tropical conditions the higher temperatures mean a more rapid breakdown of pesticides by biological, chemical and physical processes. Thus there should be less concern about environmental contamination arising from persistence in the tropics than in the temperate zone. Most pesticides in use since about 1980 are readily degraded in both temperate and tropical conditions. Nevertheless many concerns have been expressed about their effects, not least in relation to biodiversity.

The question of whether pesticides have a deleterious effect on biodiversity is not simple. The naive assumption is that because they are toxic and used to destroy some members of the biological community they must be undesirable. They certainly change the immediate structure of the community to which they are applied, but as discussed in Chapter 4 (section 4.4), removal of certain members of the community changes the ecological balance and can lead to an increase in the diversity of the community. The comparison of biological community composition at Ifugao in the Philippines, a pesticide free area, with that at IRRI, where pesticide use has been considerable, showed lower diversity in Ifugao than at IRRI (Fig. 4.17). There have been far too few quantitative studies of biodiversity for definite conclusions to be reached. While it is appropriate to show some concern at this stage, it remains an open question whether all pesticides can be considered damaging to biodiversity. Each pesticide has to be considered in terms of its overall ecological effects. Roger *et al.* (1994) have stressed the need for more research to alleviate concerns about the long-term effects of pesticides on the organisms present in rice fields.

Pesticide use has to be managed in such a way that damage to specific species is avoided. For instance some pesticides are water soluble and if allowed to contaminate water used for aquaculture may damage or destroy the fish population. If the water is used for drinking, then human health problems become a serious concern. The general population, as well as the rice farmers who have the greatest exposure, are put at risk. Concern is also frequently expressed about the loss of biodiversity within individual crop species. As far as rice is concerned, the measures which have been taken, described in Chapter 4 (section 4.4), should ensure that there is adequate preservation of biodiversity within the genus *Oryza*.

The spread of irrigation systems has meant that some water-borne diseases and their vectors have become more prevalent. This presents a further cause for concern (IRRI/PEEM, 1988), both as regards the threat to human health and to problems arising from attempts to control the vectors with chemicals.

8.7 Social and Economic Concerns

As long as rice holds an honoured position in the community its social status is assured. As long as rice farming is profitable its economic sustainability is assured. The threats to the social status of rice farming are few, although as commercial farming becomes the dominant form of rice production, as it has in Japan, so the social status gradually diminishes. In many developed countries the social status of farmers and agriculturists is much lower than that of rice farmers in Japan. As the Japanese economy increasingly develops the characteristics of Western economies the status of rice farmers and the social importance of rice may be expected to continue to fall. Changes in eating habits with rice playing a less dominant role will also tend to reduce the social importance of rice. The large subsidies given to rice production in Japan arose because of the social status of rice, and the efforts of the government to recognize the importance that the public attach to rice. The rate at which the social status of rice declines will largely determine the rate at which the subsidies decline.

The trend towards greater commercialization of rice farming may be expected to be followed throughout Asia. This trend will impact the sustainability of rice production because it will eventually diminish the readiness of governments to subsidize rice farming. It will however be many decades before that change occurs. In countries such as India and Indonesia, as well as China, per caput consumption of rice is rising, as the population become less undernourished. In time the changes in the economic status of these countries may follow that of Japan, where rice now constitutes 0.1% of gross domestic product (GDP). In India it constitutes 8.6%, in Indonesia 8.2%, and in China 10.0% (Hossain and Fischer, 1995). The significance of rice is only just showing signs of falling in Japan. The social and political status of rice and rice farmers may be expected to remain high in Japan for a decade or more, and longer in other rice-dependent countries.

The major concern about the economic sustainability of rice farming is that subsidies are necessary if the price of rice is to be kept low. The social, political and economic importance of rice in all of the major rice producing countries is such that their governments seek to ensure that sufficient rice is available at a price all can afford. For most of these countries where there is little land to be developed, and the majority of farms are less than 1 ha in size, production can only be increased by increasing inputs. As discussed in Chapter 7, the rice farmer needs to be reasonably assured of a market and a price for his rice that will make him a profit above the cost of his inputs, before he will endeavour to increase his yields. Governments are thus faced with the dilemma previously described of whether to subsidize the rice farmer to ensure adequate production, or to subsidize the price of rice to allow the rural and urban poor to obtain the rice they need to survive. Research and the development of more efficient production technology helped governments to escape this dilemma in the recent past. The extent to which this will be possible in the coming decades will be discussed in the final chapter.

9 Increasing and Sustaining Rice Production

9.1 Sources of Increased Sustainable Production

Rice production can be increased by increasing yields, or by bringing more land into use for rice production. Of these options, most major rice consuming countries have little choice but to increase yields, as there is almost no new land suitable for rice production not already in use. There is suitable land which might be developed for rice production, but not in those parts of the world where it is most needed. The production from new land can make an important contribution to rice supplies, provided that those countries where demand is greatest are able and willing to import rice. Japan, for instance, is economically well able to import rice, but prefers to be self-sufficient, or close to self-sufficient. For many other countries where rice has political and social as well as nutritional importance, there is a strong desire to maintain self-sufficiency in rice production. Several countries are reluctant to import rice for economic as well as political and social reasons. For these and others the highest importance is attached to continued yield increases.

As noted in the previous chapter there are several reasons to be concerned that the desired increases may not be realized on a sustainable basis. Greater use of water, fertilizers and pesticides may well lead to unsustainable practices and only short-term gains, unless care is taken to ensure that the additional imputs are used efficiently.

9.2 Increasing the Efficiency of the Rice Plant

An important step towards greater efficiency in input use is to improve the efficiency of the rice plant in converting the inputs to food and other useful products. Rice, in common with other crops, is a useful plant in so far as it converts water, carbon dioxide, sunlight and a number of plant nutrients into grain which can be used as food. The more efficiently the plant performs the conversions, the greater is the yield from a given area of land, and for a given level of input. The introduction of the semi-dwarf IR8 plant type was a significant advance, but there has been no further increase in yield potential since the start of the 'green revolution' in the 1960s. A major effort is therefore necessary to produce a plant type that will have a higher yield potential than the semi-dwarfs.

There appears to be little immediate prospect of increasing the photosynthetic capacity of the rice leaf or canopy (Evans, 1990). Hence any yield increase depends on more efficient use of the photosynthate, for example by increasing the number and size of the rice grains, and the time during which the grains are being filled (Kropff *et al.*, 1994). Scientists at IRRI are endeavouring to produce such a plant type, with in addition fewer but more productive tillers than existing semi-dwarfs. Such a plant type should be more efficient in the use of inputs and have a greater yield potential. Further, by making the environment around the tillers less well suited to the multiplication and spread of stem pathogens, damage due to sheath blight, and other stem pathogens, which are difficult to control, should be reduced. Progress in the development of a plant type with these features is being made at IRRI (Peng *et al.*, 1994; Khush, 1995).

Much is expected of the contribution that recent advances in biotechnology can make to improving and redesigning the rice plant (Khush and Toenniessen, 1991; Bennett *et al.*, 1994; Bennett, 1995). Improvement in pest resistance by transfer of genes from wild species to *O. sativa* is well advanced, and *O. minuta*, *australiensis* and *officinalis* have all been used for this purpose. Tissue culture techniques have been successfully employed to improve the salt tolerance of rice lines (Zapata *et al.*, 1991). Yield and morphological characteristics of the rice plant are under multigenic control, and less easy to manage by genetic modification. The most important gene loci controlling these characteristics have first to be identified, and progress is being made in this direction.

Other possibilities are also opened by the potential of biotechnology. These include the more rapid development of perennial rice, the introduction of allelopathic characteristics whereby the rice plant could inhibit the growth of weeds in its vicinity, the transfer of nitrogen-fixing genes to rice, and the development of apomictic rice lines which would enable the production of hybrid rice seed to be greatly simplified (Bennett, 1995). Perennial rice – from which *O. sativa* originated – offers advantages in the efficiency of production, but its effects on the persistence of pest problems are unknown and could be serious. Allelopathy offers considerable advantages in reducing yield losses due to weeds. Some rice varieties with allelopathic properties have already been identified (Fujui, 1992; Dilday *et al.*, 1994).

Considerable impact may be expected from research on rice biotechnology in the medium to long term. In the short term and probably until about 2025 most improvement to the rice plant will have to rely primarily on conventional plant breeding techniques, with biotechnology serving as a valuable additional tool for the plant breeder to use for the few areas where relatively rapid progress is possible.

9.3 Increasing the Efficiency of Water Use

The scope for increasing the efficiency with which water is used for rice production is considerable. As discussed in Chapter 6 (section 6.9) and illustrated in Fig. 6.2, the water actually used by the rice crop is often less than half of the water distributed in an irrigation system. Much of the waste is due to inefficient distribution systems, which fail to provide water when and where it is most needed. The problems in the management of irrigation systems have therefore come under a great deal of professional and public scrutiny.

Small communal irrigation systems managed by the farmers who use the water have generally been found to be more efficient than large schemes managed by engineers. The advantages to be obtained from involvement of the users in the management has therefore received much emphasis (Wade and Chambers, 1980; Reyes and Jopillo, 1986; Uphoff, 1986; Chambers, 1988). Farmer participation has certainly contributed to better management of many irrigation systems, although alone it is seldom sufficient to achieve a lasting and significant improvement in efficiency. Other measures, such as rehabilitation of the system through changes in design and water distribution, institutional changes in the management structure made to accommodate farmer participation, and the introduction of charges for water used, or better enforcement of the charging system when this is already in place, are also necessary to make a major impact (Rosegrant and Svendsen, 1993).

Many calculations of the area that dams are expected to serve when they are built have been found to be overly optimistic. This is probably the most common reason why national average yields of irrigated rice are well below potential yields. The miscalculation means that the supply from the system is normally inadequate to meet the needs of all the farms supposed to be served by the system. Many studies have shown that the farms which are more remote from the reservoir tend to receive less water than those near the offtake from the dam or main canal, and that their yields are correspondingly lower. Similarly farms near the offtake often receive and take more water than is necessary, reducing the efficiency of water use. Farmer involvement in water distribution and management systems can make a major impact in improvement of water use efficiency in such cases, but much further research is needed to determine the optimum management system for different irrigation schemes.

Important savings in the water needed to produce a rice crop have been shown to be possible by using intermittent irrigation, as practised in Japan (Fig. 6.1). This requires a degree of control of the water distribution system that is at present uncommon outside of Japan and a few other developed countries. To realize the advantages more widely will require considerable research to determine how best to ensure satisfactory land preparation and levelling, and management of the system.

The life of an irrigation system, and hence its long-term economic efficiency, is largely determined by the rate at which the reservoir and distribution canals become clogged by silt eroded from the catchment area of the dam, and the rate at which waterlogging and salinity develop in the service area. These problems mostly become serious in the medium term rather than short term, and to many politicians the short-term benefits may be an overriding concern. Economists and planners need to resist political pressures so that systems are designed on a realistic medium-, and preferably long-term, assessment of the costs and benefits. A more realistic appraisal of the costs may show that a proposed irrigation scheme is not economic, and so contribute to a further decline in the rate at which new dams are built. To make more water available for rice production much greater efficiency is needed in the operation of existing systems. The potential of alternative forms of irrigation must be fully explored, as well as the advantages that can be obtained by use of more efficient and shorter duration plant types, and better soil and crop management practices to produce higher yields.

Pump irrigation systems are an alternative to storage systems wherever there is a water table in reasonable proximity to the surface. They have been actively developed since about 1980, although many hand operated pumps have been used to irrigate rice for very much longer. Many pump irrigation systems are being developed under private ownership, which has the advantage that there is usually a more realistic assessment of the costs and returns than when the systems are publically owned. They have the disadvantage that it is seldom clear at the outset of development whether the groundwater resources which are being tapped are renewable or not. The sustainability of such systems is entirely dependent on the renewal of the underground water supplies. At present many such systems are withdrawing water with inadequate knowledge of the recharge of the supply, or whether withdrawal of water will allow penetration of the supply by saline groundwater or by sea water.

Much greater efficiency in water use can be obtained using sprinkler and drip irrigation systems, but the costs for their installation and operation are much greater than for flood irrigation (Stanhill, 1985). High-efficiency irrigation systems are also typically energy intensive. When water becomes excessively costly then methods of growing rice under water-deficit conditions, and adaptation of the rice plant to water stress, will need to be given much more attention. Several possibilities exist for modifying the rice plant to make it more tolerant to water deficits (O'Toole, 1982), but none of these is easy to exploit. Measures such as dry seeding of rainfed rice, to take advantage of the initial monsoon rains, also offer prospects for growing more rice with less water (Saleh and Bhuiyan, 1995). A major unsolved, or inadequately solved problem, is to control weeds when rice is dry seeded. Possibly breeding allelopathic characteristics into the rice plant will resolve this problem.

There are of course no generic solutions to the problems of assessing how the overall hydrology of a region will be affected by the introduction of an irrigation system, whether a storage or pump system. The social and economic implications of the introduction of improved irrigation facilities will also differ between regions within countries, and between countries. Rigorous studies of each catchment and each community affected are needed to assess the problems, and make rational decisions about the sustainability of many existing as well as proposed systems.

These assessments will make the real costs of dam building and irrigation system development much greater. They will be further increased for new irrigation systems if the full social costs of disruption to the communities living in the area are taken into account. Few new storage systems will be economic in relation to their agricultural value unless there is a significant increase in the price of rice, as Rosegrant and Svendsen (1993) have concluded. Thus there will need to be continued efforts to economize in the amounts of water used to produce the crop.

9.4 Enhancing and Using Nutrient Supplies More Efficiently

The nutrients that are essential for rice come from soil reserves, organic manures, sediments which are richer in nutrients than the soil, and inorganic fertilizers. Nitrogen is additionally obtained through biological conversion of atmospheric nitrogen into plant available forms. Soil reserves, while they were replenished by sediments deposited by flood waters and supplemented by biological nitrogen fixation, sustained rice production up to yields of the order of 1–2 t ha^{-1} for millennia. Yields increased when the value of organic manures was recognized, and increased further when inorganic fertilizers were used to supplement or replace organic manures.

As the discussion of the practices of outstanding rice farmers in Chapter 7 (section 7.4) shows, the better farmers will use both inorganic fertilizers and manures. To provide the needed nutrients for yields above about 4 t ha^{-1}, the amounts of manures necessary are too large to be readily incorporated in the soil. Composting can reduce the volume, and large compost heaps were often a feature of the better rice farms in many areas of China, Japan and Korea in the early days of the green revolution. However the labour involved in transporting organic material to the compost heap and then to the field has meant that the rate of use of compost, as well as of other organic manures, has fallen considerably.

As discussed in Chapter 4 (section 4.2) there are several sources of biologically fixed nitrogen which contribute to the long-term sustainability of rice farming. The largest contributions can come from green manures, including *Azolla*, but these are usually too small to produce rice yields much above 4 t ha^{-1}. As well as the contributions to biological nitrogen fixation (BNF) which may be made by *Azolla*, green manures and free-living nitrogen-fixing organisms, the possibility of enhancing nitrogen fixation on or in the vicinity of the rice root has been studied (Ladha *et al.*, 1993; Ladha and Reddy, 1995). Although several researchers have reported that they have been able to produce nodule like formations on rice roots containing nitrogen fixing organisms (Khush and Bennett, 1992), the likelihood that this will make significant contributions to the nitrogen requirements of high-yielding rice plants appears at present to be somewhat remote.

It is obviously desirable for economic as well as environmental reasons that research to maximize the contributions obtainable from BNF should be continued. However, the levels of fixation must be substantially increased before the nitrogen requirements of crops with yields in excess of 4 t ha^{-1} can be met.

Major improvements in fertilizer efficiency are likely to come from improvements in the rice plant which will enable it to gather nutrients more efficiently, and to continue to grow and use nutrients effectively when the plant is under stress due to water shortage or excess. Other improvements are possible by changing the fertilizer formulation, and the methods of application used by farmers. The advantages associated with soil incorporation as compared with broadcasting of the fertilizer in the paddy water have been demonstrated many times. The importance of additional nitrogen after heading of the rice plant has also been demonstrated. To apply fertilizer at such a late stage involves an extra awkward operation and the route for further progress may lie with research on slow-release nitrogen fertilizer formulations.

Research and extension efforts are needed to enable these opportunities for improving fertilizer efficiency to be realized. Research on integrated nitrogen management, in which nitrogen fertilizers are used in conjunction with organic manures, and in such a way that biological nitrogen fixation is not reduced, is also needed (De Datta and Buresh, 1989). The organic manures are the source of slow-release nitrogen, but alone will not be sufficient to sustain the yield levels desired.

As shown in Chapter 5, it is an inescapable fact that to sustain yields of rice of the order of 6 t ha^{-1} and more, heavy dressings of fertilizers will be needed. Otherwise the soils will be mined of nutrients, and the yields, far from increasing, will decline. Greatest efficiency in fertilizer use is obtained if the nutrients applied are adjusted to the needs of the plant, and to the levels which are supplied from the soil. In Australia and California 'precision fertilizing' is used for rice. In this system the fertilizer applied in different parts of a field is adjusted to meet the crop needs by means of computer controlled fertilizer applicators. Such techniques are a long way removed from what will be possible in the near future in most developing countries, where the immediate needs are for better characterization of soils in terms of nutrient requirement, and better control of commercial fertilizer supplies to ensure that the quality of fertilizer corresponds to what is stated on the label on the bag. Research is needed, focused on integrated nutrient management, and supported by strong extension and fertilizer quality monitoring services.

9.5 Controlling Pests More Efficiently

The prospects for controlling insect pests of rice using less chemicals are good (Heong and Sogawa, 1994; Heong *et al.*, 1995). Varietal resistance and biological control methods, and improved understanding of pest ecology, have already resulted in substantial reductions in the use of insecticides in tropical Asia, without any loss of yield. Advances in biotechnology will enable varietal resistance to be introduced in new rice varieties more quickly than in the past, and the resistance introduced from crosses using wild species should be more durable.

It is likely that control of weeds, and of some diseases, will require the continued and greater use of chemicals, until further research enables more economically efficient and environmentally friendly methods to take over. In the meantime it is essential that research to improve varietal resistance to diseases, and particularly viral diseases, is pursued vigorously. The potential for biological control of fungal diseases, recently demonstrated, must also continue to be explored. Otherwise the advantages to be obtained from a more efficient plant type, and more efficient use of water and fertilizers, may be lost.

Where and while pesticides have still to be used, all possible measures have to be taken to ensure their efficient and safe use. Improvement in the degree of responsibility shown by manufacturers and distributors regarding what is sold, and better education of farmers, should reduce the risks to farmers' health considerably. The controls exercised by governments in regard to pesticide use, and the advice available in relation to the safe use of pesticides, have improved considerably since the start of the green revolution, but scope for further improvement exists in many countries.

The increasing extent of open waterways associated with the expansion of irrigation has raised fears of rapid increases in human diseases, due to the addition of sites where water-borne vectors of disease can breed. Careful attention needs to be given to this possibility (IRRI/PEEM, 1988). However, it is interesting to note that diseases transmitted by water-borne vectors tend to be more prevalent in Africa, where there is little irrigation, than in Asia.

9.6 Finding More Land for Rice Production

There is little prospect of finding more land well suited for rice production in the major rice growing countries of Asia. However, elsewhere in the world there are considerable areas of wetlands which are highly suitable and not at present in agricultural use. In their assessment of the areas of flat wetlands of the world, and their distribution and agricultural potential, van Dam and van Diepen (1982) categorized them as highly, moderately, poorly, or not suited for rice production. Their categorization was made solely on the basis of soil and climate characteristics. In Asia, the only area of highly suitable soils not already intensively developed for rice production is in Central Asia, and much of the area drains into the Aral Sea, where irrigation development for rice and other crops has already aroused environmental concerns. There are however considerable areas of highly suitable land in Africa and South America not already in use. Whether these can in fact be developed and used for rice will depend on economic factors and social and environmental concerns.

9.6.1 Africa

Van Dam and van Diepen (1982) give a figure of 71 Mha for the area in Africa where the climate and soils are 'highly suitable' for rice. This may be compared with the 7 Mha actually used for rice in 1993. Although it is unlikely that anything approaching 71 Mha will be cultivated for rice in the immediate future, it is probable that the very high rate of increase in the rice area could continue for some time, given the rapidly growing demand. Between 1985/87 and 1991/93 the area cultivated for rice in fact expanded at 4.82% year^{-1} (Table 8.2), a much faster rate than has been recorded for any other part of the world.

The areas which van Dam and van Diepen describe as highly suitable are wetlands where rainfed lowland or irrigated rice could be grown. In addition there are many areas in Africa where cultivation of upland rice could be increased. Although there are some potential locations where irrigation could be developed economically, most of the expansion may be expected to be for rainfed and upland rice. Hence yields will be low. For Africa south of the Sahara they averaged 1.56 t ha^{-1} in 1985/87, and 1.59 t ha^{-1} in 1991/93. In most parts of Africa assured water supplies have still to be provided. The exception is Egypt, where the Nile continues to provide the necessary water. The silt which formerly nourished the land has been trapped behind the Aswan Dam, but has so far been effectively replaced by fertilizers. Yields in Egypt reached 5.8 t ha^{-1} in 1985/87 and 7.6 t ha^{-1} in 1991/93.

In so far as the green revolution has affected rice production in Sub-Saharan Africa, it has been as a result of cultivation of more land, encouraged by improvements in the rice varieties available for rainfed systems. One example is the impact that the variety ITA 257, developed at the International Institute of Tropical Agriculture (IITA), has had on rice production from the FuntuaMaigana area near Kaduna in northern Nigeria (see Box 9.1). Availability of a better rice variety and some government and World Bank support helped the area planted to rice to grow from under 10,000 ha in 1985 to 45,000 in 1987.

9.6.2 South America

In South America the situation is very different from that in Africa. The area used for rice declined by 2.4% year^{-1} between 1985/87 and 1991/93, while yields increased by 3.33% year^{-1} (Table 8.2). The driving force in the changes in South America has been the expansion of irrigated rice production, and the economic advantages of flood irrigated rice compared with upland production, even when, or particularly when, the upland rice is irrigated using overhead sprinklers, as is done in some parts of Brazil.

The largest area of rice in South America is in Brazil. Much of this area is in the previously forested areas in the Amazon basin, which after clearing have been cultivated for a few years while upland rice is grown. The advantages of flood irrigated rice production, and the pressures on Brazil to refrain from further deforestation of the Amazon region, have led to a rapid decline in the areas cultivated for upland rice. By contrast, in southeast Brazil the production of irrigated rice has been increasing. Changes elsewhere in South America have been relatively minor, although there is a substantial area which could be developed for irrigated rice in southern Brazil and the adjacent ('southern cone') countries. There has, for example, been a steady increase in the area used for irrigated rice in Uruguay, where yields between 4 and 6 t ha^{-1} have

Box 9.1. Rice production in the Funtua-Maigana region, Kaduna, Nigeria (Chaudhary, 1988).

Production of rice in the Funtua-Maigana region of northern Nigeria increased from 9000 t in 1985 to 98,536 in 1987. Most of the increase came from the planting of a much greater area to upland rice, due to the popularity of a new higher yielding variety of rice, ITA 257. The new variety was not only higher yielding, but of better eating quality than varieties traditionally grown, was resistant to blast, the most serious pest problem, and matured more quickly, and so largely escaped the drought problems which frequently afflict the varieties traditionally grown in the area. Other contributions to the rapid expansion of rice production have come from other improved varieties grown in rainfed wetlands (fadamas) in the area, and some support for the marketing of rice when the higher production caused local market prices to fall. The restriction imposed by the Nigerian Government on rice imports at that time also had a major influence in stimulating the planting of rice.

been obtained since 1970. Van Dam and van Diepen's area of highly suitable soils in South America is 55 Mha, which may be compared with the 5.9 Mha cultivated for rice in 1993.

9.6.3 Australia, USA and Europe

There are also some areas in Australia, USA and Europe which could be developed for rice production. At the prices prevailing at the end of the 20th century there are few if any of these areas where rice production could be developed economically. There are significant problems related to water availability for further irrigation, and in many areas problems related to waterlogging and salinization. Development of coastal and some other areas of wetlands for rice production is inhibited by concerns about loss of habitats important to the preservation of biodiversity. In these developed countries where much of the land well suited to rice production is already in use it is particularly important that existing areas reserved for wildlife conservation are respected.

9.7 Sustainability and the Rice Trade – Avoiding Local, Regional and National Rice Shortages

Prospects for the sustainability of future rice production do not depend only on the factors that determine yields at a particular location, and effects on the environment. Production is obtained most economically from those areas best suited to the crop, and where least degradation of resources is likely. Local, regional and national deficiencies can be alleviated by trade or transfer of rice. For examples the Malaysian Government has for many years preferred to concentrate on plantation crops which it can export, and to purchase its rice from neighbouring Thailand. India has relieved shortages in other parts of the country by transferring rice from the Punjab, where a drier climate and access to irrigation water make high yields easier to obtain on a regular basis, to eastern India and other areas where shortages occur.

In Chapter 7 Japanese and Thai government policies were presented as examples of how policies have affected rice production. The Japanese and Thai examples represent extremes – the Japanese government procurement price for rice in 1993 was \$US2000 t^{-1}, whereas the Thai government price was \$US156. For social and political reasons the Japanese government has sought to maintain internal production although it could advantageously import rice at a much lower price. By contrast Thailand has been able to produce a rice surplus at low cost, and adopted policies to maintain low production costs, so that it can export rice at highly competitive prices on the world market. Vietnam and Myanmar are following Thailand's example, and building a competitive rice export trade. Other Asian governments are seeking to maintain self-sufficiency in rice, although this objective may well change as alternative and more profitable labour opportunities arise, and the costs of rice production increase. South Korea has imported substantial quantities of rice in the early 1990s, as its other exports have grown.

At the present time there is a substantial volume of rice leaving Asia for Africa, South America and Europe. As demand in Asia increases, more of the excess production is likely to be used within Asia. China may well find that it needs to import rice as well as wheat.

Unless or until there are major advances in technology to enable higher yields to be obtained at lower input costs, the price of rice in China and other Asian countries is likely to increase significantly. While Asia is largely dependent on land-saving, high-yield, technology to increase production, Africa can increase production by using more land for rainfed rice, and South America by developing irrigation facilities. Rice production in developed countries, where economies of scale have been made possible by large farm sizes, could also expand if the price of rice rises. Global warming may also enable rice production to be developed in areas where temperatures are presently too low for rice production.

One economist (Carruthers, 1993) has already predicted that there will be a reversal of roles between western and eastern countries. As industrial development proceeds in many Asian countries, and agricultural prices increase, what are now developing countries will be able to import food from the west and export in return more cheaply produced manufactured goods.

The prospects for greater rice production in areas such as southern Brazil, Argentina and Uruguay in South America, parts of central Asia, and Australia, are good, as they are in Mozambique and several other African countries, if they can maintain political stability. Provided liberal trade policies are maintained, as they should be under the World Trade Agreement, prospects for greater world trade are considerable, and should help to limit the problems created by increasing demand.

9.8 Conclusions – The Future Sustainability of Rice Production

What then can be concluded? Can the extra rice that will be required in the course of the coming century be produced, and produced sustainably? Can it be produced economically where it will be needed, or will those who most need it be able to afford to import it?

In terms of limitations to production the difficulty posed by water availability may well be the most intractable problem. Construction of more dams and distribution systems will not be sufficient to resolve the difficulty. They may make more water available on a temporary basis, but will also lead to more problems due to waterlogging and salinity, and as with existing facilities will have lifetimes threatened by siltation. Pump systems offer extra water but will only provide a sustainable solution where they do not lead to falling water tables. Both storage and pump systems add to the problems of sustaining high yield levels because they provide water but not the silt which for many years provided the balanced supply of nutrients required to feed the rice crop.

Nutrients will need to be supplied in large amounts by using heavy dressings of inorganic fertilizers. Inefficient use of the fertilizers will lead to excessive costs, as will inefficient use of pesticides. Environmental problems will arise in areas where agrochemicals are not used efficiently. Control of environmental problems is mostly possible, but will add to the overall costs of production. This may be acceptable in developed countries, but in many developing countries governments are often likely to believe that they have higher priorities, unless it is obvious that human health is threatened more by pollution than malnutrition. Immediate problems ensuring sufficient rice supplies usually tend to be more pressing than ensuring that biodiversity is preserved. Fortunately the measures that need to be taken to produce more rice, and produce it more efficiently, largely coincide with measures that are needed to minimize environmental damage.

Continuing research and extension efforts are essential to enable more rice to be produced, and produced more efficiently and sustainably. A new rice plant type is needed with greater yield potential, and able to respond to water and nutrients more efficiently. It must be less susceptible to pest damage, and less prone to yield loss due to temporary non-flooded conditions. Existing irrigation systems need to be rehabilitated and used more efficiently. Nutrients must be supplied in the form and at the time when the plant requires them. Finally government policies must be sufficiently flexible to ensure that advantage is taken of the best production environments, even if that means importing rice from where it is produced most cheaply and efficiently.

If the trends in economic development which have characterized much of East and Southeast Asia in the closing years of the 20th century persist into the 21st century, there should be no difficulty in financing the necessary rice purchases. Governments and aid agencies must be prepared to provide special support to those areas which are disadvantaged by the greater production in favourable areas. The inhabitants of the disadvantaged areas must be supported so that economic pressures do not force them to migrate to towns and swell the ranks of the urban poor, nor extend their cropping into fragile lands unsuited to cultivation. They will need to be encouraged, and perhaps subsidized, to enable them to develop sustainable agricultural practices on their own land, or assisted with alternative opportunities for economic development.

When adequately funded, research and extension services can deliver. An internationally organized research service enabled the great challenge of the population explosion of the 20th century to be met successfully, at least until the 1990s. The greatest rate at which people have been added to the population of the world is occurring between 1990 and 2020. The green revolution has to continue through this period if the rice is to be produced to feed the additional two billion people who will be born. Without the extra rice a series of political, economic, social and biophysical problems will arise. Yet the investments in the research essential to meet the further challenges of the population explosion declined in the decade between 1980 and 1990 (Tribe, 1994). Without the research investment the rate of increase of rice yields and production will not match the rate of increase in demand. The price of rice will escalate, and conflicts over access to land and water intensify. In some areas rice production may increase in the short term, but at the expense of the environment and the long-term prospects of producing more rice.

The present store of knowledge about how rice may be produced sustainably is sufficient to avoid short-term, non-sustainable solutions, if that knowledge is applied. There is every reason to believe that the advances needed to provide sustainable solutions to the demand for more rice can be achieved, if sufficient investment is made to support the necessary research. Without that investment the world may well tumble over the Malthusian precipice into a period of pestilence and famine, and increasing conflicts over land and water.

Appendix 1: Rice yields in China during the Song, Ming and Qing Dynasties (Min Zong-dian, 1984).

Dynasty		Region	Period (Emperor Calendars)*	Yield† Ancient systems dan mu^{-1}	Modern system jin (grain) mu^{-1}	Reference
SONG	TAIHU	Suzhou	Northern Song	2–3 (rice)	360–540	*Collection of Fan Wenzheng's papers*
		Suzhou & Jiaxin	Souhern Song	2–3 (rice)	360–540	*Xu Gu Jin Kao* [Sequel to Historic Studies] vol. 16
		Zhebei & Sunan	Southern Song Xiaozong	2–3 (rice)	360–540	*Collection of Mr Zhizhai* vol. 44 *Guiyang Army's advice on agriculture*
		Huzhou	Southern Song Linzong	3 (rice)	540	Song Hui Yao Ji Gao – *Food*, Part 6 [collection]
		Zhebei & Sunan	Late Southern Song	5–6 (grain)	445–534	*Chi Tang Cun Gao* [Manuscripts kept in Chi Tang], vol. 5 *Ningguo's advice on agriculture*
		Average yield		2.5 (rice)	450	
	OTHER	Baozhou (Chingwan, Hebei Province)	Northern Song Zhengzong Tianxi	1.8–2 (grain)	160–178	*Song Hui Yao Ji Gao – Food*, Part 4 [collections]
		Ruzhou (Linru, Henan Province)	Northern Song Zhengzong Xianping	0.38 (grain)	34	*History of The Song Dynasty – Food*, Part A.4
		Ezhou (™bei ©ovince)	Southern Song Xiaozong	2–3 (grain)	178–627	*Shuang Xi Ji* [collection of papers] vol. 11 *Shanglin Ezhou Su* [Shanglin's record for Ezhou]
		Luyintian (Hubei Province)	Southern Song Xiaozong Chunxi	0.4 (grain)	71	*Song Hui Yao Ji Gao – Food*, Part 6 [collection]
		Muqu (Xiangyang, Hubei)	Southern Song Xiaozong Chunxi	0.3 (grain)	53–62	*Song Hui Yao Ji Gao – Food*, Part 6 [collection]
		Guiyang	Southern Song Xiaozong	1 (grain)	180	*Collection of Mr Zhizhai* vol. 44 *Guiyang Army's advice on agriculture*
		Hengzhou (Hebei)	Southern Song Xiaozong Chunxi	1 (rice)	180	*Sheng Zhai Ji* [collection] vol. 4 *Shi Gu Shu Yuan Tian Ji* [On farming by Shingu Institute]
		Huai River region	Southern Song Xiaozong	2 (grain)	178	*Song Hui Yao Ji Gao – Food*, Part 6 [collection]
		Xiouning	Southern Song	1.5 (rice)	270	*Lou Su Ji* [collection] vol. 10
		Yinzhou (Huizhou, Anhui Province)	Southern Song Xiaozong Chunxi	2 (rice)	360	*Xing' an Annals – Taxes* vol. 2
		Shaoxin (Zhejiang)	Zhenghe Xiaozong	2 (rice)	360	*Collection of Zhu Wen's papers* vol. 16 *On Famine Prevention methods*
		Mingzhou (Linbo, Zhejiang Province)	Northern Song Huizong Zhenghe (first eight years)	6–7 (rice)	534–623	*Song Hui Yao Ji Gao – Food*, Part 7 [collection]
		Mingzhou (Linbo, Zhejiang Province)	Northern Song Huizong Zhenghe (after the first eight years)	3–3.5 (grain)	267–312	*Song Hui Yao Ji Gao – Food*, Part 7 [collection]
		Fuzhou (Fujian Province)	Southern Song Xiaozong	2 (rice)	360	*Chun Xi's Three Mountains Annals* vol. 14 *Taxes*
		Average Yield		1.5 (rice)	270	
MING	TAIHU	Shanghai	Hongzhi	1.5–3 (rice)	435–870	*Tian Xian Jun Guo Li Bing Shu*, Part 9
		Songjiang	Wanli	2.5–3 (rice)	725–870	*Shi You Zhai Cong Shuo* [collection] vol. 14 *History* Part 10
		Suzhou	Late Ming	1–3 (rice)	290–870	*Tian Xian Jun Gu Li Bing Shu*, Part 10
		Haiyan	Tianqi	2–5 (rice)	725	*Haiyan Tu Jin* [Atlas of Haiyan]
		Huzhou	Late Ming	1.5–3 (rice)	435–870	*Shen's Agricultural Book*
		Huzhou		2 (rice)	580	*Zheng Yin Zhai's Collections*
		Average yield		2.3 (rice)	667	
	OTHER	Tianjin	Wanli	4–5 (grain)	654–705	*Nong Zhen Quan Shu* [agriculture book]
		Nanhai	Jiajing	2.5–5 (grain)	353–705	*Ho Wei Yan Jia Shun* [Collection of letters]
		Hengshire	Zhengde	2 (grain)	282	*Jun Zhi Tang's Everyday Manual*
		Average yield		2.5 (grain)	353	

Appendix 1: *continued*

Dynasty		Region	Period (Emperor Calendars)*	Yield† Ancient systems dan mu^{-1}	Yield† Modern system jin (grain) mu^{-1}	Reference
QING	TAIHU	Su Songjahu	Kangxi	1.5–2.5 (rice)	413–688	*Collections in Qie Wen Zhaj* vol. 15 *Finance* No 1
		Suhu	Qianlong	2 (rice)	650	*Huang Cao Jing Shi Wen Bian* [collection] vol. 38
		Jiangxi	Qianlong	1–3 (rice)	375–825	*Sou Shi Tong Kao* vol. 21, *Rice* Part 2
		Suzhou	Kangxi	1.5–3.6 (rice)	413–990	*Si Bian Lu Ji Yao* [collection] vol. 11
		Suzhou	Daoguang	2–3 (rice)	550–825	*The Essential Techniques for the People's Welfare*
		Songjiang	Daoguang	2–3 (rice)	550–825	*Pu Mao Nong Zi* [agricultural report]
		Songjiang	Guangxu	2 (rice)	550	*Songjian Annals*, The Year 10 of Guangxu
		Huating	Guangxu	2 (rice)	550	*Huating Annals* (Guangxu period)
		Tongxiang	Early Qing	2–3 (rice)	550–825	*Supplement to Agricultural Book*
		Jiaxing	Qianlong	2 (rice)	550	*Zhejiang Annals, Products Part 2, Jiaxing Region*
		Haiyan	Qianlong	2.4–2.5 (rice)	660–688	*Xu He De Ti Ben* (Qianlong Year 12)
		Huzhou	Early Qing	4–5 (rice)	1100–1375	*Supplement to Agricultural Book*
		Huzhou	Kangxi	2 (rice)	550	*Report of Farmer's Situation in Three Counties*
		Average yield		2 (rice)	550	
	OTHER	Tianjin	Kangxi	3–4 (grain)	405–540	*Ji Fu Tong Zhi* [annals]
		Jingjin	Yongzheng	5–7 (grain)	675–945	*Ji Fu He Dao Shu Li Cong Shu* [series on rivers and irrigation systems]
		Songxian	Qianlong	3 (grain)	405	*Song Min Zhong Tian Shuo* [on farming]
		Yidu	Qianlong	4–6 (grain)	540–810	*So Shi Tong Kao, Qingzhou's Products*
		Sichuan	Yongzheng	7 (grain)	945	*Report to the Emperor Yongzheng from the Governor of Guangdong on 4 Oct, Yongzheng 12*
		Pengxian	Guangxu	1.6–3.6 (rice)	440–990	*Annals of Pengxian County* (Guangzu Year 4)
		Hunan	Daoguang	2.8–4.8 (grain)	372–578	*Collection of Miscellaneous Papers*
		Huangmei	Guanxu	5–6 (grain)	675–810	*Annals of Huangmei County* vol. 6
		Linchuan	Guangxu	2–4 (grain)	270–550	*Fu Jun Nong Can Kao Lue* [a preliminary account on agricultural products in Fu Jun]
		Chaoxian	Daoguang	2–3 (grain)	270–405	*Annals of Chaoxian County* Supplements
		Fuzhou	Guangxu	2–4 (grain)	270–540	*Agricultural Journal* vol. 268
		Gangzhou	Kangxi	4 (grain)	550	*Guangdong Xin Yu* [Guangdong News]
		Guizhou	Qianlong	3–4 (rice)	405–675	*Annals of Guizhou* vol. 36
		Average yield		3 (grain)	405	

*Northern Song, AD 960–1126; Southern Song, AD 1127–1279; Ming, AD 1368–1644; Qing, AD 1644–1911.

†1 dan mu^{-1} = ~ 750 kg ha^{-1}; 1 jiu mu^{-1} = ~ 7.5 kg ha^{-1}.

The yields reported are probably annual yields, and so may refer to the yields obtained from 1, 2 or 3 crops.

The ratio of rice (husked and polished) to grain (whole rice) lies between 0.5 and 0.75.

Some of the problems of converting older Chinese records to modern units are discussed on p.38.

References

Abedin Mian, M.J., Blume, HP., Bhuiyan, Z.H. and Eaqub, M. (1991) Water and nutrient dynamics of a paddy soil in Bangladesh. *Zeitschrift für Pflanzenernährung und Bodenkunde*, 153, 93–99.

Abernethy, C.L. (1994) Sustainability and growth. In: Heim, F. and Abernethy, C.L. (eds) *Irrigated Agriculture in Southeast Asia beyond 2000*. International Irrigation Management Institute, Colombo, Sri Lanka, pp. 63–73.

Abrol, I.P. (1990) Fertility management of Indian soils – a historical perspective. *Transactions of the 14th International Congress of Soil Science* 5, 203–207.

Abrol, I.P. and Sehgal, J.L. (1994) Degraded lands and their rehabilitation in India. In: Greenland, D.J. and Szabolcs, I. (eds) *Soil Resilience and Sustainable Land Management*. CAB International, Wallingford, UK, pp. 129–144.

Adachi, K. and Inoue, H. (1991) The effects of puddling on the percolation control of a paddy field. In: *Soil Management for Sustainable Rice Production in the Tropics*. IBSRAM Monograph No. 2. IBSRAM, Bangkok, Thailand, pp. 331–346.

Alagh, Y.K. (1990) Agro-climatic planning and regional development. *Indian Journal of Agricultural Economics* 45, 244–268.

Alazard, D. (1985) Stem and root nodulation in *Aeschynomene* spp. *Applied and Environmental Microbiology* 50, 732–734.

Alazard, D. and Becker, M. (1987) *Aeschynomene* as green manure for rice. *Plant and Soil* 101, 141–143.

Alexandratos, N. (ed.) (1988) *World Agriculture Towards 2000: an FAO Study*. Belhaven Press, London, UK.

Alexandratos, N. (ed.) (1995) *World Agriculture: Towards 2010. An FAO Study*. FAO, Rome, Italy, and J. Wiley, Chichester, UK.

Allen, O.N. and Allen, E.K. (1981) *The Leguminosae: a Source Book of Characteristics, Uses, and Nodulation*. Univeristy of Wisconsin Press, Madison, Wisconsin, USA.

Ando, S. (1975) Effect of crude organic matter applied to the excessively-draining low-yield paddy field. *Bulletin of the Kagawa Agricultural Experiment Station* 27, 1–81.

Andriesse, J.P. (1988) *Nature and Management of Tropical Peat Soils.* FAO Soils Bulletin No. 59. FAO, Rome, Italy.

App, A., Watanabe, I., Alexander, M., Ventura, W., Daez, C., Santiago, T. and De Datta, S.K. (1980) Nonsymbiotic nitrogen fixation associated with the rice plant in flooded soils. *Soil Science* 130, 283–289.

App, A., Santiago, C., Daez, G., Menguito, C., Ventura, W., Tirol, A., Po, J., Watanabe, I., De Datta, S.K. and Roger, P. (1984) Estimation of the nitrogen balance for irrigated rice and the contribution of phototrophic nitrogen fixation. *Field Crops Research* 9, 17–27.

App, A., Watanabe, I., Ventura, T.S., Bravo, M. and Jurey, C.D. (1986) The effect of cultivated and wild rice varieties on the nitrogen balance of flooded soil. *Soil Science* 141, 448–452.

ARC (1985) Propagation and agricultural use of *Azolla* in Vietnam by Research Centre of *Azolla*, Hanoi, Vietnam. In: *Proceedings, 16th session, International Rice Commission.* FAO, Rome, Italy, pp. 257–264.

Artzy, M. and Hillel, D. (1988) A defence of the theory of progressive soil salinisation in ancient southern Mesopotamia. *Geoarchaeology* 3, 235–238.

Awan, N.M. and Latif, M. (1982) *Technical, Social, and Economic Aspects of Water Management in Salinity Control and Reclamation Project Number 1 in Pakistan.* Centre of Excellence in Water Resources Engineering, University of Engineering and Technology, Lahore, Pakistan.

Ayers, R.S. and Westcot, D.W. (1989) *Water Quality for Agriculture.* FAO, Rome, Italy.

Bancroft, J. (1884) Food of the aborigines of central Australia. *Proceedings Royal Society of Queensland* 1, 104–107.

Barker, R. and Cordova, V.G. (1978) Labor utilization in rice production. In: *Economic Consequences of the New Rice Technology.* IRRI, Los Banos, Philippines, pp. 113–136.

Barker, R. and Herdt, R.W., with Rose, B. (1985) *The Rice Economy of Asia.* Resources for the Future, Washington, DC.

Becker, M., Ladha, J.K. and Ottow, J.C.G. (1990) Growth and nitrogen fixation of two stem nodulating legumes and their effect as green manure in lowland rice. *Soil Biology and Biochemistry* 22, 1109–1119.

Becker, M., Ali, M., Ladha, J.K. and Ottow, J.C.G. (1995) Agronomic and economic evaluation of *Sesbania rostrata* green manure establishment in irrigated rice. *Field Crops Research* 40, 135–141.

Beecher, H.G. (1991) Effect of saline water on rice yields and soil properties in the Murrumbidgee Valley. *Australian Journal of Experimental Agriculture*, 31, 819–823.

Begg, C.B.M., Kirk, G.J.D., MacKenzie, A.F. and Neue, HU. (1994) Root-induced iron oxidation and pH changes in the rice rhizosphere. *New Phytologist*, 128, 469–477.

Bellwood, P. (1992) Southeast Asia before history. In Tarling, P. (ed.) *The Cambridge History of Southeast Asia.* Cambridge University Press, Cambridge, UK, pp. 55–136.

Bennett, J. (1995) Biotechnology and the future of rice production. *Geojournal* 35, 333–335.

Bennett, J., Brar, D.S., Khush, G.S., Huang, N. and Setter, T.L. (1994) Molecular approaches. In: Cassman, K.G. (ed.) *Breaking the Yield Barrier.* IRRI, Los Banos, Philippines, pp. 63–76.

Bhatti, M.A. and Kijne, J.W. (1992) Irrigation management potential of paddy rice production in Punjab of Pakistan. In: Murty, V.V.N. and Koga, K. (eds) *Soil and*

Water Engineering for Paddy Field Managment. Asian Institute of Technology, Bangkok, Thailand, pp. 355–366.

Bhuiyan, N.I. and Islam, M.M. (1986) Sulphur deficiency problems of wetland rice soils in Bangladesh agriculture. In: Portch, S. and Hussain, S.G. (eds) *Sulphur in Agricultural Soils.* Bangladesh Agricultural Research Council and The Sulphur Institute, Dacca, Bangladesh, pp. 85–116.

Bhuiyan, S.I. (1984) Groundwater use for irrigation in Bangladesh: the prospects and some emerging issues. *Agricultural Administration* 16, 181–207.

Bhuiyan, S.I. (1992) Water management in relation to crop production: case study on rice. *Outlook on Agriculture* 21, 292–300.

Bhuiyan, S.I. (1993) Irrigation sustainability in rice-growing Asia. *Canadian Water Resources Journal* 18, 39–52.

Bhuiyan, S.I. (ed.) (1994) *On-farm Reservoir Systems for Rainfed Ricelands.* IRRI, Los Banos, Philippines.

Bhuiyan, S.I. and Undan, R.C. (1986) Drainage in rice culture in the Asian humid tropics. In: *Drainage Symposium Proceedings.* ILRI, Wageningen, Netherlands.

Bhuiyan, S.I., Wickham, T.H., Sen, C.N. and Cablayan, D. (1979) Influence of water related factors on land preparation, cropping intensity, and yield of rainfed lowland rice in Central Luzon, Philippines. In: *Rainfed Lowland Rice,* IRRI, Los Banos, Philippines, pp. 215–234.

Bhuiyan S.I., Sattar, M.A. and Khan, M.A.K. (1995) Improving water use efficiency in rice irrigation through wet-seeding. *Irrigation Science* 16, 1–8.

Bouldin, D.R. (1986) The chemistry and biology of flooded soils in relation to the nitrogen economy in rice fields. In: De Datta, S.K. and Patrick, W.H. (eds) *Nitrogen Economy of Flooded Rice Fields.* Martinus Nijhoff Publishers, Dordrecht, Netherlands, pp. 1–14.

Bouwman, A.F. (ed.) (1990) *Soils and the Greenhouse Effect.* J. Wiley, Chichester, UK.

Boyden, S. (1987) *Western Civilisation in Biological Perspective: Patterns in Biohistory.* Oxford University Press, London, UK.

Brady, N.C. (1982) Rice soils of the world: their limitations, potentials and prospects. *Transactions 12th International Congress Soil Science* 1, 42–66.

Brammer, H. and Brinkman, R. (1990) Changes in soil resources in response to gradually rising sea level. In: Scharpenseel, H.W., Shomaker, M. and Ayoub, A. (eds) *Soils on a Warmer Earth.* Elsevier, Netherlands, pp. 145–156.

Bray, F. (1984) *Science and Civilisation in China, Vol. 6, Part 2, Agriculture.* Cambridge University Press, Cambridge, UK.

Bray, F. (1986) *The Rice Economies: Technology and Development in Asian Societies.* Basil Blackwell, Oxford, UK.

Brewbaker, J.L. (1985) Leguminous trees and shrubs for Southeast Asia and the South Pacific. In: Blair, G.J., Ivory, D.A. and Evans, T.R. *Forages in Southeast Asia and South Pacific Agriculture.* Australian Centre for International Agricultural Research, Canberra, Australia, pp. 43–50.

Brewbaker, J.L. and Glover, N. (1988) Woody species as green manure crops in rice-based cropping systems. In: *Green Manure in Rice Farming.* IRRI, Los Banos, Philippines, pp. 29–43.

Brewbaker, J.L. and Hutton, E.M. (1979) Leucaena, versatile tropical tree legume. In: Ritchie, G.A. (ed.) *New Agricultural Crops.* Westview Press, Boulder, Colorado, pp. 207–259.

Brohier, R.L. (1975) *Food and the People.* Lake House Investments Ltd., Colombo, Sri Lanka.

Bronson, B. (1978) Angkor, Anarudhapurar, Prambanon and Tikal: Mayan subsistence in an Asian perspective. In: Harrison, P.D. and Turner, B.L. (eds) *Pre-Hispanic Mayan Agriculture.* University of New Mexico Press, Albuquerque, New Mexico, USA, pp. 255–300.

Brundtland, G.H. and Khalid, M. (1987) *Our Common Future. Report of the World Commission on Environment and Development.* Oxford University Press, Oxford, UK.

Buck, J.L. (1937) *Land Utilization in China.* Volume 1. University of Nanjing, Nanjing, China, reprinted 1956 by Council on Economic and Cultural Affairs, New York, USA.

Buddenhagen, I.W. and Persley, G.J. (1978) *Rice in Africa.* Academic Press, London, UK.

Buresh, R.J. and De Datta, S.K. (1991) Nitrogen dynamics and management in rice–legume cropping systems. *Advances in Agronomy* 45, 1–59.

Buresh, R.J., Woodhead, T., Shepherd, K.D., Flordelis, E. and Cabangon, R.C. (1989) Nitrate accumulation and loss in a mungbean/lowland rice cropping system. *Soil Science Society America Journal* 53, 477–482.

Burford, J.R. and Bremner, J.M. (1972) Is phosphate reduced to phosphine in water-logged soils? *Soil Biology and Biochemistry* 4, 489–495.

Burley, J. (1985) *Global Needs and Problems of the Collection, Storage and Distribution of Multipurpose Tree Germplasm.* ICRAF, Nairobi, Kenya.

Buth, G.M. and Saraswath, K.S. (1972) Antiquity of rice cultivation. In: Ghouse, A.K.M. and Yunus, Mohd. (eds) *Research Trends in Plant Anatomy.* New Delhi, India, pp. 33–38.

CAAS (eds) (1986) *The Science of Rice and Rice Cultivation in Ancient China.* Agricultural Publishing House, Beijing, China.

Cai Renkui, Ni Dashu and Wang Jianguo (1995) Rice–fish culture in China: the past, present and future. In: Mackay, K. (ed.) *Rice–fish Culture in China.* International Development Research Centre (IDRC), Ottawa, Canada, pp. 3–14.

Cannell, R.Q. and Lynch, J.M. (1984) Possible adverse effects of decomposing crop residues on plant growth. In: *Organic Matter and Rice.* IRRI, Los Banos, Philippines, pp. 455–475.

Carpenter, A.J. (1978) The history of rice in Africa. In: Buddenhagen, I.W. and Persley, G.J. (eds) *Rice in Africa.* Academic Press, London, pp. 3–10.

Carruthers, I. (1992). Focus on water – look at it this way. *Outlook on Agriculture* 21, 241–245.

Carruthers, I. (1993) Going, going, gone! Tropical agriculture as we knew it. *Tropical Agriculture Association, Newsletter* 13(3), 1–5.

Carson, R. (1966) *The Silent Spring.* The Riverside Press, Cambridge, Mass., USA.

Cassman, K.G. (ed.) (1994) *Breaking the Yield Barrier.* IRRI, Los Banos, Philippines.

Cassman, K.G. and Pingali, P.L. (1995a) Extrapolating trends from long-term experiments to farmers' fields: the case of irrigated rice systems in Asia. In: Barnet, V., Payne, R. and Steiner, R. (eds) *Agricultural Sustainability: Economic, Environmental and Statistical Considerations.* J. Wiley, Chichester, UK, pp. 63–84.

Cassman, K.G. and Pingali, P.L. (1995b) Intensification of irrigated rice systems: learning from the past to meet future challenges. *GeoJournal* 35, 299–306.

Cassman, K.G., De Datta, S.K., Olk, D.C., Alcantara, J.M., Samson, M.I., Descasota, J.P. and Dizon, M.A. (1995) Yield decline and the nitrogen economy of long-term experiments on continuous, irrigated rice systems in the tropics. In: Lal, R. and Stewart, B.A. (eds) *Soil Management: Experimental Basis for Sustainability and Environmental Quality*. Lewis/CRC Publishers, Boca Raton, Florida, USA, pp. 181–222.

Catling, D. (1993) *Rice in Deepwater*. Macmillan Press, London.

Catling, H.D., Hobbs, P.R., Islam, Z. and Alam, B. (1983) Agronomic practices and yield assessments of deepwater rice in Bangladesh. *Field Crops Research* 6, 109–132.

Centre for Science and Environment (1991) *Floods, Floodplains and Environment, A Citizens Report*. Centre for Science and Environment, New Delhi, India. (Quoted by Abrol and Sehgal, 1994.)

Chabrolin, R. (1977) Rice in West Africa. In: Leakey, C.L.A. and Wills, J.B. (eds) *Food Crops of the Lowland Tropics*. Oxford University Press, London, 7–26.

Chambers, R. (1988) *Managing Canal Irrigation*. Oxford and IBH Publishing, New Delhi, India.

Chandler, D. (1983) *A History of Cambodia*. Westview Press, Boulder, Colorado.

Chandler, R. (1985) *An Adventure in Applied Science*. IRRI, Los Banos, Philippines.

Chang, T.T. (1976a) The rice cultures. *Philosophical Transactions of the Royal Society, London*, B275, 143–157.

Chang, T.T. (1976b) The origin, evolution, cultivation, dissemination and diversification of Asian and African rices. *Euphytica* 25, 425–441.

Chang, T.T. (1983) The origins and early cultures of the cereal grains and food legumes. In: Keightley, D.N. (ed.) *The Origins of Chinese Civilization*. University of California Press, Berkeley, California, USA, pp. 65–94.

Chang, T.T. (1985) Crop history and genetic conservation: rice – a case study. *Iowa State Journal of Research* 59, 425–456.

Chang, T.T. (1987) The impact of rice on human civilization and population expansion. *Interdisciplinary Science Reviews* 12, 63–69.

Chang, T.T. and collaborators (1982) The conservation and use of rice genetic resources. *Advances in Agronomy* 35, 37–91.

Chang, T.T. and Vaughan, D.A. (1991) Conservation and potentials of rice genetic resources. *Biotechnology in Agriculture and Forestry* 14, 531–552.

Chaudhary, R.C. (1988) Increased rice production in Fantua-Maigana Belt of Kaduna Agricultural Development Project: a success story. *Rural Development in Nigeria* 3, 50–57.

Chaudhuri, K.N. (1991) Ancient irrigation and water management in the Indian Ocean region. *Endeavour (New Series)* 15, 66–71.

Chen Jia-fang and Li Shi-ye (1981) Some characteristics of high fertility paddy soils. In: Institute of Soil Science, Academia Sinica (ed.) *Proceedings of Symposium on Paddy Soil*. Springer Verlag, Berlin, pp. 20–30.

Chen Li-zhi (1988) Green manure cultivation and use for rice in China. In *Green Manure in Rice Farming*. IRRI, Los Banos, Philippines, pp. 63–70.

Cheng, B.T. and Ouellette, G.J. (1971) Manganese availability in soil. *Soils and Fertilizers* 34, 589–595.

Cholitkul, W. and Sangtong, P. (1988) Paddy soil fertility and fertilizer use in Thailand. In: *Symposium on Paddy Soil Fertility, Country Reports*. IBSRAM, Bangkok, Thailand, pp. 75–92.

CIAT (1984) Upland rice in Latin America. In: *An Overview of Upland Rice Research*. IRRI, Los Banos, Philippines, pp. 93–120.

Clarke, R. (1991) *Water: the International Crisis.* Earthscan Publications, London, UK.
Conklin, H.C. (1980) *Ethnographic Atlas of Ifugao.* Yale University Press, New Haven, Conn., USA.
Coombs, J. (1984) Prospects for improving photosynthesis for increased crop productivity. *Advances in Photosynthesis Research* 4, 85–94.
Crosson, P. and Anderson, J.R. (1992) Resources and global food prospects: supply and demand for cereals to 2030. *World Bank Technical Paper* No. 184. World Bank, Washington, DC.
David, C.C. (1991) The world rice economy: challenges ahead. In: Khush, G.S. and Toenniessen, G.H. (eds) *Rice Biotechnology.* CAB International, Wallingford, UK, pp. 1–18.
David, C.C. and Otsuka, K. (1990) The modern seed-fertilizer technology and adoption of labour saving technologies: the Philippine case. *Australian Journal of Agricultural Economics* 34, 132–146.
David, C.C. and Otsuka, K. (1994) *Modern Rice Technology and Income Distribution in Asia.* Lynne Rienner, Boulder, Colorado, USA.
Davidson, B. (1972) *Africa, History of a Continent.* Spring Books, London, UK.
De Datta, S.K. (1970) Rice races raised. *Nature (London)* 227, 230.
De Datta, S.K. (1981) *Principles and Practices of Rice Production.* J. Wiley, Chichester, UK.
De Datta, S.K. (1983) Phosphorus requirements and phosphorus fertilization of lowland rice. *Proceedings 3rd International Congress on Phosphorus Compounds.* IMPHOS, Paris, France, pp. 401–417.
De Datta, S.K. (1985) Nutrient requirement for sustained high yields of rice and other cereals. *Proceedings 19th Colloquium International Potash Institute.* IPI, Berne, Switzerland, pp. 71–94.
De Datta, S.K. (1989) Rice. In: Plucknett D.L. and Sprague, H.B. (eds) *Detecting Mineral Nutrient Deficiencies in Tropical and Temperate Crops.* Westview Press, London, UK, pp. 41–55.
De Datta, S.K. and Buresh, R.J. (1989) Integrated nitrogen management in irrigated rice. *Advances in Soil Science* 10, 143–169.
De Datta, S.K. and Malabuyoc, J. (1976) Nitrogen response of lowland and upland rice in relation to tropical environmental conditions. In: *Climate and Rice.* IRRI, Los Banos, Philippines, pp. 509–539.
De Datta, S.K., Gomez, K.A. and Descolsota, J.P. (1988) Changes in yield response to major nutrients and in soil fertility under intensive rice cropping. *Soil Science* 146, 350–358.
De Datta, S.K., Trevitt, A.C.F., Freney, J.R., Obcemea, W.N., Real, J.G. and Simpson, J.R. (1989) Measuring nitrogen losses from lowland rice using bulk aerodynamic and nitrogen-15 balance methods. *Soil Science Society of America, Journal* 53, 1275–1281.
Delvert, J. (1961) *Le Paysan Cambodgien.* Mouton, Paris, France.
Dempster, J.I.M. and Brammer, H. (1992) Flood action plan – Bangladesh. *Outlook on Agriculture* 21, 301–305.
Devai, I. and Delaune, R.D. (1995) Evidence for phosphine production and emission from Louisiana and Florida marsh soils. *Organic Geochemistry* 23, 277–279.
Dexter, A.R. and Woodhead, T. (1985) Soil mechanics in relation to tillage, implements and root penetration in lowland soils. In: *Soil Physics and Rice.* IRRI, Los Banos, Philippines, pp. 261–282.

Dilday, R.H., Lin, J. and Yan, W. (1994) Identification of allelopathy in the USDA–ZARS rice germplasm collection. *Australian Journal of Experimental Agriculture* 34, 907–910.

Dobermann, A., Goovaerts, P. and Neue, H.U. (1996) Sources of soil variation in two tropical lowland rice fields. *Soil Science Society America, Journal* in press.

Dost, H. and van Breemen, N. (eds) (1982) *Proceedings of the Bangkok Symposium on Acid Sulphate Soils.* ILRI Publication No. 31. International Institute for Land Reclamation and Drainage, Wageningen, Netherlands.

Doyle, P.T., Devendra, C. and Pearce, G.R. (1986) *Rice Straw as a Feed for Ruminants.* International Development Programme of Australian Universities and Colleges, Canberra, Australia.

Dregne, H.E. (1983) *Desertification of Arid Lands.* Harwood Academic Publishers, New York.

Dreyfus, B.L. and Dommergues, Y.R. (1981) Nitrogen fixing nodules induced by *Rhizobium* on the stems of the tropical legume *Sesbania rostrata. FEMS, Microbiology Letters* 10, 313–317.

Dumanski, J. (1993) *Sustainable Land Management for the 21st Century. Vol. 1.* IBSRAM, Bangkok, Thailand and Agriculture Canada, Ottawa, Canada.

Eaglesham, A.R.J., Ellis, J.M., Evans, W.R., Feischman, D.E., Hungria, M. and Hardy, R.W.F. (1990) The first photosynthetic nitrogen fixing *Rhizobium*: characteristics. In: Gresshof, P.M., Roth, L.E., Stacey, G. and Newton, W.L. (eds) *Nitrogen Fixation: Achievements and Objectives.* Chapman and Hall, New York, USA, pp. 805–811.

Egawa, T. (1974) The use of organic fertilizers in Japan. In *Organic Material as Fertilizer. FAO Soils Bulletin* 27, pp. 253–272.

Elvin, M. (1982) The technology of farming in late-traditional China. In: Barker, R., Sinha, R. and Rose, B. (eds) *The Chinese Agricultural Economy.* Westview Press, Boulder, Colorado, USA, pp. 13–35.

EMBRAPA (1984) Upland rice in Brazil. In: *An Overview of Upland Rice Research.* IRRI, Los Banos, Philippines, pp. 121–134.

Escalada, M.M. and Heong, K.L. (1993) Communication and implementation of change in crop protection. In: *Crop Protection and Sustainable Agriculture.* Ciba Foundation, London, pp. 191–207.

Evans, L.T. (1978) Learning from Tachai: agricultural research in China. *Search* 9, 24–33.

Evans, L.T. (1990) Assimilation, allocation, explanation, extrapolation. In: Rabbinge, R., Goudriaan, J.,van Keulen, H., Penning de Vries, F.W.T. and van Laar, H.H. (eds) *Theoretical Production Ecology; Reflections and Prospects.* Pudoc, Wageningen, Netherlands, pp. 77–87.

Evans, L.T. (1993) *Crop Evolution, Adaptation and Yield.* Cambridge University Press, Cambridge, UK.

Falkenmark, M. (1994) Landscape as life-support provider: water-related limitations. In: Graham-Smith, F. (ed.) *Population, the Complex Reality.* Royal Society, London, UK, pp. 103–116.

FAO (1977) China: recycling of organic wastes in agriculture. *FAO Soils Bulletin*, 40.

FAO (1987) Consultation on irrigation in Africa. *FAO Irrigation and Drainage Paper*, No. 42. FAO, Rome, Italy.

FAO (1991) *Third Progress Report on the Programme of Action of the World Conference on Agrarian Reform and Agrarian Development (WCARRD)Document C91/19*. FAO, Rome, Italy.

FAO (1993) *Agriculture Towards 2010*. FAO, Rome, Italy.

Fillery, I.R.P., Simpson, J.R. and De Datta, S.K. (1984) Influence of field environment and fertilizer management on ammonia loss from flooded rice. *Soil Science Society of America, Journal* 48, 914–920.

Fillery, I.R.P., Simpson, J.R. and De Datta, S.K. (1986) Contributions of ammonia volatilisation to total nitrogen loss after application of urea to rice fields. *Fertilizer Research* 8, 193–201.

Firth, P., Thitipoca, H., Suthipradit, S., Wetselaar, R. and Beech, D.F. (1973) Nitrogen balance studies in the Central Plain of Thailand. *Soil Biology and Biochemistry* 5, 41–46.

Flinn, J.C. and De Datta, S.K. (1984) Trends in irrigated rice yields under intensive cropping at Philippine research stations. *Field Crops Research* 9, 1–15.

Flinn, J.C. and Duff, B. (1985) Energy analysis, rice production systems and rice research. *IRRI Research Paper Series* No. 114.

Flinn, J.C., De Datta, S.K. and Labadan, E. (1982) An analysis of long-term rice yields in a wetland soil. *Field Crops Research* 5, 201–216.

Flowers, T.J. and Yeo, A.R. (1981) Variability in the resistance to sodium chloride salinity within rice (*Oryza sativa* L.) varieties. *New Phytologist* 88, 363–373.

Fox, R.L. and Hue, N.V. (1986) Sulphur cycling in the tropics and sulphur requirements for agriculture. In: Portch, S. and Hussain, S.G. (eds) *Sulphur in Agricultural Soils*. Bangladesh Agricultural Research Council, Dacca, and The Sulphur Institute, Washington DC, USA, pp. 139–162.

Fu Ming-hua (1981) Characteristics of high-yield paddy soils in suburbs of Shanghai. In: *Paddy Soils Symposium*. Springer Verlag, Berlin, Germany, pp. 769–774.

Fujisaka, S., Dapusala, A. and Jayson, E. (1989) Hail Mary, kill the cat: a case of traditional upland crop pest control in the Philippines. *Philippine Quarterly of Culture and Society* 17, 202–211.

Fujui, Y. (1992) The potential biological control of paddy weeds with allelopathy. *Proceedings International Sumposium on Biological Control and Integrated Management of Paddy and Aquatic Weeds in Asia*. National Agricultural Research Centre, Tsukuba, Japan, pp. 305–320.

Fukui, H. (1982) Variability of rice production in tropical Asia. In: *Drought Resistance in Crops with Special Emphasis on Rice*. IRRI, Los Banos, Philippines, pp. 17–38.

Fullen, M.A., Fearnehough, W., Mitchell, D.J. and Trueman, I.C. (1995) Desert reclamation using Yellow River irrigation water in Ningxia, China. *Soil Use and Management* 11, 77–83.

Garrity, D.P. (1984a) Asian upland rice environments. In: *An Overview of Upland Rice Research*. IRRI, Los Banos, Philippines, pp. 161–184.

Garrity, D.P. (1984b) Rice environmental classifications: a comparative review. In: *Terminology for Rice Growing Environments*. IRRI, Los Banos, Philippines, pp. 11–26.

Garrity, D.P., Oldeman, L.R., Morris, R.A. and Lenka, D. (1986) Rainfed lowland rice ecosystems: characterisation and distribution. In: *Progress in Rainfed Lowland Rice*. IRRI, Los Banos, Philippines, pp. 3–23.

Gatehouse, A.M.R., Hilder, V.A. and Boulter, D. (eds) (1992) *Plant Genetic Manipulation for Crop Protection.* CAB International, Wallingford, UK.

George, T., Ladha, J.K., Buresh, R.J. and Garrity, D.P. (1992) Managing native and legume-fixed nitrogen in lowland rice-based cropping systems. *Plant and Soil* 141, 69–91.

George, T., Ladha, J.K., Buresh, R.J. and Garrity, D.P. (1993) Nitrate dynamics during the aerobic soil phase in lowland rice-based cropping systems. *Soil Science Society America Journal* 57, 1526–1532.

George, T., Ladha, J.K., Garrity, D.P. and Buresh, R.J. (1994) Legumes as nitrate catch crops during the dry-to-wet season transition in lowland rice cropping systems. *Agronomy Journal* 86, 267–273.

George, T., Ladha, J.K., Garrity, D.P. and Torres, R.O. (1995) Nitrogen dynamics of grain legume–weedy fallow–flooded rice sequences in the tropics. *Agronomy Journal* 87, 1–6.

Geyh, M.A., Kudras, H.K. and Streif, H. (1979) Sea level changes during the late Pleistocene and Holocene in the Straits of Malacca. *Nature* 278, 441–443.

Ghalang, A.L., Bhuiyan, S.I. and Hunt, E.D. (1994) Identification of potential areas for use of the on-farm reservoir system for drought alleviation. In: Bhuiyan, S.I. (ed.) *On-farm Reservoir Systems for Rainfed Ricelands.* IRRI, Los Banos, Philippines, pp. 21–34.

Ghassemi, F., Jakeman, A.J. and Nix, H.A. (1995) *Salinisation of Land and Water Resources: Human Causes, Extent, Management and Case Studies.* CAB International, Wallingford, UK.

Gibb, H.A.R. (1971) *The Travels of Ibn Battuta, AD 1325–1354. Volume 3.* Cambridge University Press, Cambridge, UK.

Giller, K.E. and Wilson, K.J. (1991) *Nitrogen Fixation in Tropical Cropping Systems.* CAB International, Wallingford, UK.

Glaszmann, J.C. (1986) A varietal classification of Asian cultivated rice (*Oryza sativa* L) based on isozyme polymorphism. In: *Rice Genetics.* IRRI, Los Banos, Philippines, pp. 83–90.

Glover, I.C. and Higham, C.F.W. (1996) New evidence for early rice-cultivation in South, Southeast and East Asia. In: Price, D. and Gebauer, B. (eds) *Last Hunters – First Farmers.* School of American Archaeological Research, Santa Fe, Texas, USA.

Gong Zi-tong (1985) Wetland soils in China. In: *Wetland Soils: Characterization, Classification and Utilization.* IRRI, Los Banos, Philippines, pp. 473–487.

Gong Zi-tong and Liu Liang-wu (1994) Soil and archaeological research. *Transactions 15th International Congress of Soil Science.* Acapulco, 6a, 369–370.

Goonasekere, K.G.A. and Amerasinghe, F.P. (1988) Planning, design and operation of irrigation schemes – their impact on mosquito-borne diseases. In: *Vector-borne Disease Control in Humans through Rice Agro-ecosystem Management.* IRRI, Los Banos, Philippines, pp. 41–50.

Graham-Bryce, I.J. (1983) Biting the hand that feeds: progress, perception and prognostication in applied chemistry. *Chemistry and Industry, London* 1, 733–739.

Graham-Smith, F. (ed.) (1994) *Population: the Complex Reality.* The Royal Society, London, UK.

Grayson, B.T., Green, M.B. and Copping, L.G. (eds) (1990) *Pest Management in Rice.* Elsevier Science Publishers, Barking, UK.

Greenland, D.J. (1974) Evolution and development of different types of shifting cultivation. In: Shifting Cultivation and Soil Conservation in Africa. *FAO Soils Bulletin* 24, 5–13.

Greenland, D.J. (1976) Bringing the green revolution to the shifting cultivator. *Science* 190, 841–844.

Greenland, D.J. (1981) Recent progress in studies of soil structure and its relation to properties and management of paddy soils. In: *Proceedings of Symposium on Paddy Soil.* Science Press, Beijing China, and Springer Verlag, Berlin, Germany, pp. 42–58.

Greenland, D.J. (1985) Physical aspects of soil management for rice-based cropping systems. In: *Soil Physics and Rice.* IRRI, Los Banos, Philippines, pp. 1–16.

Greenland, D.J. (1990) Agricultural research and third world poverty. In: Speedy, A. (ed.) *Developing World Agriculture.* Grosvenor Press International, London, UK, pp. 8–13.

Greenland, D.J. and De Datta, S.K. (1985) Constraints to rice production and wetland soil characteristics. In: *Wetland Soils: Characterization, Classification and Utilization.* IRRI, Los Banos, Philippines, pp. 23–36.

Greenland, D.J. and Lal, R. (1979) *Soil Conservation and Management in the Humid Tropics.* J. Wiley, Chichester, UK.

Greenland, D.J. and Watanabe, I. (1982) The continuing nitrogen enigma. *Transactions 12th International Congress of Soil Science* 5, 123–137.

Greenland, D.J., Wild, A. and Adams, D. (1992) Organic matter dynamics in soils of the tropics – from myth to complex reality. In: Lal, R. and Sanchez, P.A. (eds) *Myths and Science of Soils in the Tropics.* Soil Science Society of America (Special Publication, No. 29.), Madison, Wisconsin, USA, pp. 17–34.

Greer, C. (1979) *Water Management in the Yellow River Basin of China.* University of Texas Press, Austin, Texas, USA.

Griffin, K. (1975) *The Political Economy of Agrarian Change.* MacMillan, London, UK.

Grigg, D.B. (1974) *The Agricultural Systems of the World: an Evolutionary Approach.* Cambridge University Press, Cambridge, UK.

Grist, D.H. (1975) *Rice.* Longmans, London, UK.

Gu Rong-shen and Wen Qi-xiao (1981) Cultivation and application of green manure in paddy fields of China. In: *Proceedings of Symposium on Paddy Soil.* Science Press, Beijing, China, and Springer Verlag, Berlin, Germany, pp. 207–219.

Gupta, G.P. (1980) Soil and water conservation of river valley projects. *Indian Journal of Soil Conservation* 8, 1–7.

Gupta, P.C. and O'Toole, J.C. (1986) *Upland Rice: A Global Perspective.* IRRI, Los Banos, Philippines.

Harlan, J.R. (1989) The tropical African cereals. In: Harris, D.R. and Hillman, G.C. (eds) *Foraging and Farming: the Evolution of Plant Exploration.* Unwin Hyman, London, UK, pp. 335–343.

Harrington, L.W., Hobbs, P.R., Tamang, D.B. *et al.* (1992) *Wheat and Rice in the Hills: Farming Systems, Production Techniques and Research Issues for Rice–Wheat Cropping in the Mid-hills of Nepal.* CIMMYT, Mexico City DF, Mexico.

Harrington, L.W., Hobbs, P.R. and Cassaday, K.A. (eds) (1993) *Methods of Measuring Sustainability through Farmer Monitoring: Application to the Rice–Wheat Cropping Pattern in South Asia.* CIMMYT, Mexico City DF, Mexico.

Hawksworth, D.L. (ed.) (1991). *The Biodiversity of Microorganisms and Invertebrates: its Role in Sustainable Agriculture.* CAB International, Wallingford, UK.

Hayami, Y. (1975) *A Century of Agricultural Growth in Japan.* University of Tokyo Press, Tokyo, Japan.

Hayami, Y. and Kikuchi, M. (1981) *Asian Village Economy at the Crossroads: An Economic Approach to Cultural Change.* Tokyo University Press, Tokyo, Japan.

Hayami, Y. and Otsuka, K. (1994) Beyond the green revolution: agricultural development strategy into the new century. In: Anderson, J. (ed.) *Agricultural Technology: Policy Issues for the International Community.* CAB International, Wallingford, UK, pp. 15–42.

Hazell, P. and Ramasamy, C. (1991) *The Green Revolution Reconsidered. The Impact of the High-yielding Rice Varieties in South India.* John Hopkins University Press, Baltimore, Maryland, USA.

He Guiting, Zhu Xigang, Gu Huanzhang, and Zhang Jingshun (1988) The use of hybrid rice technology: an economic evaluation. In: *Hybrid Rice.* IRRI, Los Banos, Philippines, pp. 229–242.

He Kang (1985) Rice production in China. In: *Impact of Science on Rice.* IRRI, Los Banos, Philippines, pp. 69–74.

Heckman, C.W. (1979) *Ricefield Ecology in Northeastern Thailand.* (Monographiae Biologicae No. 34.) W. Junk, The Hague, Netherlands.

Heinrichs, E.A., Aquino, G.B., Chelliah, S., Valencia, S.L. and Reissig, W.H. (1982) Resurgence of *Nilaparvata lugens* (Stal) populations as influenced by method and timing of insecticide applications in lowland rice. *Environmental Entomology* 11, 78–84.

Heong, K.L. and Sogawa, K. (1994) Management strategies for key insect pests of rice:critical issues. In: Teng, P.S., Heong, K.L. and Moody, K. (eds) *Rice Pest Science and Management.* IRRI, Los Banos, Philippines, pp. 3–14.

Heong, K.L., Teng, P.S. and Moody, K. (1995) Managing rice pests with less chemicals. *Geojournal* 35, 337–349.

Herdt, R.W. (1987) A retrospective view of technological and other changes in Philippine rice farming. *Economic Development and Cultural Change* 35, 329–350.

Herdt, R.W. and Capule, C. (1983) *Adoption, Spread, and Production Impact of Modern Rice Varieties in Asia.* IRRI, Los Banos, Philippines.

Higham, C.F.W. (1984) Prehistoric rice cultivation in Southeast Asia. *Scientific American* 250(4), 100–107.

Higham, C.F.W. (1989) *The Archaeology of Mainland Southeast Asia from 10,000 BC to the Fall of Angkor.* Cambridge University Press, Cambridge, UK.

Higham, C.F.W. (1996) The transition to rice cultivation in Southeast Asia. In: Price, D. and Gebauer, B. (eds) *Last Hunters – First Farmers.* School of American Archaeology Press, Santa Fe, USA.

Hill, R.B. and Cambournac, F.J.C. (1941) Intermittent irrigation in rice cultivation, and its effect on yield, water consumption and *Anopheles* production. *American Journal of Tropical Medicine* 21, 123–144.

Hillel, D. (1991) *Out of the Earth: Civilization and the Life of the Soil.* University of California Press, Berkeley, California, USA.

Ho Ping-ti (1959) *Studies on the Population of China, 1368–1953.* Harvard University Press, Cambridge, Massachusetts, USA.

Horton, P. (1994) Limitations to crop yield by photosynthesis. In: Cassman, K.G. (ed.) *Breaking the Yield Barrier*. IRRI, Los Banos, Philippines, pp. 111–116.

Hossain, M. and Fischer, K.S. (1995) Rice research for food security and sustainable agricultural development in Asia: achievements and future challenges. *Geojournal* 35, 286–298.

Howarth, R.W. and Jorgensen, B.B. (1984) Formation of 35S labelled elemental sulphur and pyrite in coastal marine sediments (Limfjorden and Kysing Fjord, Denmark) during short-term 35SO4 reduction measurements. *Geochimica et Cosmochimica Acta* 48, 1807–1818.

Hseung, Y. and Liu, W. (1990) Improvement and utilisation of salt-affected soils. In: *Soils of China*. Science Press, Beijing, China, pp. 734–761.

Huang Yu-kun and Chen Jia-jie (1988) Sea level changes along the coast of the South China Sea since late Pleistocene. In: Aigner, J.S., Jablonski, N.C., Taylor, G., Walker, D. and Wang Ping-xian (eds) *The Palaeoenvironment of East Asia from the Mid-tertiary*. Centre for Asian Stiudies, University of Hong Kong.

Hui Lin-li (1983) The domestication of plants in China: ecogeographical considerations. In: Keightley, D. (ed.) *The Origins of Chinese Civilisation*. University of California Press, Berkeley, Ca., USA, pp. 21–63.

Huke, R.E. (1982) *Rice Area by Type of Culture: South, Southeast and East Asia*. IRRI, Los Banos, Philippines.

Huke, R.E and Huke, E.H. (1983) *Bangladesh: Rice Area Planted by Season and Culture Type*. IRRI, Los Banos, Philippines.

Huke, R.E. and Huke, E.H. (1988) *Human Geography of Rice in South Asia*. IRRI, Los Banos, Philippines.

Huke, R.E. and Huke, E.H. (1990a) *Human Geography of Rice in Southeast Asia*. IRRI, Los Banos, Philippines.

Huke, R.E. and Huke, E.H. (1990b) *Rice: Then and Now*. IRRI, Los Banos, Philippines.

Huke, R.E., Cordova, V. and Sardido, S. (1982) *San Bartolome: Beyond the Green Revolution*. IRRI Research Paper Series, No. 74. IRRI, Los Banos, Philippines.

Hutchinson, J.B. (1976) India: local and introduced crops. *Philosophical Transactions of the Royal Society of London* B275, 129–141.

Humphreys, E., van der Lely, A., Muirhead, W.A. and Hoey, D. (1994) The development of on-farm restrictions to minimise recharge from rice in New South Wales. *Australian Journal of Soil and Water Conservation* 7, 11–20.

Hyams, E. (1976) *Soil and Civilization*. Harper and Row, New York.

IBSRAM (1991) *Soil Management for Sustainable Rice Production in the Tropics*. IBSRAM Monograph No. 2. IBSRAM, Bangkhen, Thailand.

IITA (1984) Upland rice in Africa. In: *An Overview of Upland Rice Research*. IRRI, Los Banos, Philippines, pp. 3–20.

Institute of Soil Science, Academia Sinica (1981) *Proceedings, Paddy Soils Symposium*. Springer Verlag, Berlin.

IRRI (1963) *The Rice Blast Disease*. IRRI, Los Banos, Philippines, and John Hopkins University Press, Baltimore, Maryland, USA.

IRRI (1977) *Cropping Systems Research and Development for the Asian Rice Farmer*. IRRI, Los Banos, Philippines.

IRRI (1978) *Soils and Rice*. IRRI, Los Banos, Philippines.

IRRI (1979a) *Highlights*. IRRI, Los Banos, Philippines.

IRRI (1979b) *Rainfed Lowland Rice*. IRRI, Los Banos, Philippines.

IRRI (1979c) *Brown Planthopper – Threat to Rice Production in Asia.* IRRI, Los Banos, Philippines.
IRRI (1980) *Irrigation Water Management – Report of a Planning Workshop.* IRRI, Los Banos, Philippines.
IRRI (1981a) *A Continuous Rice Production System – the Rice Garden.* IRRI, Los Banos, Philippines.
IRRI (1981b) *Research Highlights.* IRRI, Los Banos, Philippines.
IRRI (1982a) *Drought Resistance in Crops, with Emphasis on Rice.* IRRI, Los Banos, Philippines.
IRRI (1982b) *Cropping Systems Research in Asia.* IRRI, Los Banos, Philippines.
IRRI (1982c) *A Plan for IRRI's Third Decade.* IRRI, Los Banos, Philippines.
IRRI (1982d) *Rice Research Strategies for the Future.* IRRI, Los Banos, Philippines.
IRRI (1982e) *Research Highlights.* IRRI, Los Banos, Philippines.
IRRI (1983a) *Weed Control in Rice.* IRRI, Los Banos, Philippines.
IRRI (1983b) *Research Highlights.* IRRI, Los Banos, Philippines.
IRRI (1984a) *Terminology for Rice Growing Environments.* IRRI, Los Banos, Philippines.
IRRI (1984b) *Research Priorities in Tidal Swamp Rice.* IRRI, Los Banos, Philippines.
IRRI (1984c) *An Overview of Upland Rice Research.* IRRI, Los Banos, Philippines.
IRRI (1984d) *Organic Matter and Rice.* IRRI, Los Banos, Philippines.
IRRI (1984e) *Judicious and Effective Use of Insecticides on Rice.* IRRI, Los Banos, Philippines.
IRRI (1985a) *International Rice Research: 25 years of Partnership.* IRRI, Los Banos, Philippines.
IRRI (1985b) *Rice Improvement in Eastern, Southern and Central Africa.* IRRI, Los Banos, Philippines.
IRRI (1985c) *Wetland Soils: Characterization, Classification and Utilization.* IRRI, Los Banos, Philippines.
IRRI (1985d) *Soil Physics and Rice.* IRRI, Los Banos, Philippines.
IRRI (1985e) *Genetic Evaluation for Insect Resistance in Rice.* IRRI, Los Banos, Philippines.
IRRI (1985f) *Insights of Outstanding Rice Farmers.* IRRI, Los Banos, Philippines.
IRRI (1985g) *Annual Report.* IRRI, Los Banos, Philippines, pp. 240–247.
IRRI (1986a) *Progress in Rainfed Lowland Rice.* IRRI, Los Banos, Philippines.
IRRI (1986b) *Progress in Upland Rice Research.* IRRI, Los Banos, Philippines.
IRRI (1986c) *Rice Genetics.* IRRI, Los Banos, Philippines.
IRRI (1987a) *Azolla Utilization.* IRRI, Los Banos, Philippines.
IRRI (1987b) *Efficiency of Nitrogen Fertilizers for Rice.* IRRI, Los Banos, Philippines.
IRRI (1987c) *Helpful Insects, Spiders and Pathogens: Friends of the Rice Farmer.* IRRI, Los Banos, Philippines.
IRRI (1987d) *World Rice Statistics.* IRRI, Los Banos, Philippines.
IRRI (1988) *Green Manure in Rice Farming.* IRRI, Los Banos, Philippines.
IRRI (1989) *Hybrid Rice.* IRRI, Los Banos, Philippines.
IRRI (1993a) *1993–1995 IRRI Rice Almanac.* IRRI, Los Banos, Philippines.
IRRI (1993b) *Rice Research in a Time of Change.* IRRI, Los Banos, Philippines.
IRRI (1994) *Rice: The Challenges of Research.* (Pamphlet) IRRI, Los Banos, Philippines.
IRRI (1995a) Feeding 4 Billion People: the Challenge for Rice Research in the 21st. Century. *GeoJournal* 35(3), pp. 253–388.
IRRI (1995b) *Water, a Looming Crisis.* IRRI, Los Banos, Philippines.
IRRI (1995c) *World Rice Statistics, 1993/94.* IRRI, Los Banos, Philippines.

IRRI/PEEM (1988) *Vector-borne Disease Control in Humans through Rice Agrosystem Management.* IRRI in collaboration with the WHO/FAO/UNEP Panel of Experts on Environmental Management for Vector Control (PEEM). IRRI, Los Banos, Philippines.

Ishii, Y. (ed.) (1978) *Thailand, a Rice-growing Society.* University of Hawaii Press, Honolulu, USA.

Ishikawa, M. (1988) Green manure in rice: the Japan experience. In: *Green Manure in Rice Farming.* IRRI, Los Banos, Philippines, pp. 45–62.

Islam, A. and Islam, W. (1973) Chemistry of submerged soils and growth and yield of rice. I. Benefit from submergence. *Plant and Soil* 39, 555–565.

Ives, J. (1991) Floods in Bangladesh: who is to blame? *New Scientist* 13 April, 34–37.

Jackson, M.T. (1995) Protecting the heritage of rice biodiversity. *GeoJournal* 35, 267–274.

Jiang Ci Mao and Dai Ge (1995) Economic research in rice–fish farming. In: Mackay, K. (ed.) *Rice–fish Culture in China.* International Development Research Centre (IDRC), Ottawa, Canada, pp. 253–258.

Joffe, J.S. (1955) Green manuring viewed by a pedologist. *Advances in Agronomy* 3, 142–186.

Jones, U.S., Katyal, J.C., Mamaril, C.P. and Park, C.S. (1982) Wetland rice nutritional deficiencies other than nitrogen. In: *Rice Research Strategies for the Future.* IRRI, Los Banos, Philippines, pp. 327–378.

Jones, P.D., Wigley, T.M.L., Folland, C.K., Parker, D.E., Angell, J.K., Lebedeff, D. and Hansen, J.E. (1988) Evidence for global warming in the past decade. *Nature* 332, 790.

Joshi, P.K. and Jha, D. (1991) Farm-level effects of soil degradation in Sharda Sahayak irrigation project. *IFPRI Working Paper.* International Food Policy Research Institute, Washington, DC, USA.

Juliano, B.O. and Bechtel, D.B. (1985) The rice grain and its gross composition. In: Juliano, B.O. (ed.) *Rice: Chemistry and Technology.* American Association of Cereal Chemists, St. Paul, Minnesota, pp. 17–57.

Kanazawa, N. (1984) Trends and economic factors affecting organic matter in Japan. In: *Organic Matter and Rice.* IRRI, Los Banos, Philippines, pp. 557–567.

Kanno, I. (1962) A new classification system for rice soils in Japan. *Pedologist* 6, 2–10.

Kanno, I., Honyo, Y., Arimura, S. and Takudome, S. (1964) Genesis and characterisation of rice soils developed in polder lands of Shiroishi area, Kyushu. *Soil Science and Plant Nutrition* 10, 1–20.

Katayama, T.C. (ed.) (1987) *Studies on Distribution and Ecotypic Differentiation of Wild and Cultivated Rice Species in Africa.* Research Centre for the South Pacific, Kagoshima University, Kagoshima, Japan.

Kawaguchi, K. and Kyuma, K. (1977) *Paddy Soils in Tropical Asia: their Material Nature and Fertility.* University of Hawaii Press, Honolulu, Hawaii, USA.

Kawai, K. (1988) Sulphur in Japanese Agriculture. In: Jae Song-chin (ed.) *Sulphur in Korean Agriculture.* Korean Society of Soil Science and The Sulphur Institute, Washington, DC, pp. 35–41.

Kenmore, P.E. (1991) *Indonesia's Integrated Pest Management: a Model for Asia.* FAO, Rome, Italy.

Kenmore, P.E., Carino, F.O., Perez, C.A., Dyck, V.A. and Gutierrez, A.P. (1984) Population regulation of the rice brown planthopper (*Nilaparvata lugens* Stal) within rice fields in the Philippines. *Journal of Plant Protection in the Tropics* 1, 19–38.

Kenmore, P.E., Litsinger, J.A., Banday, J.P., Santiago, A.C. and Salac, M.M. (1987) Philippine rice farmers and insecticides: thirty years of growing dependency and options for change. In: Tait, J. and Napompeth, B. (eds) *Management of Pests and Pesticides: Farmers' Perceptions and Practices.* Westview Press, Boulder, Colorado, USA.

Khush, G.S. (1977) Disease and insect resistance in rice. *Advances in Agronomy* 29, 265–341.

Khush, G.S. (1978). Genetics of and breeding for resistance to brown plant hopper. In: *Brown Plant Hopper: Threat to Rice Production in Asia.* IRRI, Los Banos, Philippines, pp. 320–334.

Khush, G.S. (1984) Terminology for rice growing environments. In: *Terminology for Rice Growing Environments.* IRRI, Los Banos, Philippines, pp. 5–10.

Khush, G.S. (1987) Development of rice varieties suitable for double cropping. *Tropical Agricultural Research Series* No. 20. Tropical Agricultural Research Centre, Tokyo, Japan, pp. 235–246.

Khush, G.S. (1989) Multiple disease and insect resistance for increased yield stability in rice. In: *Progress in Irrigated Rice Research.* IRRI, Los Banos, Philippines, pp. 79–92.

Khush, G.S. (1990) Rice breeding – Accomplishments and challenges. *Plant Breeding Abstracts* 60, 461–469.

Khush, G.S. (1995) Breaking the yield frontier of rice. *Geojournal* 35, 329–332.

Khush, G.S. and Bennett, J. (eds) (1992) *Nodulation and Nitrogen Fixation in Rice: Potential and Prospects.* IRRI, Los Banos, Philippines.

Khush, G.S. and Toenniessen, G.H. (1991) *Rice Biotechnology.* IRRI, Los Banos, Philippines, and CAB International, Wallingford, UK.

Kikuchi, M. (1975) Irrigation and rice technology in agricultural development: a comparative history of Taiwan, Korea and the Philippines. PhD Thesis, University of Hokkaido, Sapporo, Japan.

Kirk, G.J.D. (1990) Diffusion of inorganic materials in soil. *Philosophical Transactions of the Royal Society of London* B 329, 331–342.

Kirk, G.J.D. and Saleque, M.A. (1995) Solubilisation of phosphate by rice plants growing in reduced soil: prediction of the amount solubilised and the resultant increase in uptake. *European Journal of Soil Science* 46, 247–256.

Kirk, J.G.D., Nye, P.H. and Ahmad, A.R. (1990) Diffusion and oxidation of iron in the rice rhizosphere. *Transactions, 14th International Congress Soil Science* 2, 153–157.

Klinge, H. (1977) Fine litter production and return to the soil in three natural forest stands in East Amazonia. Geo Eco Trop 1, 159–167.

Kobayashi, T., Noguchi, Y., Hiwada, T., Kanayama, K. and Maruoka, N. (1973) Studies on the arthropod associations in paddy fields, with particular reference to insecticidal effects on them. Part 1. *Kontyu* 41, 359–373.

Koenigs, F.F.R. (1950) A "sawah" profile near Bogor, Indonesia. *Transactions, 4th International Congress of Soil Science* 1, 297–300.

Koenigs, F.F.R. (1963) The puddling of clay soils. *Netherlands Journal of Agricultural Science* 11, 145–156.

Kohno, E. (1992) Site conditions of paddy fields and characteristics of paddy soil. In: Murty, V.V.N. and Koga, K. (eds) *Soil and Water Engineering for Paddy Soil Management.* Asian Institute of Technology, Bangkok, Thailand, pp. 286–294.

Konishi, M. and Ito, Y. (1973) Early entomology in East Asia. In: Smith, R.F., Mittler, T.E. and Smith, C.N. (eds) *History of Entomology*. Annual Reviews Inc., California, USA, pp. 1–20.

Koyama, T. (1981) The transformation and balance of nitrogen in Japanese rice fields. *Fertilizer Research* 2, 261–278.

Koyama, T. and App, A. (1979) Nitrogen balance in flooded rice soils. In: *Nitrogen and Rice*. IRRI, Los Banos, Philippines, pp. 95–104.

Kropff, M.J., Cassman, K.G., Peng, S., Matthews, R.B. and Setter, T.L. (1994) Quantitative understanding of yield potential. In: Cassman, K.G. (ed.) *Breaking the Yield Barrier*. IRRI, Los Banos, Philippines, pp. 21–38.

Kumar, T.T. (1988) *History of Rice in India: Mythology, Culture and Agriculture*. Gian Publishing House, Delhi, India.

Kuo Ching-hui (1956) The silt of the Yellow River and its erosive action. In: Ti-li Chih-shih. *Geographic Knowledge*. Geographic Society of China (In Chinese, quoted by Greer, 1979.)

Kyuma, K. (1991) Paddy soils in Japan in comparison with those in tropical Asia. In: *Soil Management for Sustainable Rice Production in the Tropics*. IBSRAM Monograph No.2, 33–46.

Ladha, J.K. and Garrity, D.P. (eds) (1994) *Green Manure Production Systems for Asian Ricelands*. IRRI, Los Banos, Philippines.

Ladha, J.K. and Reddy, P.M. (1995) Extension of nitrogen fixation to rice – necessity and possibilities. *GeoJournal* 35, 363–372.

Ladha, J.K. and Watanabe, I. (1987) Biochemical basis of *Azolla–Anabaena* symbiosis. In: *Azolla Utilisation*. IRRI, Los Banos, Philippines, pp. 47–58.

Ladha, J.K., George, T. and Bohlool, B.B. (1992a) *Biological Nitrogen Fixation for Sustainable Agriculture*. Kluwer Academic Publishers, Dordrecht, Netherlands.

Ladha, J.K., Pareek, R.P. and Becker, M. (1992b) Stem-nodulating legume–*Rhizobium* symbiosis and its agronomic use in lowland rice. *Advances in Soil Science* 20, 147–192.

Ladha, J.K., Tirol-Padre, A., Reddy, C.K. and Ventura, W. (1993) Prospects and problems of biological nitrogen fixation in rice production: a critical assessment. In: Palacios, R., Mora, J. and Newton, W.E. (eds) *New Horizons in Nitrogen Fixation*. Kluwer, Dordrecht, Netherlands, pp. 677–682.

Laing, D.R., Posada, R., Jennings, P.R., Martinez, C.P. and Jones, P.G. (1984) Upland rice in Latin America: Environment, constraints and potential. In: *An Overview of Upland Rice Research*. IRRI, Los Banos, Philippines, pp. 197–228.

Lambert, D.H. (1985) *Swamp Rice Farming: the Indigenous Pahang Malay Agricultural System*. Westview Press, Boulder, Colorado, USA.

Leeden, F. van der, Troise, F.L. and Todd, D.K. (1990) *The Water Encyclopoedia*, 2nd edn. Lewis Publishers, Chelsea, Michigan, USA.

Lefroy, R.D.B., Mamaril, C.P., Blair, G.J. and Gonzales, P.J. (1992) Sulphur cycling in rice wetlands. In: Howarth, R.W., Stewart, J.W.B. and Ivanov, M.B. (eds) *Sulphur Cycling on the Continents*. J. Wiley, Chichester, UK, pp. 279–300.

Levine, G. (1971) Water environment and crop production. In: Turk, K.L. (ed.) *Some Issues Emerging from Recent Breakthroughs in Food Production*. Cornell University Press, Ithaca, New York, USA.

Li Zhuo-xin (1982) Nitrogen fixation by *Azolla* in ricefields in China. *Transactions 12th International Congress of Soil Science* 2, 83–95.

Lim, R.P. (1980) Population changes of some aquatic invertebrates in ricefields. In: *Tropical Ecology and Development*. International Society of tropical Ecology, Kuala Lumpur, Malaysia, pp. 971–980.

Lin Bao (1985) Effect and management of potassium fertilizer on wetland rice in China. In: *Wetland Soils: Characterization, Classification and Utilization*. IRRI, Los Banos, Philippines, pp. 285–292.

Lin, Q.M. (1992) Study on the effect of magnesium fertilizer on rice and the diagnostic indices of magnesium nutrition of rice. In: Portch, S. (ed.) *International Symposium on the Role of Sulphur, Magnesium and Micronutrients in Balanced Plant Nutrition*. The Sulphur Institute, Washington DC, USA, pp. 234–239.

Lin Shih-Cheng and Yuan Loung-Ping (1980) Hybrid rice breeding in China. In: *Innovative Approaches to Rice Breeding*. IRRI, Los Banos, Philippines, pp. 35–52.

Lipton, M. (1979) The technology, the system and the poor: the case of the new cereal varieties. In: Institute of Social Studies. *Developing Societies: the Next 25 Years*. Martinus Nijhoff, The Hague, Netherlands, pp. 121–135.

Lipton, M. and Longhurst, R. (1985) *Modern Varieties, International Agricultural Research and the Poor*. Study Paper No. 2, CGIAR, World Bank, Washington DC, USA.

Liu Chung Chu (1979) Use of *Azolla* in rice production in China. In: *Nitrogen and Rice*. IRRI, Los Banos, Philippines, pp. 375–394.

Liu Chung Chu (1988) Integrated use of green manure in ricefields in South China. In: *Green Manure in Rice Farming*. IRRI, Los Banos, Philippines, pp. 319–332.

Liu Chung Chu and Zheng Weiwen (1989) *Azolla in China*. Agricultural Publishing House, Beijing, China.

Liu Liang-wu (1990) The application of 14C dating to soil research. *Quaternary Glaciology and Quaternary Geology* (Special issue on 14C dating) 6, 200–205. Geological Publishing House.

Liu Zhiyu, Shi Weiming and Fan Xiaohui (1990) The rhizosphere effects of phosphorus and iron in soils. *Transactions 14th International Congress Soil Science* 2, 147–152.

Liu Zhongqun (1986) Preliminary study of soil sulphur and sulphur fertilization efficiency in China. In: Portch, S. and Hussain, S. G. (eds) *Sulphur in Agricultural Soils*. Bangladesh Agricultural Research Council, Dacca, and The Sulphur Institute, Washington DC, USA, pp. 371–387.

Lu Ru-kun (1981) The fertility and fertilizer use of the important paddy soils of China. In: *Symposium on Paddy Soils*. Springer Verlag, Berlin, pp. 160–170.

Lumpkin, T.A. and Plucknett, D.L. (1980) Azolla: Botany, physiology and use as a green manure. *Economic Botany* 34, 111–153.

Lumpkin, T.A. and Plucknett, D.L. (1982) *Azolla as a Green Manure: Use and Management in Crop Production*. Westview Press, Boulder, Colorado, USA.

MacKay, K.T. (1995) *Rice–Fish Culture in China*. International Development Research Centre, Ottawa, Canada.

McNeill, W.H. and Sedlar, J.W. (1970) *Classical China*. Oxford University Press, London, UK.

MAFF (1970) *Modern Farming and the Soil*. HMSO, London, UK.

Maglinao, A.R., Belen, E.M., Domingo, S.N. and Predicala, N.J. (1994) *Technology promotion and transfer: the Philippine experience with small farm reservoir technology. Proceedings, International Workshop on Conservation Farming for Sloping Uplands in Southeast Asia: Challenges, Opportunities and Prospects*. IBSRAM, Bangkok, Thailand, pp. 12.

Maloney, B.K. (1991) Rice agricultural origins: recent advances. *Geojournal* 23, 121–124.

Maloney, B.K., Higham, C.F.W. and Bannanurag, R. (1989) Early rice cultivation in Southeast Asia: archaeological and palynological evidence from the Bong Pakong Valley, Thailand. *Antiquity* 63, 363–370.

Malthus, T.R. (1798) *Of the Checks to Population in the less civilized Parts of the World, and in Past Times.* Book 1 of An Essay on Population, first published anonymously. Reprinted 1989, with editorial commentary by Patricia James. Cambridge University Press, Cambridge, UK.

Mamaril, C.P. and Gonzales, P.B. (1988) Response of lowland rice to sulphur in the Philippines. In: *Proceedings, International Symposium on Sulphur for Korean Agriculture.* Korean Society of Soil Science, and The Sulphur Institute, Washington DC, USA, pp. 72–76.

Masicat, P., De Vera, M.V. and Pingali, P.L. (1990) Philippine irrigation infrastructure: degradation trends for Luzon, 1966–1989. *IRRI Social Science Division Paper* 90-03. IRRI, Los Banos, Philippines.

Matthews, R.B., Kropff, M.J. and Bachelet, D. (1994a) Climate change and rice production in Asia. *Entwicklung und Landlischer Raum*, January, 16–19.

Matthews, R.B., Kropff, M.J., Bachelet, D., Horie, T., Lee, M.H., Centeno, H.G., Shin, J.C., Mohandass, S., Singh, S. and Defeng, Z. (1994b) Modeling the impact of climate change on rice production in east and southeast Asia. *Proceedings, International Symposium on Climate Change and Rice.* IRRI, Los Banos, Philippines.

Matthews, R.B., Kropff, M.J., Bachelet, D. and Van Laar, H.H. (1995) *Modelling the Impact of Climate Change on Rice Production in Asia.* CAB International, Wallingford, UK.

Mew, T.W., Rosales, A.M. and Maningas, G.V. (1994) Biological control of *Rhizoctonia* sheath blight and blast of rice. In: Ryder, M.H., Stephens, P.M. and Bowen, G.D. (eds) *Improving Plant Productivity with Rhizosphere Bacteria.* CSIRO, Melbourne, Australia, pp. 9–13.

Mikkelsen, D.S. and Kuo, S. (1977) Zinc fertilization and behaviour in flooded soils. *Commonwealth Bureau of Soils, Special Publication* No. 5. CAB International, Wallingford, UK.

Miao, Q.Y. (1982) Translation of '*The Essential Techniques for the People's Welfare*, by S.S. Jia.' Agricultural Publishing House, Beijing, China. Quoted by Chu Chong Chu and Zheng Wei-wen, 1989.

Min Zong-dian (1984) Study of rice yield in the Taihu region in the Song, Ming and Qing Dynasties. *Chinese Agricultural History* No. 3.

Miranda, S.M. and Levine, G. (1978) Effects of physical water control parameters on lowland irrigation water management. In; *Irrigation Policy and Management in Southeast Asia.* IRRI, Los Banos, Philippines, pp. 77–92.

Miura, K., Badayos, R.B. and Briones, A.M. (1995) *Pedological Characterization of Lowland Areas in the Philippines.* Dept. Soil Science, University of the Philippines at Los Banos, Los Banos, Philippines.

Moncharoen, L. (1985) Major characteristics and classification of wet inceptisols and histosols. In: *Wetland Soils: Characterization, Classification and Utilization.* IRRI, Los Banos, Philippines, pp. 393–405.

Momtaz, A. (1989) Research and management strategies for increased rice production in Egypt. In: *Rice Farming Systems: New Directions.* IRRI, Los Banos, Philippines, pp. 55–70.

Moormann, F.R and Juo, A.S.R. (1986) Present land-use and cropping systems in Africa. In: Juo, A.S.R. and Lowe, J.A. (eds) *The Wetlands and Rice in Subsaharan Africa.* IITA, Ibadan, Nigeria, pp. 187–194.

Moormann, F.R. and van Breemen, N. (1978) *Rice: Soil, Water, Land.* IRRI, Los Banos, Philippines.

Morishima, H. (1986) Wild progenitors of cultivated rice and their population dynamics. In: *Rice Genetics.* IRRI, Los Banos, Philippines, pp. 3–14.

Morris, R.A. and Meelu, O.P. (1985) Soil fertility and fertilizer management of rice-based cropping sequences. In: *Wetland Soils: Characterization, Classification and Utilization.* IRRI, Los Banos, Philippines, pp. 321–346.

Morris, R.A., Pantanothai, A., Syarifuddin, A. and Carangal, V.R. (1986) Cropping systems for rainfed lowland rice environments. In: *Progress in Rainfed Lowland Rice.* IRRI, Los Banos, Philippines, pp. 59–76.

Motomura, S., Lapid, E.M. and Yokoi, H. (1970) Soil structural development in the Ariake polder soils in relation to iron forms. *Soil Science and Plant Nutrition* 16, 47–54.

Munck, L. and Rexen, F. (1985) Increasing income and employment in rice farming areas – role of whole plant utilization and min-rice refineries. In: *Impact of Science on Rice.* IRRI, Los Banos, Philippines, pp. 271–280.

Murty, V.V.N. and Koga, K. (1992) *Soil and Water Engineering for Paddy Field Management.* Asian Institute of Technology, Bangkok, Thailand.

Nambiar, K.K.M. (1989) *Annual Report, 1985/86–1986/87, All-India Coordinated Research Project on Long-term Fertilizer Experiments.* Indian Agricultural Research Institute, New Delhi, India.

Nambiar, K.K.M. and Ghosh, A.B. (1984) *Highlights of Research of a Long-term Fertilizer Experiment in India (1972–82).* LTFE Bulletin No. 1. Indian Council of Agricultural Research, New Delhi, India.

Narayana, D.V.V. and Ram Babu (1983) Estimation of soil erosion in India. *Journal of Irrigation and Drainage Engineering* 109, 419–434.

Neue, H.U. (1985) Organic matter dynamics in wetland soil. In: *Wetland Soils: Characterization, Classification and Utilization.* IRRI, Los Banos, Philippines, pp. 109–122.

Neue, H.U. (1991) A holistic view of the chemistry of flooded soils. In: *Soil Management for Sustainable Rice Production in the Tropics.* IBSRAM, Bangkok, Thailand, pp. 5–32.

Neue, H.U. and Scharpenseel, H.W. (1984) Gaseous products in the decomposition of organic matter in submerged soils. In: *Organic Matter and Rice.* IRRI, Los Banos, Philippines, pp. 311–328.

Neue, H.U., Ziska, L.H., Matthews, R.B. and Dai, Qiujie (1995) Reducing global warming – the role of rice. *Geojournal* 35, 351–362.

Norland, I., Cederroth, S. and Gerdin, P. (1986) *Rice Societies: Asian Problems and Prospects.* Curzon Press, London, UK.

Norton, G.A. and Way, M.J. (1990) Rice pest management systems: past and future. In: Grayson, B.T., Green, M.B. and Copping, L.G. (eds) *Pest Management in Rice.* Elsevier Applied Science, Barking, London, UK, pp. 1–14.

NRC (1989) *Alternative Agriculture.* National Research Council, National Academy Press, Washington, DC, USA.

Nye, P.H. (1961) Organic matter and nutrient cycling under tropical forest. *Plant and Soil* 13, 333–346.

Nye, P.H. and Greenland, D.J. (1960) *The Soil under Shifting Cultivation*. CAB International, Wallingford.

Oka, H.I. (1988) *Origin of Cultivated Rice*. Elsevier, Amsterdam, Netherlands.

Oka, H.I. and Morishima, H. (1971) The dynamics of plant domestication: cultivation experiments with *Oryza perennis* and its hybrid with *Oryza sativa*. *Evolution* 25, 356–364.

Oldeman, L.R. (1994) The global extent of soil degradation. In: Greenland, D.J. and Szabolcs, I. (eds) *Soil Resilience and Sustainable Land Use*. CAB International, Wallingford, UK, pp. 99–118.

Oldeman, L.R. and Suardi, D. (1977) Climatic determinants in relation to cropping patterns. In: *Cropping Systems Research and Development for the Asian Rice Farmer*. IRRI, Los Banos, Philippines, pp. 61–82.

Olk, D.C., Cassman K.G. and Fan, T.W.M. (1995) Characterization of two humic acid fractions from a calcareous vermiculitic soil: implications for the humification process. *Geoderma* 65, 195–208.

Ooi, P.A.C. and Waage, J.K. (1994) Biological control in rice: applications and research needs. In: Teng, P.S., Heong, K.L. and Moody, K. (eds) *Rice Pest Science and Management*. IRRI, Los Banos, Philippines, pp. 209–216.

O'Toole, J.C. (1982) Adaptation of rice to drought-prone environments. In: *Drought Resistance in Crops with Special Emphasis on Rice*. IRRI, Los Banos, Philippines, pp. 195–216.

Ottow, J.C.G., Benckiser, G., Watanabe, I. and Santiago, S. (1983) Multiple nutritional soil stress as the prerequisite for iron toxicity of wetland rice (*Oryza sativa* L.) *Tropical Agriculture, (Trinidad)* 60, 102–106.

Pagliai, M. and Painuli, D.K. (1991) The physical properties of paddy soils and their effects on post-rice cultivation for upland crops. *IBSRAM Monograph* No. 2, IBSRAM, Bangkok, Thailand, pp. 391–415.

Pal, A.R., Rathore, A.L. and Pandey, V.K. (1994) On-farm rainwater storage for improving riceland productivity in eastern India: opprtunities and challenges. In: Bhuiyan, S.I. (ed.) *On-farm Reservoir Systems for Rainfed Ricelands*. IRRI, Los Banos, Philippines, pp. 105–125.

Parish, D.H. (1993) *Agricultural Productivity, Sustainability, and Fertilizer Use*. International Fertilizer Development Center (IFDC), Muscle Shoals, Alabama, USA.

Parry, J.T. (1992) The investigative role of Landsat™ in the examination of pre- and proto-historic water management sites in Northeast Thailand. *Geocarto International* 7(4), 5–24.

Patnaik, S. (1978) Natural sources of nutrients in rice soils. In: *Soils and Rice*. IRRI, Los Banos, Philippines, pp. 501–520.

Patnaik, S. and Sahoo, R. (1971) Some considerations on the improvement of saline soils. *Oryza* 8 (Suppl.), 309–312.

Patrick, W.H. and Reddy, C.N. (1977) Chemical changes in rice soils. In: *Soils and Rice*. IRRI, Los Banos, Philippines, pp. 361–379.

Paul, W.R.C. (1945) *Paddy Cultivation in Ceylon*. Government Press, Colombo, Ceylon (Sri Lanka.)

Pearce, F. (1991) The rivers that won't be tamed. *New Scientist* 13 April, 38–41.

Peng, S., Khush, G.S. and Cassman, K.G. (1994) Evolution of the new plant ideotype for increased yield potential. In: Cassman, K.G. (ed.) *Breaking the Yield Barrier.* IRRI, Los Banos, Philippines, pp. 5–20.

Pereira, H.C. (1989) *Policy and Practice in the Management of Tropical Watersheds.* Westview Press, Boulder, Colorado, USA.

Pingali, P.L. (1994) Technological prospects for reversing the declining trend in Asia's rice productivity. In: Anderson, J.R. (ed.) *Agricultural Technology: Policy Issues for the International Community.* CAB International, Wallingford, pp. 384–401.

Pingali, P. and Hossain, M. (1997) *Asian Rice Bowls: from Crisis to Complacency and Back to Crisis?* CAB International, Wallingford, UK (in press).

Pingali, P. and Roger, P. (eds) (1994) *Impact of Pesticides on Farmer Health and the Rice Environment.* Kluwer Academic Publishers, Dordrecht, Netherlands.

Pingali, P.L. and Rosegrant, M.W. (1993) Confronting the environmental consequences of the green revolution in Asia. Paper to *AAEA International Preconference on 'Post Green Revolution Agricultural Development Strategies in the Third World: What Next?'* Department of Economics, University of Florida, Gaines, Florida, USA.

Pingali, P.L. and Rosegrant, M.W. (1995) Agricultural commercialization and diversification: processes and policies. *Food Policy* 20, 171–185.

Pinstrup-Andersen, P. (1993) World food trends and how they may be modified. *Paper to CGIAR International Centers' Week.* World Bank, Washington, DC.

Plucknett, D.L. (1993) Science and agricultural transformation. *IFPRI Lecture Series.* International Food Policy Research Institute, Washington DC.

Ponnamperuma, F.N. (1972) The chemistry of submerged soils. *Advances in Agronomy* 24, 29–96.

Ponnamperuma, F.N. (1984a) Effects of flooding on soils. In: Koslowski, T.T. (ed.) *Flooding and Plant Growth.* Academic Press, New York, USA, pp. 10–46.

Ponnamperuma, F.N. (1984b) Straw as a source of nutrients for wetland rice. In: *Organic Matter and Rice.* IRRI, Los Banos, Philippines, pp. 117–136.

Ponnamperuma, F.N. (1985) Chemical kinetics of wetland rice soils relative to soil fertility. In: *Wetland Soils: Characterization, Classification and Utilization.* IRRI, Los Banos, Philippines, pp. 71–89.

Portères, R. (1950) Vieilles agricultures africaines avant le XVIe siècle. Berceaux d'agriculture et centres de variation. *L'Agronomie Tropicale* 5, 489–507.

Portères, R. (1956) Taxonomie agrobotanique des riz cultivés. *Journal d'Agriculture Tropicale et de Botanique* 3, 341–384, 541–580, 629–700, 821–856.

Portères, R. (1976) African cereals. In: Harlan, J.R., de Wet, J.M.J. and Stemmler, A.B.L. (eds) *Origins of African Plant Domestication.* Mouton, The Hague, Netherlands, pp. 441–452.

Prihar, S.S., Ghildyal, B.P., Painuli, D.K. and Sur, H.S. (1985) Physical properties of mineral soils affecting rice-based cropping systems. In: *Soil Physics and Rice.* IRRI, Los Banos, Philippines, pp. 57–70.

Qi-xiao Wen (1984) Utilization of organic materials in rice production in China. In: *Organic Matter and Rice.* IRRI, Los Banos, Philippines, pp. 45–56.

Quick, G.R. (1990) *Rodents and Rice.* IRRI, Los Banos, Philippines.

Rabesa Zafera Antoine (1985) Rice production in Madagascar. In: *Impact of Science on Rice.* IRRI, Los Banos, Philippines, pp. 127–136.

Ramakrishnan, P.S. (1992) *Shifting Agriculture and Sustainable Development.* UNESCO, Paris, France.

Randhawa, M.S. (1980) *The History of Indian Agriculture*, Volumes 1 to 3. Indian Council of Agricultural Research, New Delhi, India.

Randhawa, N.S. and Abrol, I.P. (1990) Sustaining agriculture: the Indian scene. In: Edwards, C.A., Lal, R., Madden, P., Miller, R.H. and House, G. (eds) *Sustainable Agricultural Systems*. Soil and Water Conservation Society, Ankeny, Iowa, USA, pp. 439–452.

Randhawa, N.S., Sinha, M.K. and Takkar, P.N. (1978) Micronutrients. In: *Soils and Rice*. IRRI, Los Banos, Philippines, pp. 581–604.

Rashid, M., Bajwa, M.I., Hussain, R., Naeem-ud-Din, M. and Rehman, F. (1992) Rice response to sulphur in Pakistan. *Sulphur in Agriculture* 16, 3–5.

Reddy, S.T.S. (1989) Declining groundwater levels in India. *Water Resources Development* 5, 183–190.

Reddy, K.R. and Patrick, W.H. (1975) Effect of alternate aerobic and anaerobic conditions on redox potential, organic matter decomposition and nitrogen loss in a flooded soil. *Soil Biology and Biochemistry* 7, 87–94.

Reddy, K.R. and Patrick, W.H. (1986) Denitrification losses in flooded rice fields. In: De Datta, S.K. and Patrick, W.H. (eds) *Nitrogen Economy of Flooded Rice Soils*. Martinus Nijhoff, Dordrecht, Netherlands, pp. 99–116.

Reissig, W.H., Heinrichs, E.A., Litsinger, J.A., Moody, L.F., Mew, T.W. and Barrion, A.T. (1986) *Illustrated Guide to Integrated Pest Management in Rice in Tropical Asia*. IRRI, Los Banos, Philippines.

Rerkasem, B. and Rerkasem, K. (1991) A decline of soil fertility under intensive cropping in northern Thailand. In: *Soil Management for Sustainable Rice Production*. IBSRAM Monograph no. 2. IBSRAM, Bangkok, Thailand, pp. 315–328.

Revelle, R. (1963) Water. *Scientific American* 209(3), 92–109.

Reyes, R. de los and Jopillo, S. (1986) *An Evaluation of the Philippine Participatory Communal Irrigation Programme*. Institute of Philippine Culture, Ateneo de manila University, Quezon City, Philippines.

Rhoades, J.D. (1974) The problem of salt in agriculture. In: *Yearbook of Science and the Future*. Encyclopaedia Britannica, Chicago, Illinois, USA.

Rinaudo, G., Alazard, D. and Mondiongui, A. (1988) Stem nodulating legumes as green manure for rice. In: *Green Manure in Rice Farming*. IRRI, Los Banos, Philippines, pp. 97–109.

Robinson, W.O. (1930) Some chemical phases of submerged soil conditions. *Soil Science* 30, 197–217.

Robison, D.M. and McKean, S.J. (1992) *Shifting Cultivation and Alternatives: an Annotated Bibliography, 1972–1989*. CAB International, Wallingford, UK.

Roger, P.A. (1991) Reconsidering the utilization of blue-green algae in wetland rice cultivation. In: Dutta, S.K. and Sloger, C. *Biological Nitrogen Fixation Associated with Rice Production*. Oxford and IBH Publications, New Delhi, India, pp. 119–141.

Roger, P.A. and Bhuiyan, S.I. (1990) Ricefield ecosystem management and its impact on disease vectors. *Water Resource Development* 6, 1–18.

Roger, P.A. and Kulasooriya, S.A. (1980) *Blue-green Algae and Rice*. IRRI, Los Banos, Philippines.

Roger, P.A. and Kurihara, Y. (1991) The floodwater biology of tropical wetland rice fields. In: *Soil Management for Sustainable Rice Production in the Tropics*. IBSRAM, Bangkok, Thailand, pp. 211–234.

Roger, P.A. and Ladha, J.K. (1990) Estimation of biological N fixation and its estimation in wetland rice fields. *Transactions 14th International Congress Soil Science* 3, 128–133.

Roger, P.A. and Ladha, J.K. (1992) Biological nitrogen fixation in wetland rice fields: estimation and contribution to nitrogen balance. *Plant and Soil* 141, 41–55.

Roger, P.A. and Watanabe, I. (1984) Algae and aquatic weeds as a source of organic matter and plant nutrients for wetland rice. In: *Organic Matter and Rice.* IRRI, Los Banos, Philippines, pp. 147–168.

Roger, P.A. and Watanabe, I. (1986) Technology for utilising biological nitrogen fixation in wetland rice: potentialities, current usage and limiting factors. *Fertilizer Research,* 9, 39–77.

Roger, P.A., Grant, I.F., Reddy, P.M. and Watanabe, I. (1987) The photosynthetic aquatic biomass in wetland rice fields and its effect on nitrogen dynamics. In: *Efficiency of Nitrogen Fertilizers for Rice.* IRRI, Los Banos, Philippines, pp. 43–68.

Roger, P.A., Heong, K.L. and Teng, P.S. (1991) Biodiversity and sustainability of wetland rice production: role and potential of microorganisms and invertebrates. In: Hawksworth, D.L. (ed.) *The Biodiversity of Microorganisms and Invertebrates; Its Role in Sustainable Agriculture.* CAB International, Wallingford, UK, pp. 117–136.

Roger, P.A., Simpson, I., Oficial, R., Ardales, S. and Jimenez, R. (1994) Effect of pesticides on soil and water microflora and mesofauna in wetland rice fields. *Australian Journal of Experimental Agriculture* 34, 1057–1068.

Rola, A.C. and Pingali, P.L. (1993) *Pesticides, Rice Productivity and Farmers' Health: an Economic Assessment for the Philippines.* IRRI, Los Banos, Philippines.

Rose, B. (1985) *Appendix to The Rice Economy of Asia.* Resources for the Future, Washington, DC, USA.

Rosegrant, M.W. and Binswanger, H.P. (1994) Markets in tradeable water rights: potential for efficiency gains in developing country water resource allocation. *World Development* 22, 1613–1625.

Rosegrant, M.W. and Svendsen, M. (1993) Asian food production in the 1990's: irrigation investment and management policy. *Food Policy* February, 13–33.

Rowell, D.L. (1981) Oxidation and reduction. In: Greenland, D.J. and Hayes, M.H.B. (eds) *The Chemistry of Soil Processes.* J. Wiley, Chichester, UK, pp. 401–462.

Runsheng, Du (1984) China's countryside under reform. *Beijing Review* 27, 17.

Russell, E.W. (1988) In: Wild, A. (ed.) *Soil Conditions and Plant Growth,* 11th edn. Longmans, Harlow, UK.

Ruthenberg, H. (1971) *Farming Systems in the Tropics.* Oxford University Press, London, UK.

Saleh, A.F.M. and Bhuiyan, S.I. (1995) Crop and rainwater management strategies for increasing productivity of rainfed lowland rice systems. *Agricultural Systems* 49, 259–276.

Salokhe, V.M. and Shirin, A.K.M. (1992) Effect of puddling on soil properties: a review. In: Murty, V.V.N. and Koga, K. (eds) *Soil and Water Engineering for Paddy Soil Management.* Asian Institute of Technology, Bangkok, Thailand, pp. 276–285.

Sanchez, P.A. (1973a) Puddling tropical rice soils. 1. Growth and nutritional aspects. *Soil Science* 115, 149–158.

Sanchez, P.A. (1973b) Puddling tropical soils. 2. Effects on water losses. *Soil Science* 115, 303–308.

Sanchez, P.A. (1976) *Properties and Management of Soils in the Tropics*. J. Wiley, New York, USA.

Sanchez, P.A. (1981) Soil management in the oxisol savannahs and ultisol jungles of tropical South America. In: Greenland, D.J. (ed.) *Characterization of Soils in Relation to their Classification and Management: Examples from Some Areas of the Humid Tropics*. Oxford University Press, Oxford, UK, pp. 214–252.

Sanchez, P.A. and Benites, J.R. (1987) Low-input cropping for acid soils of the humid tropics. *Science* 238, 1521–1527.

Sarma, P.B.S. (1986) Water resources and their role in food production. In: Swaminathan, M.S. and Sinha, S.K. (eds) *Global Aspects of Food Production*. Tycooly International, Oxford and IRRI, Los Banos, Philippines, pp. 171–198.

Sass, R.L. (1995) Mitigation of methane emissions from irrigated rice agriculture. *International Geophysical and Biophysical Programme (IGBP) Newsletter* No. 22, pp. 4–5.

Sattar, M.A. (1988) Fertility status of paddy soils of Bangladesh. In: *Symposium on Paddy Soil Fertility*, Country Reports. IBSRAM, Bangkok, Thailand, pp. 99–119.

Schiere, J.B. and Ibrahim, M.N.M. (1989) *Feeding of Urea-Ammonia Treated Rice Straw: A Compilation of Miscellaneous Reports Produced by the Straw Utilization Project (Sri Lanka)*. PUDOC, Wageningen, Netherlands.

Seshu, D.V. (1986) Rice varietal testing in the rainfed lowlands of south and southeast Asia. In: *Progress in Rainfed Lowland Rice*. IRRI, Los Banos, Philippines, pp. 77–101.

Setter, T.L., Peng, S., Kirk, J.G.D., Virmani, S.S., Kropff, M.J. and Cassman, K.G. (1994) Physiological considerations and hybrid rice. In: Cassman, K.G. (ed.) *Breaking the Yield Barrier*. IRRI, Los Banos, Philippines, pp. 39–62.

SFI (1985) Utilization of *Azolla* in China, by Soil and Fertilizer Institute, Hangzhou, China. In: *Proceedings 16th Session, International Rice Commission*. FAO, Rome, Italy, pp. 265–274.

Shao, Q., Yi, H. and Chen, Z. (1986) New findings concerning the origin of rice. In: *Rice Genetics*, IRRI, Los Banos, Philippines, pp. 53–58.

Sharma, R.D. (1987) *The Story of Rice*. IRRI, Los Banos, Philippines.

Sharma, P.K. and De Datta, S.K. (1985a) Effects of puddling on soil physical properties and processes. In: *Soil Physics and Rice*. IRRI, Los Banos, Philippines, pp. 217–234.

Sharma, P.K. and De Datta, S.K. (1985b) Puddling influence on soil, rice development and yield. *Soil Science Society America Journal* 49, 1451–1457.

Sharma, P.K. and De Datta, S.K. (1986) Physical properties and processes of puddled rice soils. *Advances in Soil Scence* 5, 139–178.

Sharma, P.K. and De Datta, S.K. (1992) Response of wetland rice to percolation under tropical conditions. In: Murty, V.V.N. and Koga, K. (eds) *Soil and Water Engineering for Paddy Soil Management*. Asian Institute of Technology, Bangkok, Thailand, pp. 109–117.

Sharma, G.R., Mishra, V.D., Mandal, D., Mishra, B.B. and Paul, J.N. (1980) *Beginnings of Agriculture: (Epi-paleolithic to Neolithic: Excavations at Choponi-Mando, Mahadaha and Mahagara)*. Abinash Prakashan, Allahabad, India.

Sharma, P.K., De Datta, S.K. and Redulla, C.A. (1989) Effect of percolation rate on nutrient kinetics and rice yield in tropical rice soils. I Role of soil organic matter. II Role of straw amendments. *Plant and Soil*, 119, 111–126.

Shen Lian (1990) Evaluation of potential yields in grain production of the rice soils in Taiwan. In: *Maximum Yield Symposium. 14th International Congress of Soil Science*, Kyoto, Japan, pp. 104–109.

Shen, F. and Wolter, W. (1992) *Managing the 'Four Waters', an innovative concept of modernization.* ITPRID/World Bank, Washington, DC. (Quoted in Alexandratos, 1995.)

Shepherd, K.D., Gregory, P.J., Woodhead, T., Pandey, R.K. and Magbujos, E.C. (1988) Growth of soybean (*Glycine max*) and mungbean (*Vigna radiata*) in the post-monsoon season after upland rice. *Experimental Agriculture* 24, 433–441.

Sheudzhen, A. K. (1991) Effect of trace elements on rice yield. *Khimizatsiya Sel'skogo Khozyaistva* 6, 71–73. (*Soils and Fertilizers*, 1993, 55, abstract 7895.)

Shi-ye Li (1984) Azolla in the paddy fields of Eastern China. In: *Organic Matter and Rice.* IRRI, Los Banos, Philippines, pp. 169–178.

Shin, J.S. and Han, K.H. (1988) Sulphur balance in Korean agriculture. In: *Proceedings, International Symposium on Sulphur for Korean Agriculture.* Korean Society of Soil Science and The Sulphur Institute, Washington DC, USA, pp. 29–34.

Shorrocks, V.R. (1972) Mineral nutrition, growth and nutrient cycle of *Hevea brasiliensis. Journal, Rubber Research Institute Malaysia* 19, 32–61.

Simpson, I.C., Roger, P.A., Oficial, R. and Grant, I.F. (1993) (a) Impacts of agricultural practices on aquatic oligochaete populations in ricefields. (b) Density and composition of aquatic oligochaete populations in different farmers' ricefields. *Biology and Fertility of Soils* 16, 27–33 and 34–40.

Simpson, I.C., Roger, P.A., Oficial, R. and Grant, I.F. (1994) Effects of nitrogen fertilizer and pesticide management on floodwater ecology in a wetland ricefield. II. Dynamics of microcrustaceans and dipteran larvae. *Biology and Fertility of Soils* 17, 138–146.

Small, L.E. and Carruthers, I. (1991) *Farmer-financed Irrigation: the Economics of Reform.* Cambridge University Press, Cambridge, UK.

Smith, J. and Gascon, F.E. (1981) The effect of the new rice technology on family labour utilisation in Laguna, Philippines. *IRRI Research Paper Series* No. 42. IRRI, Los Banos, Philippines.

Solheim, W.G.II (1972) An earlier agricultural revolution. *Scientific American* 226(4), 34–41.

Sorensen, P. (1986) On the problem of early rice in southeast Asia. In: Norland, I., Cederroth, S. and Gerdin, I. (eds) *Rice Societies: Asian Problems and Prospects.* Curzon Press, London, UK, pp. 267–279.

Spendjian, G. (1991) Economic, social and policy aspects of sustainable land use. In: Dumanski, J., Pushparajah, J., Latham, M. and Myers, R. (eds) *Evaluation for Sustainable Land Management in the Developing World, Vol. 2.* IBSRAM Proceedings No. 12. IBSRAM, Bangkok, Thailand, pp. 415–436.

Stangel, P. (1976) World fertilizer reserves in relation to future demand. In: Wright, M.J. (ed.) *Plant Adaptation to Mineral Stress in Problem Soils.* Cornell University Agricultural Experiment Station, Special Publication, Cornell University, Ithaca, New York, USA, pp. 31–54.

Stanhill, G. (1985) The water resource for agriculture. *Philisophical Transactions of the Royal Society of London*, B310, 161–173.

Svendsen, M. and Rosegrant, M. (1994) Will the future be like the past? In: Heim, F. and Abernethy, C. (eds) *Irrigated Agriculture in Southeast Asia beyond 2000.* International Irrigation Management Institute, Colombo, Sri Lanka, pp. 75–90.

Tabbal, D.F., Undan, R.C., Alagcan, M.M., Lampayan, R., Bhuiyan, S.I. and Woodhead, T. (1993) Farm-level water management for rice-based farming systems in the

Philippines. In: Miranda, S.M. and Maglinao, A.R. (eds) *Irrigation Water Management for Rice-based Farming Systems: Proceedings of a Tri-country Workshop.* IIMI, Colombo, Sri Lanka, pp. 151–173.

Tainter, J.A. (1988) *The Collapse of Complex Societies.* Cambridge University Press, Cambridge, UK.

Takahashi, J. (1965) Natural supply of nutrients in relation to plant requirements. In: IRRI (ed.) *The Mineral Nutrition of the Rice Plant.* Johns Hopkins University Press, Baltimore, Maryland, pp. 271–294.

Takahashi, R.M., Miura, T. and Wilder, W.H. (1982) A comparison between the area sampler and two other sampling devices for aquatic fauna in ricefields. *Mosquito News* 44, 211–216

Takajima, Y. (1964) Studies on organic acids in paddy field soils with reference to their inhibitory effects on the growth of rice plants: 2. *Soil Science and Plant Nutrition* 10, 212–219.

Tandon, H.L.S. (1992) Sulphur in Indian agriculture: Update, 1992. *Sulphur in Agriculture* 16, 20–23.

Tang, A.M. and Stone, B. (1980) *Food Production in the People's Republic of China.* Research Report 15, International Food Policy Research Institute, Washington DC, USA.

Tang Li-hua (1981) The status of microelements in relation to crop production in paddy soils of China. II. Molybdenum. In: *Paddy Soils Symposium.* Springer Verlag, Berlin, Germany, pp. 832–836.

Tang Shen-xiang (1994) Rice remains from Neolithic age excavated in China. *International Rice Research Notes* 19, 1.

Tanji, K.K., Lauchli, A. and Meyer, J. (1986) Selenium in the San Joaquin Valley. *Environment* 28, 6–11 and 34–39.

Thompson, G.V. (1992) Archaeological investigations at Kok Phanom Di, Central Thailand. Doctoral dissertation. Australian National University, Canberra, Australia

Thompson, G.B. (1995) *The Excavation of Khok Phanom Di, Central Thailand. Volume 4. Subsistence and Environment: the Botanical Evidence.* Reports of the Research Committee, Society of Antiquaries of London. Oxbow Books, London, UK.

Thorp, J. (1936) *Geography of the Soils of China.* The National Geological Survey of China, Nanjing, China.

Tisdale, S.L. and Nelson, W. (1966) *Soil Fertility and Fertilizers.* 2nd edn. MacMillan, New York, USA.

Tribe, D. (1994) *Feeding and Greening the World: the Role of International Agricultural Research.* CAB International, Wallingford, UK.

Trolldenier, G. (1981) Iron toxicity in rice plants and nitrogenase activity in the rhizosphere as related to potassium application. In: *Paddy Soils Symposium.* Agricultural Press, Beijing, China and Springer Verlag, Berlin, Germany, pp. 375–380.

Trolldenier, G. (1988) Visualisation of the oxidising power of rice roots, and possible partnership of bacteria in iron deposition. *Zeitschrift für Pflanzenernährung, Düngung Bodenkunde* 151, 117–121.

Tsutsuki, K. (1984) Volatile products and low molecular weight phenolic products of the anaerobic decomposition of organic matter. In: *Organic Matter and Rice.* IRRI, Los Banos, Philippines, pp. 329–343.

Uehara, G., Nishina, M.S. and Tsuji, G.Y. (1974) *The Composition of Mekong River Silt and its Possible Role as a Source of Plant Nutrient in Delta Soils.* Department of Agronomy and Soil Science, College of Tropical Agriculture, University of Hawaii, Honolulu, Hawaii, USA.

Uhlig, H. (1978) Geoecological controls on high-altitude rice cultivation in the Himalayas and mountain regions of Southeast Asia. *Arctic and Alpine Research* 10, 519–529.

UN (1991) World population prospects, 1990. *Population Studies* No. 120. United Nations, New York, USA.

Uphoff, N. (1986) *Improving International Irrigation Management with Farmer Participation: Getting the Process Right.* Westview Press, Boulder, Colorado, USA.

Vacharotayan, S. and Takai, Y. (1983) *Paddy Nitrogen Economy.* Nodai Research Institute, Tokyo Agricultural University, Tokyo, Japan.

van Dam, A.J. and van Diepen, C.A. (1982) *Polders of the World. The Soils of the Flat Wetlands of the World – their distribution and their agricultural potential.* International Soil Museum (ISRIC), Wageningen, Netherlands.

Vaughan, D.A. and Stich, L.A. (1991) Gene flow from the jungle to farmers: wild-rice genetic resources and their uses. *Bioscience* 41, 22–28.

Vijayalakshmi, K. and Mathan, K.K. (1991) Influence of graded doses of magnesium and potassium fertilization on dry matter production and yield of rice. *Journal of Potassium Research* 7, 221–225.

Virmani, S.S. (1994) *Heterosis and Hybrid Rice Breeding.* IRRI, Los Banos, Philippines and Springer Verlag, Berlin, Germany.

Vishnu-Mittre (1974) Paleobotanical evidence in India. In: Hutchinson, J.B. (ed.) *Evolutionary Studies on World Crops.* Cambridge University Press, Cambridge, pp. 3–30.

Vishnu-Mittre (1975) The early domestication of plants in South and Southeast Asia: a critical review. *Paleobotanist* 22, 83–88.

Von Uexkull, H.R. (1985) Availability and management of potassium in wetland soils. In: *Wetland Soils: Characterization, Classification and Utilization.* IRRI, Los Banos, Philippines, pp. 293–305.

Waage, J.K. (1991) Biodiversity as a resource for biological control. In: Hawksworth, D.L. (ed.) *The Biodiversity of Microorganisms and Invertebrates: its Role in Sustainable Agriculture.* CAB International, Wallingford, UK, pp. 149–164.

Waage, J.K. and Greathead, D.J. (1988) Biological control: challenges and opportunities. *Philosophical Transactions of the Royal Society of London* B318, 111–128.

Wade, R. and Chambers, R. (1980) *Managing the Main System: Canal Irrigation's Blind Spot.* Institute of Development Studies, University of Sussex, Brighton, UK.

WARDA (1984) Upland Rice in West Africa. In: *An Overview of Upland Rice Research.* IRRI, Los Banos, Philippines, pp. 21–43.

Warrick, R.A. and Barrow, E.M. (1990) Climate and sea level change: a perspective. *Outlook on Agriculture* 19, 5–8.

Warrick, R.A., Barrow, E.M. and Wigley, T.M.L. (1993) *Climate and Sea Level Change: Observations, Projections and Implications.* Cambridge University Press, Cambridge, UK.

Watanabe, I. (1984) Anaerobic decomposition of organic matter in flooded rice soils. In: *Organic Matter and Rice.* IRRI, Los Banos, Philippines, pp. 237–258.

Watanabe, I. (1987) Fertility and fertilizer evaluation for rice. In: *Azolla Utilisation.* IRRI, Los Banos, Philippines, pp. 197–206.

Wetselaar, R., Simpson, J.R. and Rosswall, T. (eds) (1981) *Nitrogen Cycling in South-East Asian Wet Monsoonal Ecosystems.* Australian Academy of Science, Canberra, Australia.

White, W.R. (1990) Reservoir sedimentation and flushing. In: *Hydrology in Mountainous Regions. Part II. Artificial Reservoirs; Water and Slopes.* IAHS Publication, No. 194. International Association of Hydrological Science, Wallingford, UK, pp. 129–139.

Whitton, B.A. and Rother, J.A. (1988) Environmental features of deepwater rice fields in Bangladesh during the flood season. *Proceedings, 1987 International Deepwater Rice Workshop,* Bangkok, Thailand. IRRI, Los Banos, Philippines, pp. 47–54.

Whitton, B.A., Karim, N.H. and Rother, J.A. (1988) Nutrients and nitrogen in deepwater ricefields in Bangladesh. *Proceedings, International Deepwater Rice Workshop,* Bangkok, Thailand. IRRI, Los Banos, Philippines, pp. 1–8.

Wickham, T.H. and Sen, C.N. (1978) Water management for lowland rice: water requirements and yield response. In: *Soils and Rice.* IRRI, Los Banos, Philippines, pp. 649–670.

Widjaja-Adhi, I.P.G. (1988) Physical and chemical characteristics of peat soils of Indonesia. *Indonesian Agricultural Research and Development Journal* 10, 59–64.

Wiens, T.B. (1984) Microeconomic study of organic fertilizer use in intensive farming in Jiangsu Province, China. In: *Organic Matter and Rice.* IRRI, Los Banos, Philippines, pp. 534–556.

Wood, A. (1950) *The Groundnut Affair.* Bodley Head, London, UK.

Woodburn, A. (1993) Rice – the Crop and its Agrochemicals Market. Allan Woodburn Associates Ltd., Midlothian, UK, and Managing Resources Ltd, Cambridge, UK.

Wu Hui (1985) *Study of Grain Yield in Ancient China.* Agriculture Publishing House, Beijing, China.

Xing Jia-ming (1988) The relationship between environment changes and human activities since the late quarternary in North China. In: Aigner, J.S., Jablonski, N.G., Taylor, G., Walker, D. and Wang Ping-xian (eds) *The Palaeoenvironment of East Asia from the Mid-tertiary.* Centre of Asian Studies, University of Hong Kong, 2, pp.1076–1083.

Xu, J.X. and Dong, W.R. (1989) Copper status of permanently and long-term waterlogged paddy soils and response of rice to copper fertilizer. *Acta Pedologica Sinica* 26, 149–158.

Yamanaka, M. (1979) Palynological studies of quaternary sediments in Northeast Japan. IV. Tsukimino Moor in Aomori Prefecture, with special reference to the history of the rice crop. *Ecological Review* 19, 113–121.

Yan, W. (1991) China's earliest rice agricultural remains. *Bulletin Indo-Pacific Prehistory Association* 10, 118–126.

Yang Huai-jen and Xie Zhi-ren (1984) Sea level changes over the past 20,000 years. In: Whyte, R.O. (ed.) *The Evolution of the East Asian Environment.* Centre for Asian Studies, University of Hong Kong, 1, pp.288–308.

Yao, B. and Chen, Q. (1983) South–north water transfer project plans. In: Biswas, A.K., Dakang, Z., Nickum, J.E. and Changming, L. (eds) *Long-distance Water Transfer: a Chinese Case Study and International Experiences.* Tycooly International, Dublin, Ireland, pp. 127–149.

Yap, C.L. (1992) A comparison of the cost of producing rice in selected countries. *FAO Economic and Social Development Paper* No. 101. FAO, Rome, Italy.

Yatazawa, M. (1977) Agroecosystems in Japan. In: *Cycling of Mineral Nutrients in Agro-ecosystems* 4, 167–179.

Yoshida, S. (1981) *Fundamentals of Rice Crop Science*. IRRI, Los Banos, Philippines.

Yoshida. S. (1983) Rice. In: *Potential Productivity of Field Crops under Different Environments*. IRRI, Los Banos, Philippines, pp. 103–127.

Yoshida, S. and Oka, H.I. (1982) Factors influencing rice yield, production potential and stability. In: *Rice Research Strategies for the Future*. IRRI, Los Banos, Philippines, pp. 51–70.

Yoshida, S., Parao, F.T. and Beachell, H.M. (1972) A maximum annual rice production trial in the Philippines. *International Rice Commision Newsletter* 21(3), 27–32.

Yoshida, S., Ogawa, M., Suenaga, K. and He, Chun-ye (1983) Induction and selection of salt-tolerant mutant rices by tissue culture – recent progress at IRRI. In: *Cell and Tissue Culture Techniques for Cereal Crop Improvement*. Science Press, Beijing, China, and IRRI, Los Banos, Philippines, pp. 237–254.

You, X.L. (1995) *History of Rice Cultivation in China*. Series on History of Chinese Culture. Agricultural Publishing House, Beijing, China.

Yu, C.L. and Buckwell, A. (1991) *Chinese Grain Economy and Policy*. CAB International, Wallingford.

Yu, Tian-ren (1981) Oxidation and reduction properties of paddy soils. In: *Symposium on Paddy Soils*. Springer Verlag, Berlin, Germany, pp. 95–106.

Yu, Tian-ren (1985) *Physical Chemistry of Paddy Soils*. Springer Verlag, Berlin.

Yu Ye-fei (1980) The statistics of rice yield in ancient China. *Journal of Chongqing Normal University* No. 3.

Yuan Loung-Ping and Virmani, S.S. (1988) Status of hybrid rice research and development. In: *Hybrid Rice*. IRRI, Los Banos, Philippines, pp. 7–24.

Yukawa, K. (1989) Historical development of rice cultivation and its water management. *Irrigation Engineering and Rural Planning* 16, 60–70.

Zandstra, H.G., Price, E.C., Litsinger, J.A. and Morris, R.A. (1981) *A Methodology for On-farm Cropping Systems Research*. IRRI, Los Banos, Philippines.

Zapata, F.J., Alejar, M.S., Torrizo, L.B., Novero, A.U., Singh, V. and Senadhira, D. (1991) Field performance of anther-culture derived lines from F1 crosses of indica rices under saline and non-saline conditions. *Theoretical and Applied Genetics* 83, 6–11.

Zeven, A.C. (1973) Dr. Th. H. Engelbrecht's view on the origin of cultivated plants. *Euphytica* 22, 279–286.

Zhao Qiguo (1990) Soil and water management in farming systems with flooded rice. *Transactions 14th International Congress of Soil Science*. Plenary Papers, pp. 13–29.

Zheng, H.K. and Huang, Z.Q. (1986) Effects of copper on paddy rice and the application technique. *Fujian Agricultural Science and Technology* No. 4, 21–22.

Zoschke, E. (1990) Yield losses in tropical rice as influenced by the composition of weed flora and the timing of elimination. In: Grayson, B.T., Green, M.B. and Copping, L.G. (eds) *Pest Management in Rice*. Elsevier Applied Science, London, UK, pp. 300–313.

Index